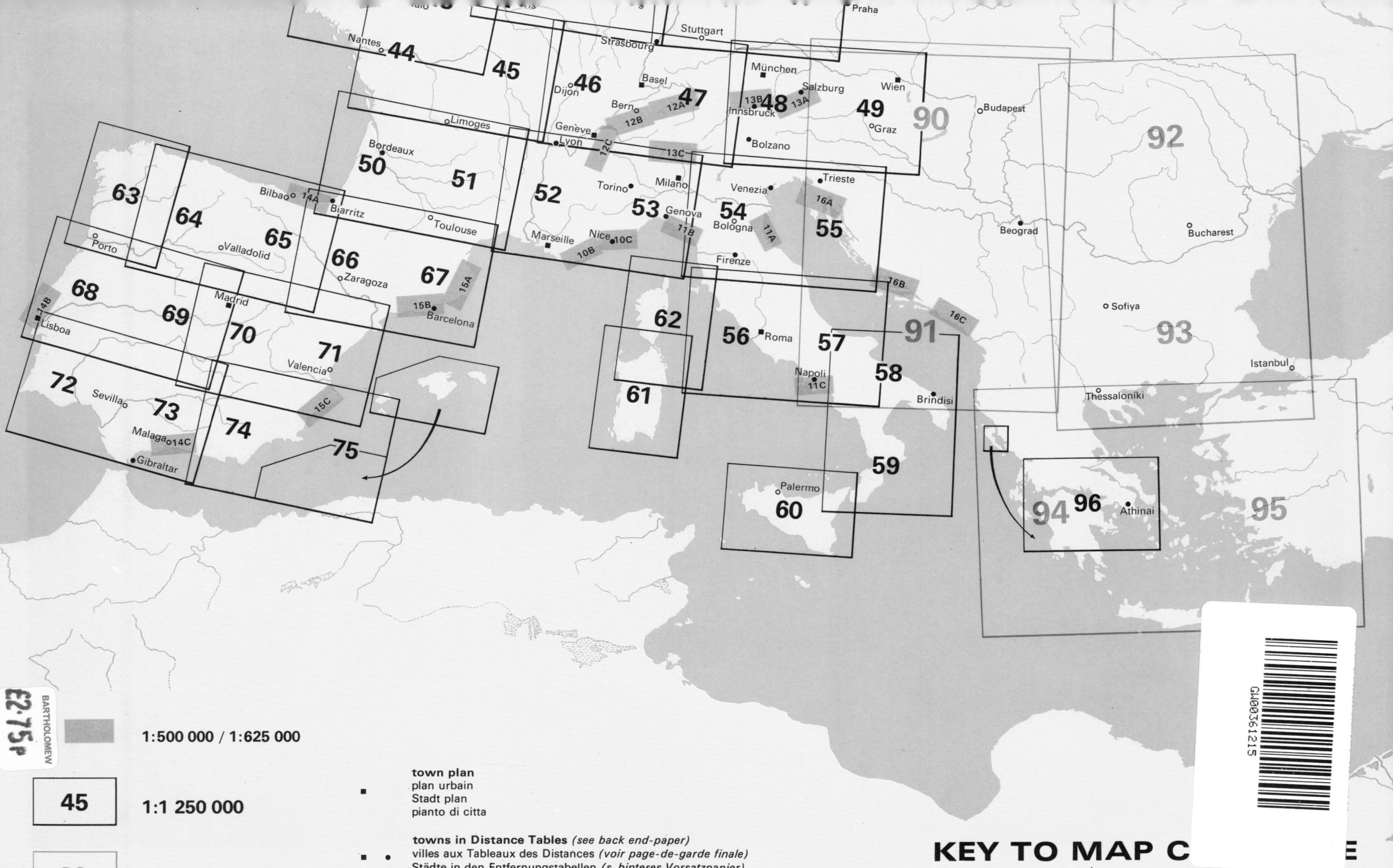

KEY TO MAP C
TABLEAU D'ASSEMBLAGE / KARTENVERTEILUNG / INDICE DELLE TAVOLE

GW00361215

town plan
plan urbain
Stadt plan
pianto di citta

towns in Distance Tables *(see back end-paper)*
villes aux Tableaux des Distances *(voir page-de-garde finale)*
Städte in den Entfernungstabellen *(s. hinteres Vorsatzpapier)*
le citta con Distanze Stradali *(cf. l'ultima pagina)*

1:500 000 / 1:625 000

1:1 250 000

1:2 500 000

45

89

BARTHOLOMEW

£2·75p

ROAD ATLAS EUROPE
BARTHOLOMEW

CONTENTS

6513

RESORT AREAS 10 – 16

motorway
autoroute
Autobahn
Autostrada

motorway under construction
autoroute en construction
Autobahn im Bau
Autostrada in costruzione

main road
route principale
Hauptstraße
strada principale

other roads
autres routes
übrige Straßen
altre strade

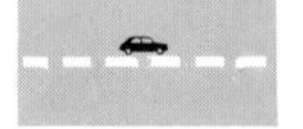

ferry
bac
Fähre
traghetto

mountain railway
funiculaire
Bergbahn
funicolare

cable car
téléphérique
Luftseilbahn
funivia

viewpoint
belvédère
schöne Aussicht
bel panorama

airport
aéroport
Flughafen
aeroporto

shrine
sanctuaire
Heiligtum
santuario

RUINAS

ancient site
site antique
antikes Baudenkma
antichita o rovine

height in metres
altitude en mètres
Hohe in Metern
altitudine in metri

▲ 1654

lighthouse
phare
Leuchtturm
faro

port
port
Hafen
porto

forest
forêt
Forst
foresta

ROAD ATLAS EUROPE

18 - 75   84 - 95   KEY CONTINUED

airport
aéroport
Flughafen
aeroporto

international boundary
frontiére national
Staatsgrenze
confine di stato

regional boundary
limite régionale
Ländesgrenze
confine di regione

CH   GR   **identification letters**
plaques nationales
Landeszeichen
sigla nazionale

•1654   ≥ 965   **height in metres**
altitude en mètres
Hohe in Metern
altitudine in metri

canal
canal
Kanal
canal

marsh
marais
Marsch
palude

RUINAS   **ancient site**
site antique
antikes Baudenkmal
antichita o rovine

white   **overlap of adjoining pages**
blanc   recouvrement des pages avoisinantes
weiß   Ubereinandergreifen der anliegenden Seiten
bianco   risvolto delle pagine aggiunte

see pages 76-83
voyez pages 76-83
siehe Seiten 76-83
vedere pagine 76-83

18 - 75   84 - 95   KEY

motorway with junction
autoroute et accès
Autobahn mit Kreuzung
autostrada con stazione

motorway under construction/projected
autoroute en construction/en projet
Autobahn im Bau/geplant
autostrada in costruzione/in progetto

trunk road
grand itinéraire
Fernverkehrstraße
Strada di grande comunicazione

main road
route principale
Hauptstraße
strada principale

other roads
autres routes
ubrige Straßen
altre Strade

40   40   **distances in kilometres**
14   26   distances en kilomètres
Entfernungen in kilometern
distanze chilometriche

E 25   E 3   **road numbering**
N IV   N 430   numérotage des routes
Straßennumerierung
numerazione delle strade

TOLL   **toll road**
route à péage
taxpflichtige Straße
strada a pedaggio

car ferry
bac pour autos
Autofähre
autotraghetto

transport of cars by ship
transport d'autos par bateau
Autoverladung per Schiff
trasporto auto per via mare

railway
chemin de fer
Eisenbahn
ferrovia

pass   **tunnel**
col   tunnel
Paß   Tunnel
passo   galleria

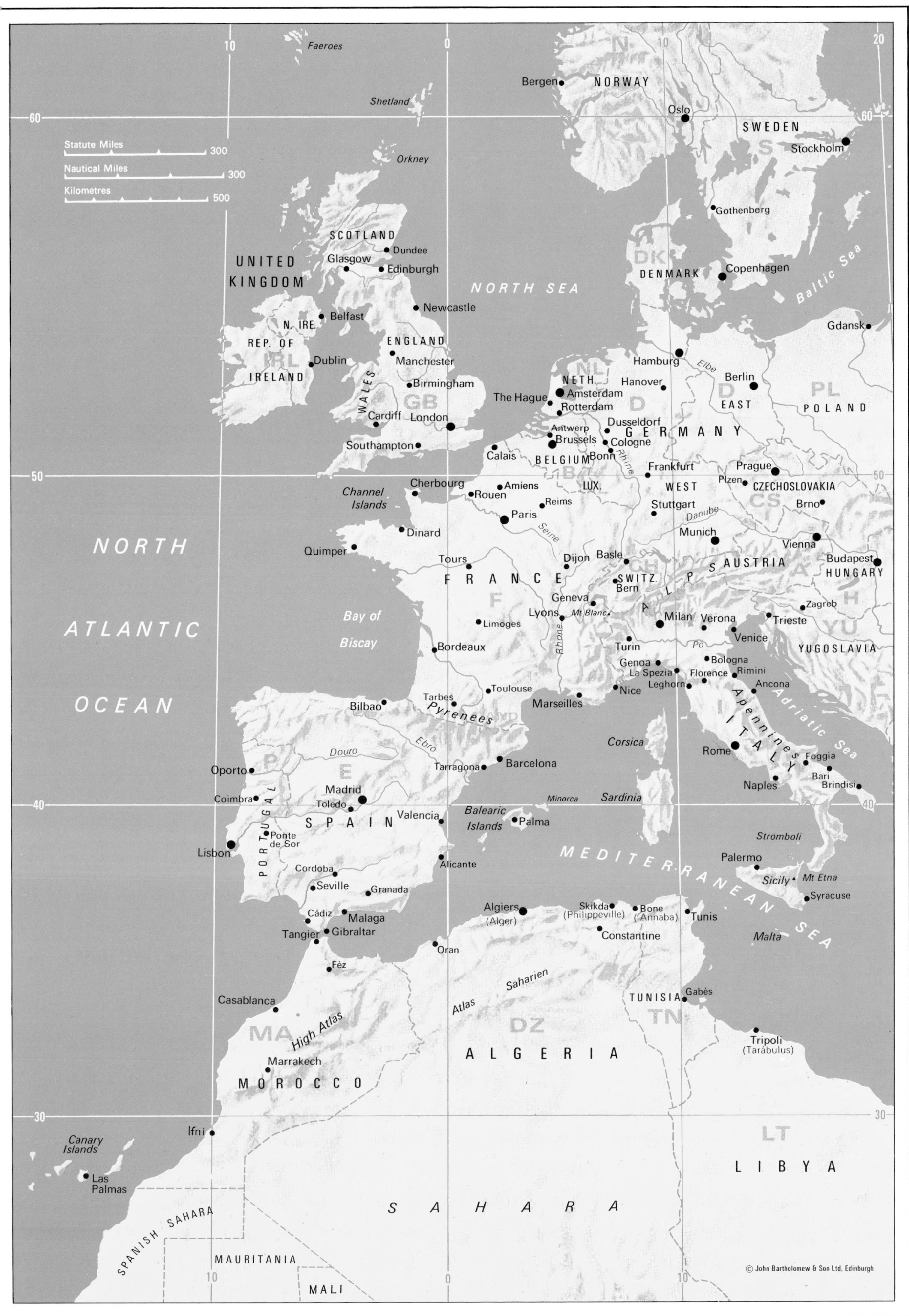

Faeroes

Shetland

60

Statute Miles 300
Nautical Miles 300
Kilometres 500

Orkney

Bergen   NORWAY

N

Oslo

SWEDEN

S

Stockholm

60

Gothenberg

SCOTLAND
Dundee
Glasgow   Edinburgh

NORTH SEA

DK

DENMARK

Copenhagen

Baltic Sea

Gdansk

UNITED
KINGDOM

Newcastle

N. IRE. Belfast

REP.
OF
IRELAND   Dublin

ENGLAND
Manchester

Birmingham

Hamburg   Elbe   Berlin

NL   Hanover   D   EAST

NETH.

The Hague   Amsterdam
Rotterdam

PL
POLAND

WALES

GB

Cardiff   London

Southampton

Dusseldorf   GERMANY

Antwerp   Cologne   Rhine

Brussels

Calais   BELGIUM   Bonn

LUX.   Frankfurt   Prague

WEST   Plzen   CZECHOSLOVAKIA

B

50

Cherbourg

Channel
Islands

Dinard

Quimper

NORTH

ATLANTIC

OCEAN

Amiens
Rouen   Reims

Paris

Tours

FRANCE

F

Seine

Dijon   Basle

Limoges

Bay of
Biscay

Bordeaux

Geneva
Lyons   Mt Blanc

Rhone

Stuttgart

Danube

Munich

SWITZ.
Bern   A   L   P   S

Turin

Po

Milan   Verona

Genoa
La Spezia   Bologna
Florence   Rimini
Leghorn   Ancona

50

CS   Brno

Vienna

Budapest
HUNGARY

Zagreb

Trieste

Venice   YU
YUGOSLAVIA

Tarbes
Bilbao   Pyrenees   Ebro   Nice

Douro

Oporto   P

Coimbra

PORTUGAL

Lisbon   Ponte
de Sor

Cordoba

Seville

Cádiz
Tangier   Malaga
Gibraltar
Fèz

Toulouse

Marseilles

Corsica

Rome

Tarragona   Barcelona

E

Madrid
Toledo   SPAIN   Valencia

Minorca   Sardinia

Balearic
Islands   Palma

Alicante

Apennines

I T A L Y

Foggia
Bari
Brindisi

Naples

Adriatic Sea

40   Oran

MEDITERRANEAN SEA

Stromboli

Palermo

Sicily   Mt Etna
Syracuse

40

Casablanca

Marrakech

MA

High Atlas

MOROCCO

Atlas

Saharien

DZ

ALGERIA

Algiers
(Alger)

Skikda   Bone
(Philippeville) (Annaba)   Tunis

Constantine

TUNISIA   Gabès

Malta

TN

Tripoli
(Tarábulus)

LT

30

Canary
Islands

Ifni

Las
Palmas

SPANISH SAHARA

MAURITANIA

MALI

S A H A R A

L I B Y A

30

© John Bartholomew & Son Ltd, Edinburgh

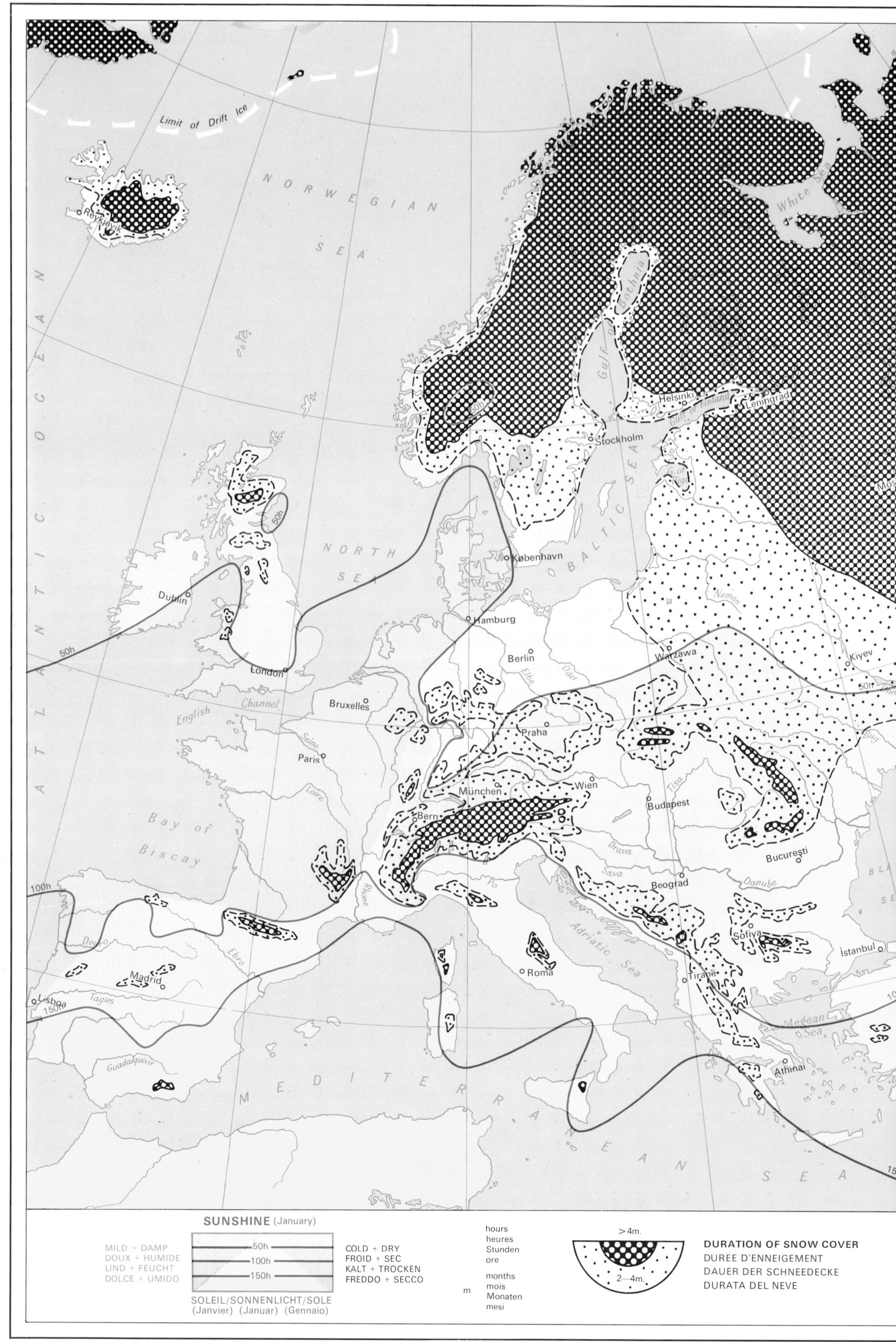

SUNSHINE (January)

MILD + DAMP
DOUX + HUMIDE
LIND + FEUCHT
DOLCE + UMIDO

————— 50h —————
————— 100h —————
————— 150h —————

COLD + DRY
FROID + SEC
KALT + TROCKEN
FREDDO + SECCO

SOLEIL/SONNENLICHT/SOLE
(Janvier) (Januar) (Gennaio)

hours
heures
Stunden
ore

months
mois
Monaten
mesi

m

>4m.

2—4m.

DURATION OF SNOW COVER
DUREE D'ENNEIGEMENT
DAUER DER SCHNEEDECKE
DURATA DEL NEVE

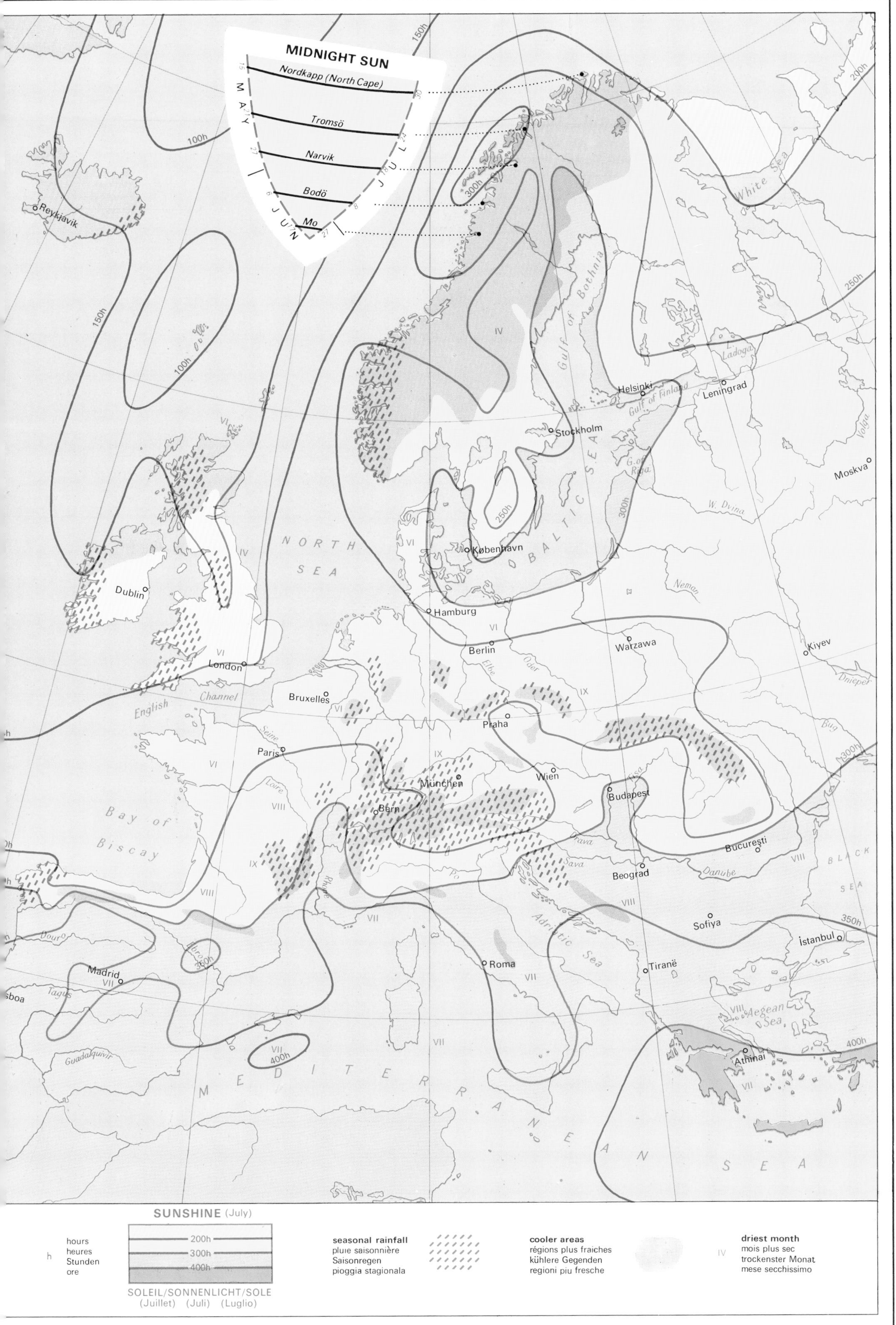

MIDNIGHT SUN

Nordkapp (North Cape)

Tromsö

Narvik

Bodö

Mo

SUNSHINE (July)

| hours |
| heures |
| h Stunden |
| ore |

200h

300h

400h

seasonal rainfall
pluie saisonnière
Saisonregen
pioggia stagionala

cooler areas
régions plus fraîches
kühlere Gegenden
regioni piu fresche

driest month
IV mois plus sec
trockenster Monat
mese secchissimo

SOLEIL/SONNENLICHT/SOLE
(Juillet) (Juli) (Luglio)

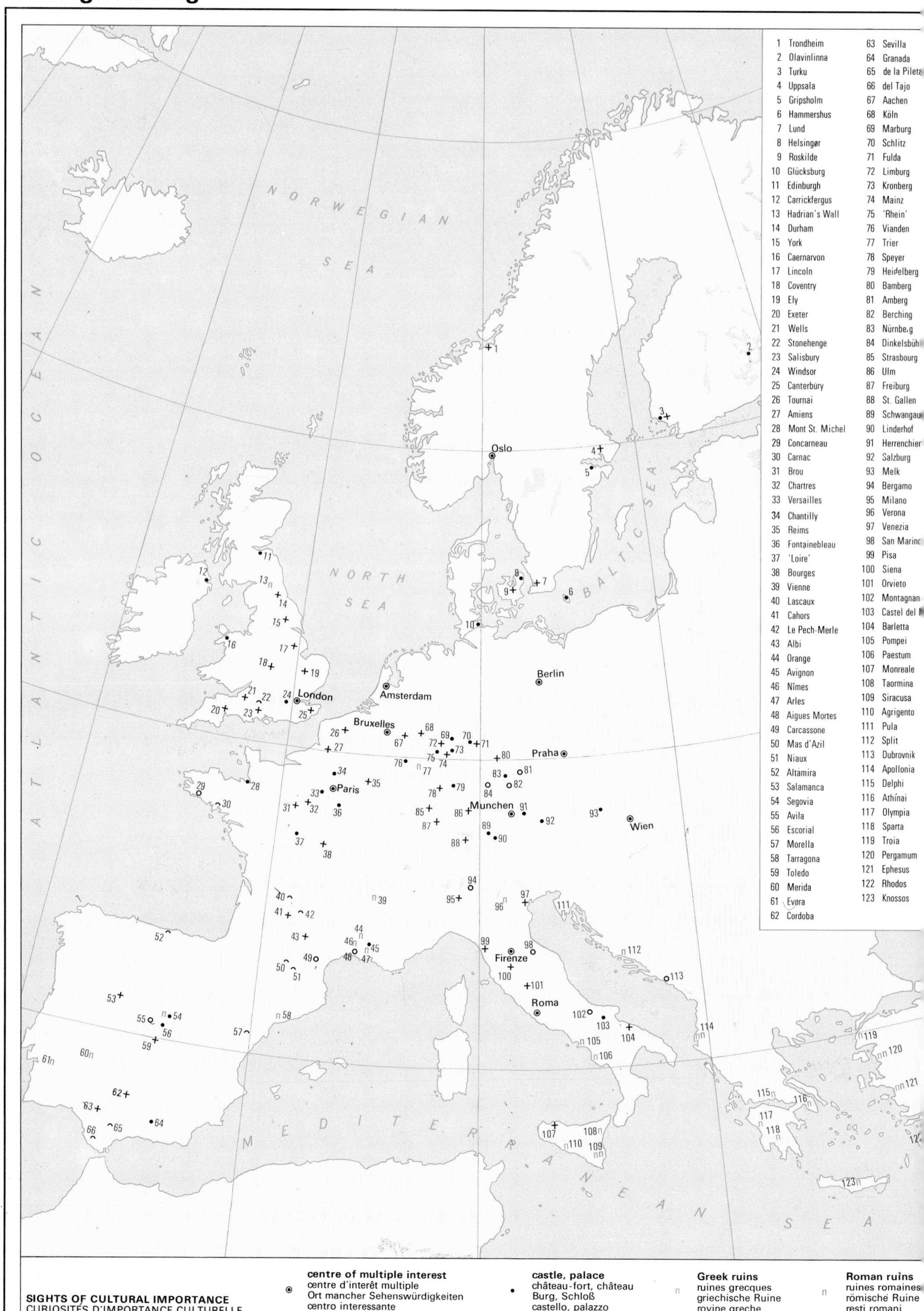

NORWEGIAN SEA

NORTH SEA

BALTIC SEA

BLACK SEA

MEDITERRANEAN SEA

Bergen

Helsinki

Stockholm

Moskva

Iona

Edinburgh

York

København

Berlin

Warszawa

Cork

Walsingham
Bath   Aldeburgh   'Holland'   Arnhem
Glastonbury   London   Canterbury
Flanders'   Koblenz   Oberhausen   Leipzig

Kraków

Wiesbaden   Praha

Lisieux   Mannheim   Heidelberg   Bayreuth
Chartres   Strasbourg
Oberammergau   München   Wien
Zürich   Salzburg   Graz
Besançon   Bregenz
Luzern
Montreux

Venézia

Bordeaux
Lyon
La Salette
Santander   Aix-en   Firenze
Santiago de   Provence   Perugia   Loreto
Compostela   Bilbao   San   Lourdes   Stes   Cannes   Assisi   Dubrovnik
Sebastian   Maries   Spoleto
Montserrat   Roma
Fátima   Barcelona
Lisboa

Sevilla
Granada

Athínai

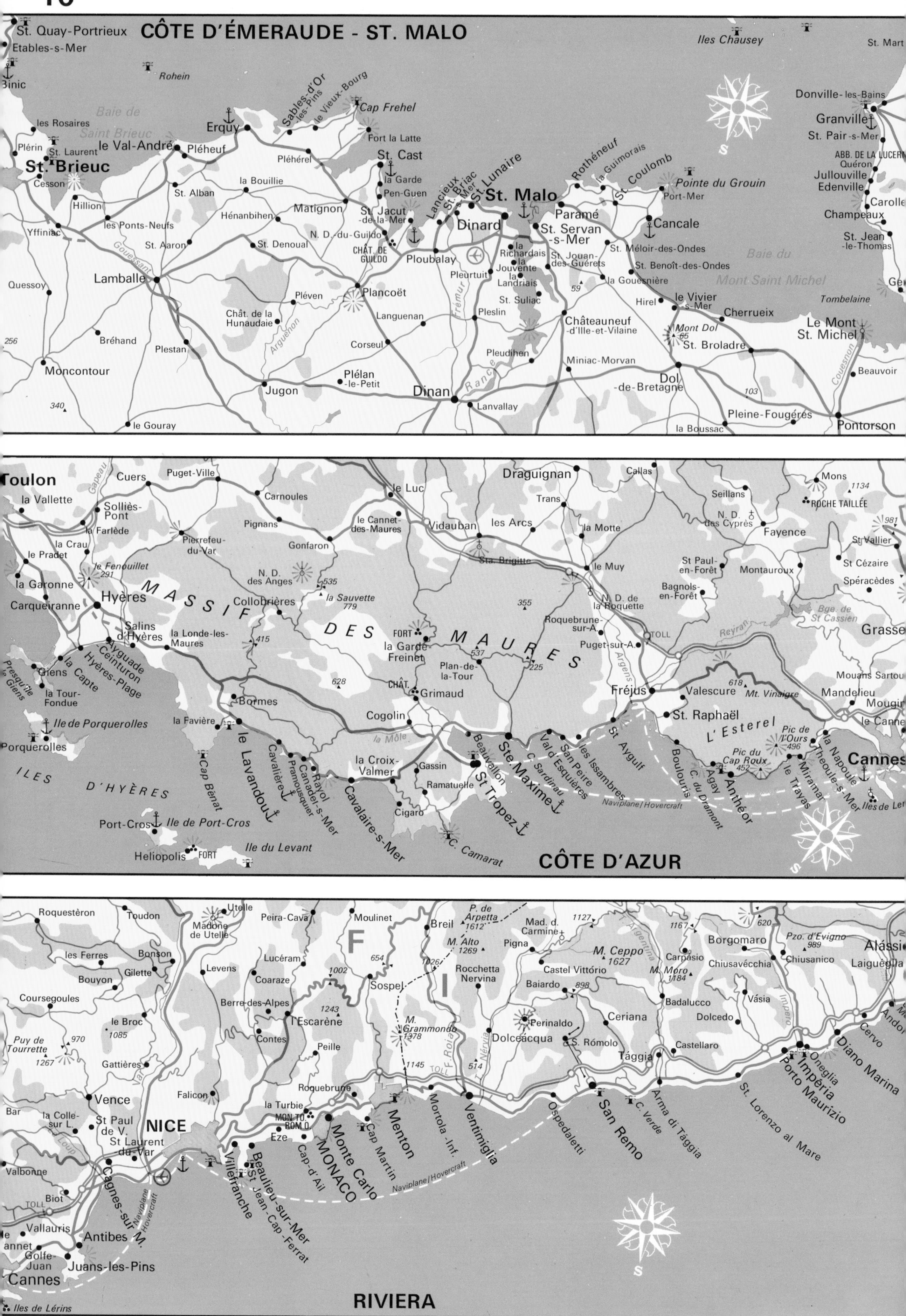
CÔTE D'ÉMERAUDE - ST. MALO
St. Quay-Portrieux
Etables-s-Mer
Binic
Rohein
Baie de Saint Brieuc
les Rosaires
Plérin
St. Laurent
St. Brieuc
Cesson
le Val-André
Pléneuf
Erquy
Sables-d'Or-les-Pins
le Vieux-Bourg
Cap Frehel
Fort la Latte
St. Cast
la Garde
Pen-Guen
Pléhérel
Hillion
Yffiniac
les Ponts-Neufs
St. Alban
la Bouillie
Hénanbihen
Matignon
St. Jacut-de-la-Mer
N. D.-du-Guildo
CHÂT. DE GUILDO
St. Aaron
St. Denoual
Gouessant
Lamballe
Quessoy
Pléven
Plancoët
Châť. de la Hunaudaie
Arguenon
Languenan
Corseul
Bréhand
Plestan
256
Moncontour
340
Jugon
Plélan-le-Petit
Dinan
Lanvallay
le Gouray
Lancieux
St. Briac-s-Mer
St. Lunaire
Dinard
St. Malo
Ploubalay
Pleurtuit
Frémur
la Richardais
la Jouvente
la Landriais
St. Suliac
Pleslin
Rance
Pleudihen
Rothéneuf
la Guimorais
St. Coulomb
Pointe du Grouin
Port-Mer
Paramé
St. Servan-s-Mer
St. Jouan-des-Guérets
Cancale
St. Méloir-des-Ondes
St. Benoît-des-Ondes
la Gouesnière
59
Hirel
le Vivier-s-Mer
Châteauneuf-d'Ille-et-Vilaine
Mont Dol
65
St. Broladre
Cherrueix
Miniac-Morvan
Dol-de-Bretagne
103
Pleine-Fougérés
la Boussac
Pontorson
Iles Chausey
St. Mart
Donville-les-Bains
Granville
St. Pair-s-Mer
ABB. DE LA LUCERNE
Quéron
Jullouville
Edenville
Carolles
Champeaux
St. Jean-le-Thomas
Baie du Mont Saint Michel
Tombelaine
Le Mont St. Michel
Couesnon
Beauvoir
Toulon
la Vallette
Cuers
Puget-Ville
Gapeau
Solliès-Pont
la Farlède
la Crau
le Pradet
la Garonne
Carqueiranne
le Fenouillet
291
Pierrefeu-du-Var
Carnoules
Pignans
Gonfaron
N. D. des Anges
535
la Sauvette
779
Hyères
MASSIF DES MAURES
Collobrières
Salins d'Hyères
la Londe-les-Maures
415
Ayguade
Ceinturon
Hyères-Plage
la Capte
Giens
Presqu'île de Giens
la Tour-Fondue
Ile de Porquerolles
Porquerolles
ILES D'HYÈRES
Bormes
la Favière
le Lavandou
Cap Bénat
Cavalière
Pramousquier
Canadel-s-Mer
Rayol
Cavalaire-s-Mer
Port-Cros
Ile de Port-Cros
Heliopolis
FORT
Ile du Levant
628
le Luc
le Cannet-des-Maures
Vidauban
Draguignan
Trans
les Arcs
la Motte
Sta. Brigitte
le Muy
355
FORT
la Garde-Freinet
537
Plan-de-la-Tour
225
CHÂT.
Grimaud
Cogolin
la Môle
la Croix-Valmer
Gassin
Ramatuelle
Cigaro
C. Camarat
Beauvallon
St Tropez
Ste Maxime
C. Sardinaux
Val d'Esquières
San Peire
les Issambres
St Aygulf
Callas
Seillans
N. D. des Cyprès
Fayence
Mons
1134
ROCHE TAILLÉE
981
St Vallier
St Cézaire
Spéracèdes
St Paul-en-Forêt
Montauroux
Bagnols-en-Forêt
N. D. de la Roquette
Bge. de St Cassien
Roquebrune-sur-A.
TOLL
Reyran
Grasse
Puget-sur-A.
Argens
Fréjus
Valescure
618
Mt. Vinaigre
St. Raphaël
L'Esterel
Mouans Sartoux
Mandelieu
Mougins
le Cannet
Pic de l'Ours
496
Pic du Cap Roux
452
Boulouris
C. du Dramont
Agay
Anthéor
le Trayas
Miramar
Théoule-s-Mer
la Napoule
Cannes
Iles de Lérins
Naviplane / Hovercraft
CÔTE D'AZUR
Roquestèron
Toudon
Utelle
Madone de Utelle
Peira-Cava
Moulinet
Breil
P. de Arpetta
1612
Mad. d. Carmine
1127
1167
620
Pzo. d'Evigno
989
Alássio
Laiguéglia
les Ferres
Bonson
Gilette
Bouyon
Levens
Lucéram
Coaraze
1002
654
Sospel
F
I
M. Alto
1269
Pigna
M. Ceppo
1627
Argentina
Carpásio
Borgomaro
Chiusavécchia
Chiusanico
Rocchetta Nervina
Castel Vittório
M. Moro
1184
Baiardo
898
Coursegoules
Berre-des-Alpes
1243
l'Escarène
Badalucco
Vásia
Impero
le Broc
1085
970
Puy de Tourrette
1267
Contes
M. Grammondo
1378
Perinaldo
Ceriana
Dolcedo
Dolceacqua
S. Rómolo
Castellaro
Peille
Gattières
Nérvia
F. Roia
1145
514
TOLL
Tággia
Vence
Var
Falicon
Roquebrune
la Turbie
MON. TO. ROM. O.
Eze
Bar
la Colle-sur-L.
St Paul de V.
NICE
St Laurent du-Var
Loup
Valbonne
Biot
TOLL
Cagnes-sur-M.
Naviplane Hovercraft
Vallauris
Golfe-Juan
Antibes
Juans-les-Pins
Cannes
Iles de Lérins
Villefranche
Beaulieu-sur-Mer
St Jean-Cap-Ferrat
Cap-d'Ail
MONACO
Monte Carlo
Cap Martin
Menton
Mortola-Inf.
Ventimiglia
Naviplane / Hovercraft
Ospedaletti
San Remo
C. Verde
Arma di Tággia
St. Lorenzo al Mare
Porto Maurizio
Impéria
Oneglia
Diano Marina
Cervo
Andora
RIVIERA
S

1:500 000

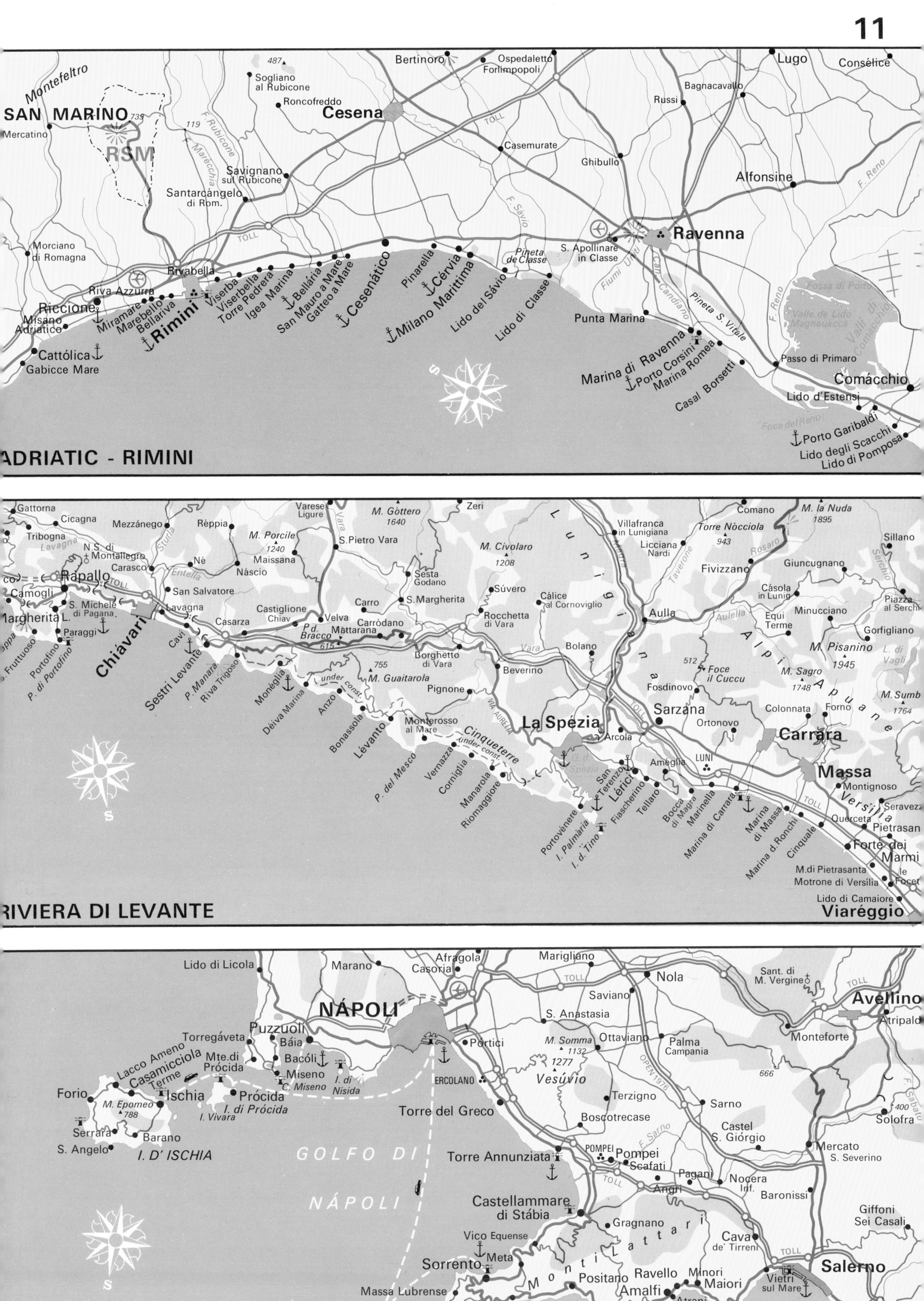

ADRIATIC - RIMINI

Bertinoro
Ospedaletto
Forlimpopoli
Lugo
Consélice
Sogliano
al Rubicone
487
Russi
Bagnacavallo
Montefeltro
Roncofreddo
Cesena
Casemurate
Alfonsine
SAN MARINO 739
Savignano
sul Rubicone
Ghibullo
F. Reno
Mercatino
119
F. Rubicone
S. Apollinare
in Classe
Ravenna
RSM
Santarcángelo
di Rom.
Marécchia
TOLL
F. Savio
Fossa di Porto
Morciano
di Romagna
Rivabella
Pineta
de Classe
Fiumi Uniti
Valle de Lido
di Mugnauacca
Riva Azzurra
Punta Marina
Pineta S. Vitale
Valli di Comácchio
Viserba
Pinarella
Riccione
Miramare
Viserbella
Bellária
Cérvia
Lido del Sávio
Canale Candiana
Mísano
Adriático
Marebello
Torre Pedrera
San Mauro a Mare
Milano Maríttima
Lido di Classe
F. Reno
Bellariva
Rimini
Igea Marina
Cesenático
Marina di Ravenna
Passo di Primaro
Cattólica
Gatteo a Mare
Porto Corsini
Gabicce Mare
Marina Romea
Comácchio
Casal Borsetti
Lido d'Estensi
Foce del Reno
Porto Garibaldi
Lido degli Scacchi
Lido di Pomposa

RIVIERA DI LEVANTE

Gattorna
Varese
Ligure
Zeri
Comano
M. la Nuda
1895
Cicagna
M. Góttero
1640
Villafranca
in Lunigiana
Torre Nócciola
943
Sillano
Mezzánego
Réppia
M. Porcile
1240
S.Pietro Vara
Licciana
Nardi
Tribogna
N.S. di
Montallegro
Né
Maissana
Fivizzano
Giuncugnano
Rapallo
Carasco
Náscio
M. Civolaro
1208
Cásola
in Lunig
Piazza
al Serchi
co
Camogli
San Salvatore
Sesta
Godano
Cálice
al Cornoviglio
Aulla
Minucciano
Gorfigliano
S. Michele
di Pagana
Lavagna
Castiglione
Chiav.
Velva
Carro
S. Margherita
Rocchetta
di Vara
M. Pisanino
1945
Margherita L.
Paraggi
Casarza
P. d.
Bracco
Mattarana
Carródano
Fosdinovo
Equi
Terme
L. di
Vagli
Portofino
P. di Portofino
Chiávari
615
755
Borghetto
di Vara
Bolano
512
Foce
il Cuccu
M. Sagro
1748
M. Sumb
1764
Fruttuoso
Cavi
Sestri Levante
P. Manara
Riva Trigoso
M. Guaitarola
Pignone
Beverino
Sarzana
Colonnata
Forno
P
Monéglia
under const.
La Spézia
Ortonovo
Carrara
Déiva Marina
Anzo
Monterosso
al Mare
Arcola
LUNI
Bonassola
Cinqueterre
under const.
G. di
Spézia
Améglia
Massa
Lévanto
Vernazza
San
Terenzo
Montignoso
P. del Mesco
Corniglia
Lérici
Seravezza
Manarola
Portovénere
Fiaschérino
Marinella
Versília
Querceta
Riomaggiore
I. Palmária
I. d. Tino
Tellaro
Bocca
di Magra
Marina di Carrara
Marina
di Massa
Pietrasan
Marina d. Ronchi
Cinquale
Forte dei
Marmi
M. di Pietrasanta
Motrone di Versília
le
Focet
Lido di Camaiore

RIVIERA DI LEVANTE
Viaréggio

NÁPOLI - SALERNO

Lido di Licola
Marano
Afragola
Casoria
Marigliano
Nola
Sant. di
M. Vergine
Avellino
Torregáveta
Puzzuoli
Báia
NÁPOLI
Saviano
S. Anastasia
Atripald
Lacco Ameno
Mte.di
Prócida
Pórtici
M. Somma
1132
Ottaviano
Palma
Campania
Monteforte
Casamícciola
Terme
Bacóli
Ischia
Miseno
I. di
Nísida
ERCOLANO
1277
Vesúvio
OPEN 1916
666
Forio
M. Epomeo
788
Prócida
C. Miseno
I. di Prócida
I. Vivara
Torre del Greco
Terzigno
Sarno
Castel
S. Giórgio
Solofra
Serrara
Boscotrecase
Mercato
S. Severino
S. Angelo
Barano
I. D'ISCHIA
GOLFO DI
Torre Annunziata
POMPEI
Pompei
Scafati
Pagani
Nocera
Inf.
400
NÁPOLI
Angri
Baronissi
F. Sarno
Castellammare
di Stábia
Gragnano
Cava
de' Tirreni
Giffoni
Sei Casali
Vico Equense
TOLL
Salerno
Meta
Monti
Lattari
Vietri
sul Mare
Sorrento
Positano
Ravello
Minori
Massa Lubrense
Amalfi
Maiori
I. DI CAPRI
Grotta Azzurra
Termini
S. Agata
s. due Golfi
Praiano
Atrani
Conca
dei Marini
C. d'Orso
Anacapri
Capri
P. Campanella
Golfo di Salerno
Faraglioni

NÁPOLI - SALERNO

0   5   10   15   20 Miles
0   5   10   15   20   25   30 Kilometres

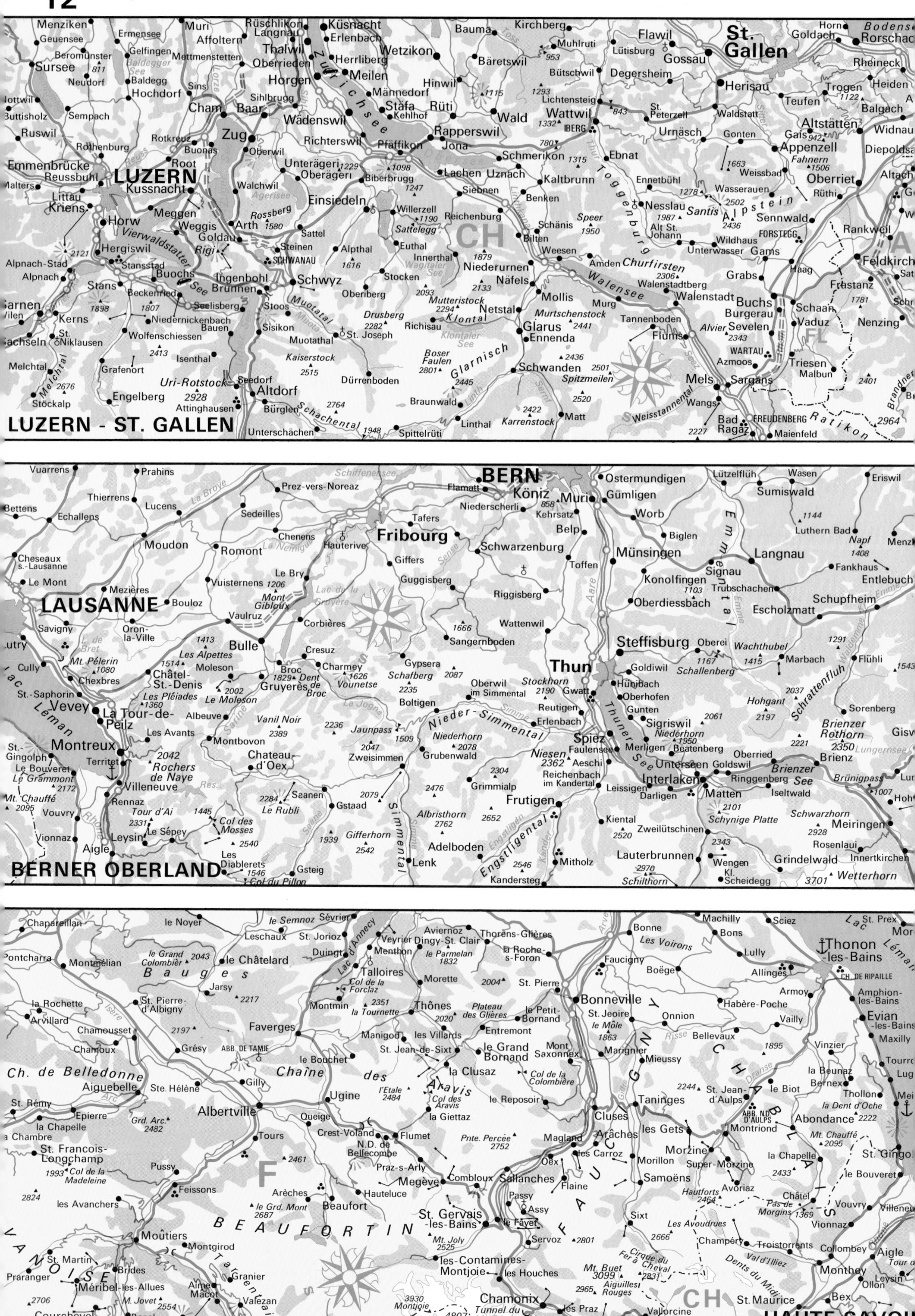

12

LUZERN - ST. GALLEN

BERNER OBERLAND

HAUTE SAVOI

1:625 000

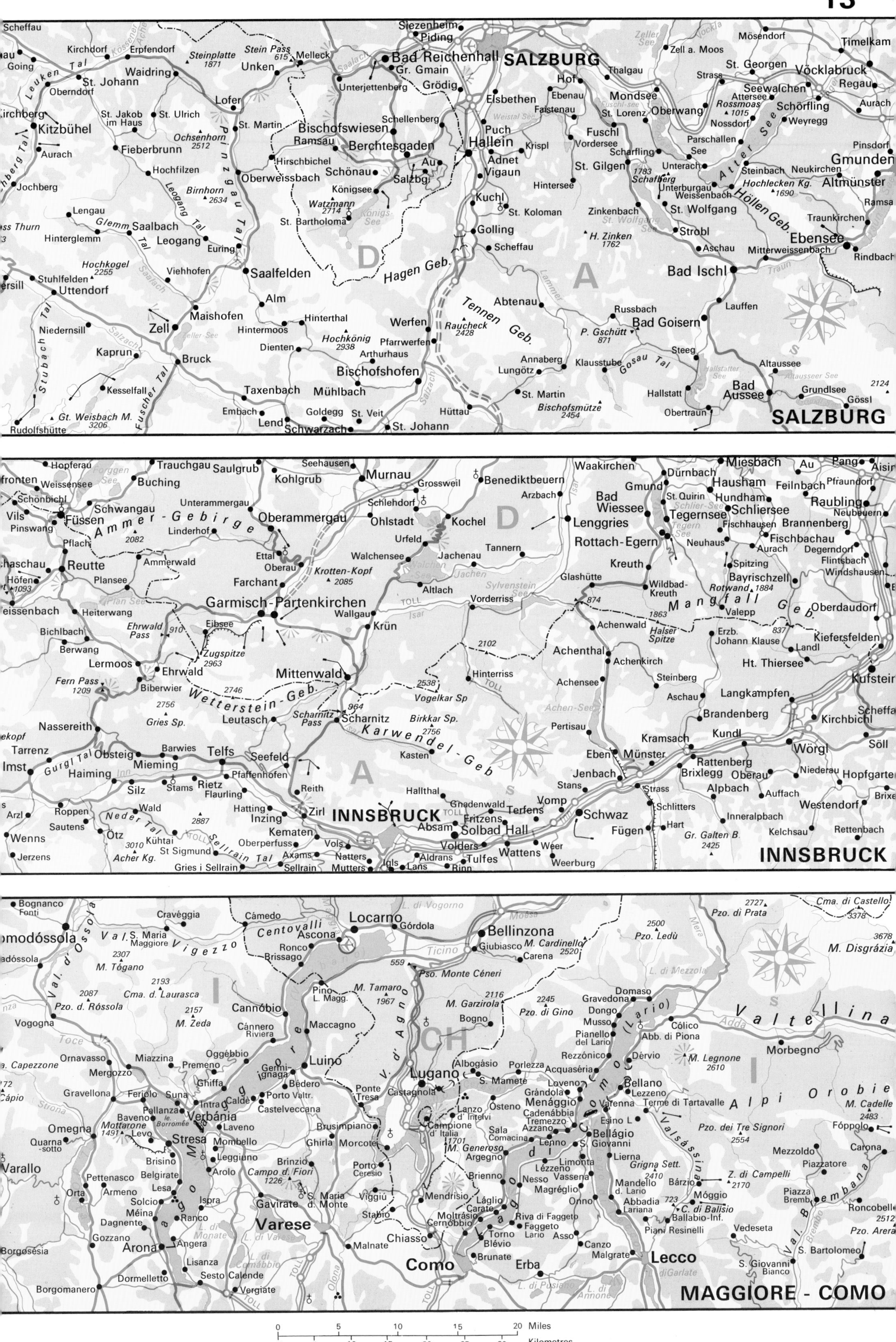

SALZBURG
Scheffau
Kirchdorf
Erpfendorf
Steinplatte
1871
Stein Pass
615
Melleck
Siezenheim
Piding
Bad Reichenhall
Gr. Gmain
Unterjettenberg
Grödig
Going
Leuken Tal
St. Johann
Oberndorf
Waidring
Unken
Lofer
Kirchberg
Kitzbühel
St. Jakob im Haus
St. Ulrich
St. Martin
Ochsenhorn
2512
Pinzgau Tal
Bischofswiesen
Ramsau
Berchtesgaden
Schellenberg
Elsbethen
Hallein
Puch
Adnet
Vigaun
Krispl
Kuchl
St. Koloman
Golling
Scheffau
Fieberbrunn
Aurach
Hochfilzen
Jochberg
Oberweissbach
Hirschbichel
Schönau
Au
Salzbg.
Königsee
Watzmann
2714
St. Bartholomä
Königs-See
Birnhorn
2634
Leogang Tal
Lengau
Glemm
Saalbach
Hinterglemm
Leogang
Euring
Hochkogel
2255
Viehhofen
Saalfelden
Stuhlfelden
Uttendorf
Mittersill
Maishofen
Zell
Zeller See
Alm
Hintermoos
Hinterthal
Dienten
Hochkönig
2938
Arthurhaus
Pfarrwerfen
Werfen
Raucheck
2428
Bischofshofen
Mühlbach
Kaprun
Bruck
Niedernsill
Stubach Tal
Fuscher Tal
Kesselfall
Taxenbach
Embach
Lend
Goldegg
St. Veit
Schwarzach
St. Johann
Hüttau
Gt. Weisbach M.
3206
Rudolfshütte
Hagen Geb.
Tennen Geb.
D
A
Abtenau
Annaberg
Lungötz
St. Martin
Bischofsmütze
2454
Zeller See
Zell a. Moos
Mösendorf
Timelkam
Thalgau
Hof
Ebenau
Faistenau
Mondsee
St. Lorenz
Oberwang
Fuschl
Vordersee
Scharfling
Hintersee
St. Gilgen
Schafberg
1783
Unterach
Unterburgau
Weissenbach
Strass
Attersee
Rossmoas
1015
Nossdorf
Parschallen
See
St. Georgen
Seewalchen
Vöcklabruck
Regau
Schörfling
Weyregg
Aurach
Atter See
Pinsdorf
Gmunden
Altmünster
Steinbach
Neukirchen
Hochlecken Kg.
1690
Höllen Geb.
Ramsau
Traunkirchen
Ebensee
Rindbach
Mitterweissenbach
St. Wolfgang
Strobl
Aschau
Zinkenbach
St. Wolfgang See
H. Zinken
1762
Bad Ischl
Lauffen
Russbach
Bad Goisern
P. Gschütt
871
Gosau Tal
Klausstube
Steeg
Hallstatter See
Hallstatt
Obertraun
Altaussee
Altausseer See
Bad Aussee
Grundlsee
Gössl
2124
SALZBURG
Hopferau
Trauchgau
Saulgrub
Seehausen
Murnau
Benediktbeuern
Waakirchen
Miesbach
Au
Pang
Aising
Pfronten
Weissensee
Buching
Kohlgrub
Grossweil
Dürnbach
Gmund
Hausham
Feilnbach
Pfraundorf
Schönbichl
Unterammergau
Schlehdorf
Arzbach
Bad Wiessee
St. Quirin
Hundham
Raubling
Vils
Füssen
Schwangau
Oberammergau
Ohlstadt
Kochel
D
Lenggries
Tegernsee
Schliersee
Neubeuern
Pinswang
Ammer-Gebirge
Linderhof
2082
Urfeld
Rottach-Egern
Fischhausen
Brannenberg
Pflach
Ettal
Walchensee
Jachenau
Tannern
Neuhaus
Fischbachau
Reutte
Ammerwald
Oberau
Krotten-Kopf
2085
Kreuth
Aurach
Degerndorf
Plansee
Farchant
Altlach
Glashütte
Spitzing
Flintsbach
Höfen
1093
Bayrischzell
Windshausen
Garmisch-Partenkirchen
Vorderriss
Wildbad-Kreuth
Rotwand
1884
Weissenbach
Heiterwang
Wallgau
Krün
Mangfall Geb.
Oberdaudorf
Achenwald
1863
Valepp
Bichlbach
Ehrwald Pass
910
Eibsee
Halser Spitze
837
Erzb. Johann Klause
Kiefersfelden
Berwang
Zugspitze
2963
2102
Achenthal
Landl
Lermoos
Ehrwald
Achenkirch
Ht. Thiersee
Mittenwald
Hinterriss
Steinberg
Kufstein
Fern Pass
1209
Biberwier
2746
2538
Vogelkar Sp
Achensee
Aschau
Langkampfen
Wetterstein-Geb.
964
Scheffau
2756
Gries Sp.
Leutasch
Scharnitz Pass
Scharnitz
Birkkar Sp.
2756
Brandenberg
Kirchbichl
Nassereith
Karwendel-Geb
Pertisau
Kundl
Tarrenz
Obsteig
Barwies
Telfs
Seefeld
Kramsach
Söll
Imst
Gurgl Tal
Mieming
Kasten
Eben
Münster
Wörgl
Haiming
Pfaffenhofen
A
Jenbach
Rattenberg
Brixlegg
Oberau
Niederau
Hopfgarten
Silz
Stams
Rietz
Flaurling
Reith
Hallthal
Stans
Strass
Alpbach
Auffach
Brixen
Arzl
Roppen
Wald
Hatting
Zirl
Inzing
INNSBRUCK
Gnadenwald
Terfens
Vomp
Schwaz
Schlitters
Westendorf
Wenns
Sautens
Neder Tal
2887
Ötz
Kühtai
Kematen
Absam
Fritzens
Solbad Hall
Fügen
Hart
Inneralpbach
Kelchsau
Rettenbach
Jerzens
3010
Acher Kg.
St Sigmund
Sellrain Tal
Oberperfuss
Axams
Vols
Natters
Volders
Wattens
Weer
Gr. Galten B.
2425
Gries i Sellrain
Sellrain
Mutters
Igls
Lans
Aldrans
Tulfes
Rinn
Weerburg
INNSBRUCK
Bognanco Fonti
Crevéggia
Cámedo
Centovalli
Locarno
Górdola
L. di Vogorno
Bellinzona
2727
Pzo. di Prata
Cma. di Castello
3378
Domodóssola
Val S. Maria Maggiore Vigezzo
Ascona
Ronco
Brissago
Giubiasco
M. Cardinello
2520
Carena
Ticino
2500
Pzo. Ledù
3678
M. Disgrázia
2307
M. Tógano
559
Pso. Monte Céneri
L. di Mezzola
Val. d'Ossola
2193
Cma. d. Laurasca
2087
Pzo. d. Róssola
Pino
L. Magg.
M. Tamaro
1967
2116
M. Garzirola
2245
Pzo. di Gino
Domaso
Gravedona
Dongo
Valtellina
Vogogna
2157
M. Zeda
Cannóbio
Maccagno
V. d'Agno
Bogno
Musso
Pianello del Lario
Cólico
Abb. di Piona
Adda
Toce
Cánnero Riviera
CH
Rezzónico
Dérvio
M. Legnone
2610
Morbegno
Ornavasso
Miazzina
Oggébbio
Luino
Albogásio
Porlezza
Acquaséria
Mergozzo
Premeno
Germignaga
Bédero
Lugano
S. Mamete
Loveno
Bellano
I
Gravellona
Feriolo
Suna
Ghiffa
Intra
Caldè
Porto Valtr.
Ponte Tresa
Castagnola
Grándola
Menággio
Lezzeno
Alpi Orobie
Baveno
Pallanza
Verbánia
Castelveccana
Lanzo d'Intelvi
Osteno
Cadenábbia
Varenna
Terme di Tartavalle
M. Cadelle
2483
Omegna
Mottarone
1491
Levo
Laveno
Brusimpiano
Campione d'Italia
Sala Comacina
Azzano
Tremezzo
Esino L.
Valsássina
Pzo. dei Tre Signori
Fóppolo
Quarna-sotto
Stresa
Mombello
Ghirla
Morcote
1701
M. Generoso
Argegno
Lenno
Bellágio
S. Giovanni
2554
Carona
Varallo
Brisino
Leggiuno
Brinzio
Porto Ceresio
Limonta
Lierna
Grigna Sett.
2410
Mezzoldo
Piazzatore
Pettenasco
Belgirate
Arolo
Campo d. Fiori
1226
Brienno
Nesso
Vassena
Mandello d. Lario
Bárzio
Z. di Campelli
2170
Orta
Armeno
Lesa
Viggiù
Mendrisio
Magréglio
Móggio
Piazza Brembana
Solcio
Méina
Ispra
Gavirate
S. Maria d. Monte
Stabio
Láglio
Carate
Onno
Abbadia Lariana
723
C. di Balisio
Ballabio-Inf.
Roncobello
Dagnente
Ranco
Varese
Moltrásio
Cernobbio
Riva di Faggeto
Faggeto Lario
Piani Resinelli
Vedeseta
Val. Brembana
2512
Pzo. Arera
Gozzano
L. di Monate
L. di Varese
Chiasso
Torno
Blévio
Asso
Canzo
Arona
Angera
Malnate
Brunate
Malgrate
Lecco
S. Bartolomeo
Borgosésia
Lisanza
L. di Comábbio
Como
Erba
S. Giovanni Bianco
Dormelletto
Sesto Calende
L. di Pusiano
L. di Annone
L. di Garlate
Borgomanero
Vergiate
Olona
TOLL
MAGGIORE - COMO
0 5 10 15 20 Miles
0 5 10 15 20 25 30 Kilometres

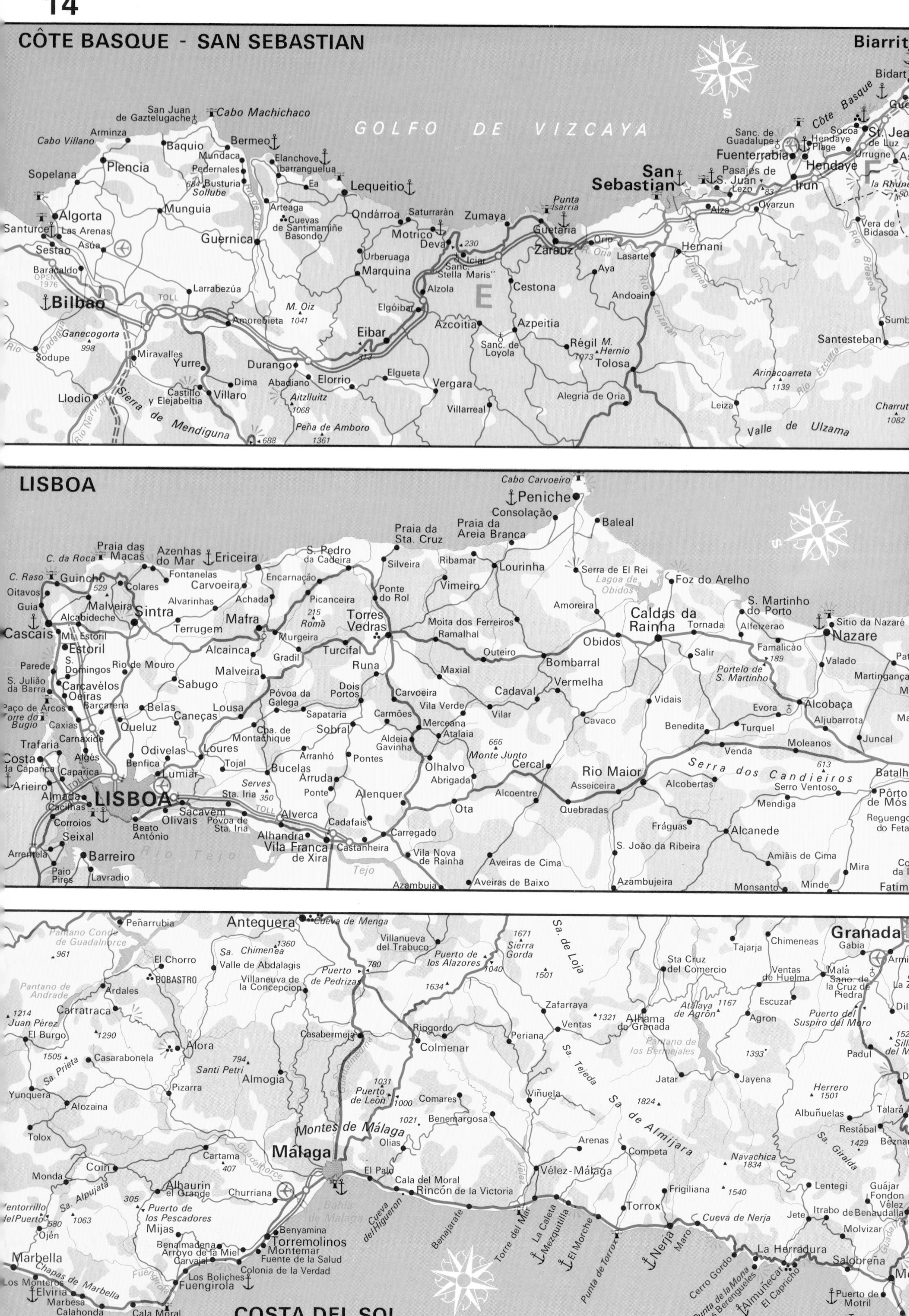

14

CÔTE BASQUE - SAN SEBASTIAN

Biarrit

GOLFE DE VIZCAYA

Cabo Villano
San Juan de Gaztelugache
Arminza
Cabo Machichaco
Baquio
Bermeo
Mundaca
Pedernales
Busturia
Ibarranguelua
Elanchove
Sollube
Ea
Sopelana
Plencia
Munguia
Arteaga
Lequeitio
Cuevas de Santimamiñe Basondo
Ondárroa
Saturrarán
Zumaya
Guetaria
Punta Isarria
Socoa
St. Jea
Côte Basque
Hendaye Plage
la Luz
Urrugne
Bidart
Gué

Algorta
Las Arenas
Asúa
Santurce
Guernica
Motrico
Deva
Zarauz
Orio
Lasarte
Hernani
San Sebastian
Sanc. de Guadalupe
Fuenterrabia
Irun
Hendaye
la Rhune

Sestao
Baracaldo OPEN 1976
Larrabezúa
Amorebieta
M. Oiz 1041
Urberuaga
Marquina
Iciar Sanc. "Stella Maris"
Cestona
Aya
Andoain
Oyarzun
Alza
Pasajes de S. Juan Lezo
Vera de Bidasoa
Santesteban
Sumbi

Bilbao
Ganecogorta 998
Miravalles
Yurre
Durango
Elorrio
Eibar
Elgóibar
Azcoitia
Azpeitia
Sanc. de Loyola
Régil
M. Hernio 1073
Tolosa
Arinacarreta 1139
Charrut 1082

Sodupe
Llodio
Castillo y Elejabeltia
Dima
Abadiano
Aitzlluitz 1068
Villaro
Vergara
Elgueta
Villarreal
Alegria de Oria
Leiza
Valle de Ulzama

Sierra de Mendiguna
Peña de Amboro 1361
688

LISBOA

Cabo Carvoeiro
Peniche
Consolação
Baleal
Praia da Sta. Cruz
Praia da Areia Branca
Ribamar
Lourinha
Serra de El Rei
Foz do Arelho
S. Martinho do Porto
Sitio da Nazaré
Nazare

C. da Roca
Praia das Maçãs
Azenhas do Mar
Ericeira
S. Pedro da Cadeira
Silveira
Vimeiro
Amoreira
Lagoa de Obidos
Alfeizerao
Famalicão 189
Valado
Martingança
M

C. Raso
Guincho
Colares
Fontanelas
Carvoeira
Achada
Picanceira
Ponte do Rol
Moita dos Ferreiros
Ramalhal
Caldas da Rainha
Tornada
Alcobaça

Oitavos
Guia
Malveira
Sintra
Alvarinhas
Encarnação
Romã 215
Torres Vedras
Outeiro
Obidos
Salir
Portelo de S. Martinho
Aljubarrota
Evora
Juncal

Cascais
Alcabideche
Mt. Estoril
Terrugem
Alcainca
Gradil
Turcifal
Runa
Maxial
Bombarral
Vermelha
Vidais
Benedita
Turquel
Moleanos
Venda

Estoril
S. Domingos
Rio de Mouro
Malveira
Sabugo
Póvoa da Galega
Dois Portos
Carvoeira
Cadaval
Cavaco
Serra dos Candieiros
613
Batalh

Parede
Carcavelos
Oeiras
Belas
Caneças
Lousa
Sapataria
Sobral
Carmões
Mercena
Vila Verde
Vilar
Rio Maior
Alcobertas
Mendiga
Pôrto de Mós

Paço de Arcos
Torre do Bugio
Caxias
Queluz
Odivelas
Loures
Tojal
Cba. de Montachique
Arranhó
Pontes
Aldeia Gavinha
Atalaia
Monte Junto 666
Cercal
Asseiceira
Serro Ventoso
Reguengo do Fetal

Trafaria
Algés
Carnaxide
Benfica
Lumiar
Bucelas
Árruda
Olhalvo
Abrigada
Alcoentre
Quebradas
Fráguas
Alcanede
Amiãis de Cima

Costa da Caparica
Caparica
Almada
Cacilhas
LISBOA
Serves
Sta. Iria 350
Olivais
Sacavém
Póvoa de Sta. Iria
Alverca
Ponte
Alenquer
Ota
S. João da Ribeira
Mira
Co da I

Arieiro
Corroios
Seixal
Beato António
Alhandra
Cadafais
Castanheira
Vila Nova de Rainha
Aveiras de Cima
Azámbujeira
Monsanto
Minde
Fatim

Arrentela
Paio Pires
Lavradio
Barreiro
Vila Franca de Xira
Rio Tejo
Tejo
Azambuja
Aveiras de Baixo

Costa del Sol · GRANADA

Peñarrubia
Antequera
Cueva de Menga
Pantano Conde de Guadalcorce 961
Sa. Chimenea 1360
Villanueva del Trabuco
1671
Sierra Gorda
Sa. de Loja
Tajarja
Chimeneas
Granada
Gabia

El Chorro
BOBASTRO
Valle de Abdalagis
Villanueva de la Concepcion
Puerto de Pedrizas 780
Puerto de los Alazores 1040
1501
Sta Cruz del Comercio
Ventas de Huelma
Malá
Sano. de la Cruz de Piedra
La Z

Pantano de Andrade 1214
Ardales
Casabermeja
1634
Zafarraya
Ventas
Alhama de Granada
Atalaya de Agrón 1167
Escuzar
Puerto del Suspiro del Moro
Dila

Juan Pérez
Carratraca
El Burgo 1290
Alora 794
Riogordo
Periana
1321
Agron
1393
Padul
152 Sill del M

1505
Casarabonela
Santi Petri
Almogia
Colmenar
Sa. Tejeda
Jatar
Jayena
Herrero 1501
Albuñuelas
Talara

Yunquera
Alozaina
Pizarra
1031
Puerto de León 1000
Comares
Viñuela
1824
Sa. de Almijara
Restábal
1429
Béznar

Tolox
1021
Benamargosa
Arenas
1393
Navachica 1834
Lentegi
Guájar Fondon
Vélez Molvizar

Monda
Coin
Cartama 407
Montes de Málaga
Olias
Competa
Sa. de Giralda
Itrabo
Benaudalla

Marbella
Alhaurin el Grande
Churriana
Málaga
El Palo
Cala del Moral
Vélez-Málaga
Frigiliana 1540
Lentegi
M

Chapas de Marbella
Benyamina
Benalmadena
Arroyo de la Miel
Torremolinos
Montemar
Fuente de la Salud
Rincón de la Victoria
Torrox
Cueva de Nerja
La Herradura
Salobreña

Los Monteros
Elvira
Marbesa
Calahonda
Cala Moral
Los Boliches
Fuengirola
Colonia de la Verdad
Bahia de Málaga
Cueva del Higueron
Benajarafe
Torre del Mar
La Caleta
Mezquitilla
El Morche
Nerja
Maro
Cerro Gordo
Punta de la Mona
Los Berengueles
Almuñecar
El Capricho
Jete
Puerto de Motril
Torrenueva
M

Ventorrillo del Puerto 580
Sa. Alpujata 305
1063
Puerto de los Pescadores
Mijas
Ojén
Sa. Prieta
Punta de Torrox

COSTA DEL SOL

1:625 000

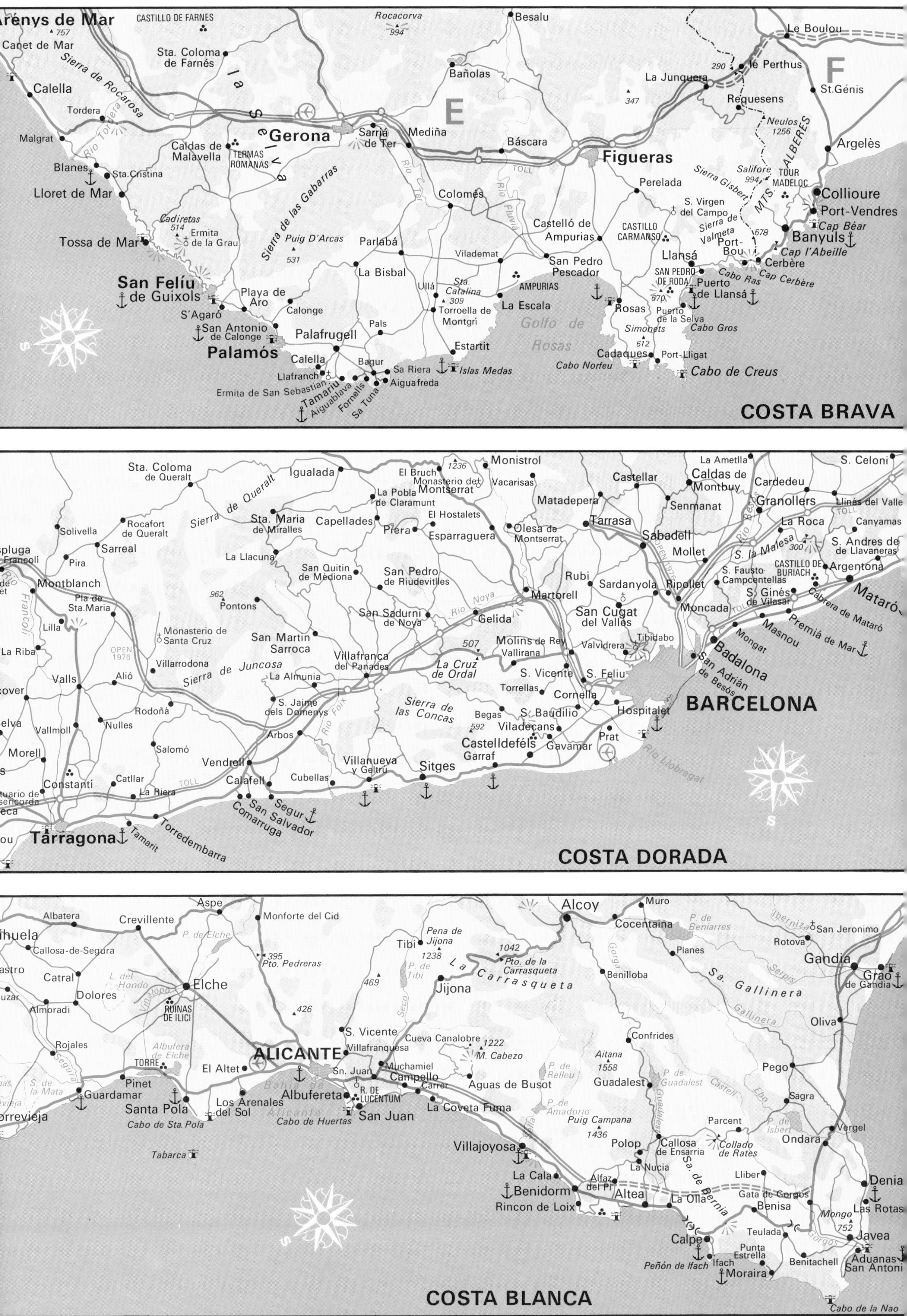

15

COSTA BRAVA

Arenys de Mar
Canet de Mar
Calella
Tordera
Malgrat
Blanes
Sta. Cristina
Lloret de Mar
Tossa de Mar
San Felíu de Guixols
S'Agaró
San Antonio de Calonge
Palamós
Calella
Llafranch
Ermita de San Sebastian
Tamariu
Aiguablava
Fornells
Sa Tuna
Aiguafreda
Sa Riera
Bagur

CASTILLO DE FARNES
Sta. Coloma de Farnés
Sierra de Rocacorsa
Caldas de Malavella
TERMAS ROMANAS
Gerona
Sarriá de Ter
Mediña
la Selva
Cadiretas 514
Ermita de la Grau
Sierra de las Gabarras
Puig D'Arcas 531
Calonge
Playa de Aro
Palafrugell
Pals
Bagur
La Bisbal
Parlabá
Ullá
Torroella de Montgri
Sta. Catalina 309
Estartit

Rocacorva 994
Besalu
Bañolas
Colomés
Vilademat
Castelló de Ampurias
San Pedro Pescador
AMPURIAS
La Escala
Islas Medas
Golfo de Rosas
Cabo Norfeu

Le Boulou
le Perthus 290
St.Génis
La Junquera
Requesens
Neulos 1256
ALBERES
Argelès
Figueras
347
Perelada
Salifore 994
TOUR MADELOC
Sierra Gisberti
S. Virgen del Campo
Sierra de Valmeta
Port-Bou
678
CASTILLO CARMANSO
SAN PEDRO DE RODA
Llansá
Puerto de Llansá
Puerto de la Selva
Rosas
Simonets 612
Cadaques
Port-Lligat
Cabo Gros
Cabo de Creus

Collioure
Port-Vendres
Cap Béar
Banyuls
Cap l'Abeille
Cerbère
Cap Cerbère
Cabo Ras

COSTA DORADA

Sta. Coloma de Queralt
spluga de Francolí
Solivella
Pira
Rocafort de Queralt
Sarreal
Montblanch
Pla de Sta.Maria
La Riba
Lilla
Valls
cover
Selva
Vallmoll
Morell
Nulles
Constantí
tuario de sericorda
seca
Tarragona
Tamarit
Torredembarra

Sierra de Queralt
La Llacuna
San Quitin de Mediona
Monasterio de Santa Cruz
Villarrodona
Alió
Rodoñá
Salomó
Catllar
La Riera
OPEN 1976
962
Pontons
San Martin Sarroca
La Almunia
S. Jaime dels Domenys
Arbos
Vendrell
Calafell
San Salvador
Comarruga
Segur
La Riera

Igualada
El Bruch
Monasterio de Montserrat
La Pobla de Claramunt
Sta. Maria de Miralles
Capellades
Piera
Esparraguera
El Hostalets
San Pedro de Riudevitlles
San Sadurni de Noya
Villafranca del Panades
Sierra de Juncosa
La Cruz de Ordal
Sierra de las Concas 592
Villanueva y Geltrú
Cubellas
Sitges
Garraf

1236
Monistrol
Vacarisas
Olesa de Montserrat
Matadepera
Tarrasa
Sabadell
Mollet
Rubi
Sardanyola
Ripollet
Moncada
Martorell
Rio Noya
Gelida
Molins de Rey
Vallirana
Valvidrera
Tibidabo
S. Vicente
S. Feliu
Torrellas
Cornella
Begas
S. Baudilio
Viladecans
Castelldefels
Gavamar
Prat
Hospitalet
507
Rio Llobregat

La Ametlla
Caldas de Montbuy
Castellar
Senmanat
Cardedeu
Llinás del Valle
Granollers
La Roca
Canyamas
S. Andres de de Llavaneras
S. Fausto Campcentellas
S. la Malesa 300
CASTILLO DE BURIACH
Argentona
Mataró
S. Ginés de Vilasar
Cabrera de Mataró
Masnou
Premiá de Mar
Mongat
Badalona
San Adrián de Besós
BARCELONA

COSTA BLANCA

ihuela
Albatera
Crevillente
Aspe
Callosa-de-Segura
Catral
astro
juzar
Dolores
Almoradi
Rojales
nas
S. de la Mata
rrevieja
Guardamar
Pinet
Santa Pola
TORRE
El Altet
Los Arenales del Sol
Cabo de Sta. Pola
Tabarca

Monforte del Cid
P. de Elche
L. del Hondo
Elche
RUINAS DE ILICI
395
Pto. Pedreras
426
S. Vicente
Villafranquesa
ALICANTE
R. DE LUCENTUM
Sn. Juan
Albufereta
San Juan
Cabo de Huertas
Albufera de Elche
Bahía de Alicante

Tibi
Pena de Jijona 1238
P. de Tibi
469
Jijona
Cueva Canalobre
Muchamiel
Campello
Carrer
La Coveta Fuma
Seco
1042
Pto. de la Carrasqueta
La Carrasqueta
1222
M. Cabezo
Aguas de Busot
Puig Campana 1436
Villajoyosa
La Cala
Benidorm
Rincon de Loix
Alfaz del Pi
Altea
La Olla

Alcoy
Muro
Cocentaina
Planes
Benilloba
Confrides
Aitana 1558
P. de Relleu
Guadalest
P. de Amadorio
Polop
Callosa de Ensarria
La Nucia
Iberniza
P. de Beniarres
San Jeronimo
Rotova
Gandia
Grao de Gandia
Sa. Gallinera
Gorga
Serpis
Ebro
Oliva
Pego
Sagra
Parcent
Collado de Rates
Lliber
Gata de Gorgos
La Olla
P. de Isbert
Benisa
Teulada
Ondara
Vergel
Denia
Las Rotas
Mongo 752
Javea
Aduanas
San Antoni
Calpe
Peñón de Ifach
Ifach
Moraira
Punta Estrella
Benitachell
Cabo de la Nao
Sa. de Bernia
Guadalest
Castell

0   5   10   15   20 Miles
0 5 10 15 20 25 30 Kilometres

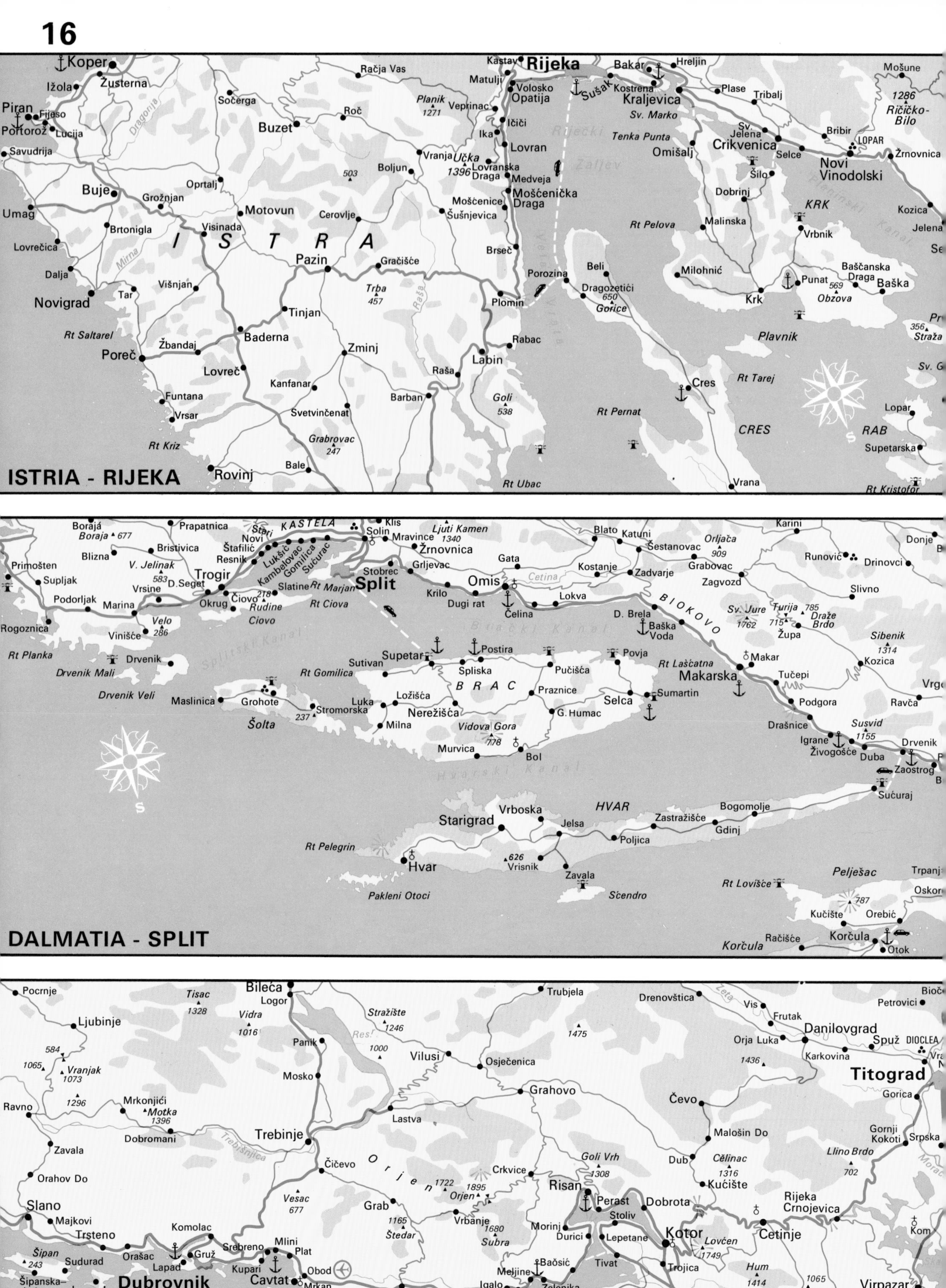

16

ISTRIA - RIJEKA

Koper
Izola
Žusterna
Piran
Fijeso
Portorož
Lucija
Savudrija
Buje
Grožnjan
Umag
Brtonigla
Lovrečica
Dalja
Novigrad
Tar
Višnjan
Rt Saltarel
Žbandaj
Poreč
Lovreč
Funtana
Vrsar
Rt Kriz
Rovinj
Bale

Sočerga
Roč
Buzet
Planik
1271
Motovun
Visinada
Cerovlje
Pazin
Tinjan
Zminj
Baderna
Kanfanar
Svetvinčenat
Barban
Grabrovac
247

Račja Vas
Matulji
Veprinac
Vranja
Ika
Lovranska
Draga
Medveja
Moščenice
Moščenička
Draga
Šušnjevica
Brseč
Gračišće
Trba
457
Raša
Rabac
Labin
Raša
Goli
538

Kastav
Rijeka
Volosko
Opatija
Ičiči
Lovran
Učka
1396
Porozina
Beli
Dragozetiči
650
Gorice
Plomin
Rt Ubac
Rt Pernat
Rt Tarej
Cres
CRES
Rt Kristofor

Bakar
Hreljin
Sušak
Kostrena
Kraljevica
Sv. Marko
Sv.
Jelena
Tenka Punta
Omišalj
Krk
Milohnić
Malinska
Dobrinj
Šilo
Punat
569
Obzova
356
Straža
Vran
Plase
Tribalj
Bribir
Crikvenica
Selce
Plavnik
Sv. G
Mošune
1286
Ričičko-
Bilo
LOPAR
Žrnovnica
Novi
Vinodolski
KRK
Kozica
Jelena
Baščanska
Draga
Baška
RAB
Lopar
Supetarska

DALMATIA - SPLIT

Borajá
Boraja 677
Prapatnica
Bristivica
Blizna
V. Jelinak
583
Primošten
Supljak
Podorljak
Marina
Vrsine
Rogoznica
Vinišće
Rt Planka
Drvenik Mali
Drvenik Veli
Maslinica
Grohote
Šolta
237
Stromorska
Luka
Milna
Murvica

Stari
Novi
Štafilić
Resnik
Lukšić
Kambelovac
Gomilica
Sućurac
Trogir
D.Seget
Okrug
Ćiovo
278
Rudine
Ćiovo
Rt Ciova
Slatine
Rt Marjan
Velo
286
Rt Gomilica
Sutivan
Supetar
Spliska
Postira
BRAC
Ložišća
Nerežišća
Vidova Gora
778
Bol

KASTELA
Solin
Klis
Mravince
Žrnovnica
Stobrec
Split
Gata
Grljevac
Krilo
Dugi rat
Čelina
Bračcki Kanal
Pučišća
Praznice
G. Humac
Selca
Sumartin
Hvarski Kanal

Ljuti Kamen
1340
Omiš
Cetina
Lokva
Kostanje
Zadvarje
D. Brela
Baška
Voda
Povja
Rt Laščatna
Makarska
Tučepi
Podgora
Drašnice
Igrane
Živogošće
Duba
Zaostrog
Sućuraj

Blato
Katuni
Šestanovac
Orljača
909
Zagvozd
BIOKOVO
Sv. Jure
1762
Furija
715
785
Draže
Brdo
Makar
Župa
Sibenik
1314
Kozica
Ravča
Susvid
1155
Drvenik

Karini
Runović
Donje
Drinovci
Slivno
Vrg
Pelješac
Trpanj

Starigrad
Vrboska
Jelsa
HVAR
Zastražišće
Bogomolje
Gdinj
Rt Pelegrin
Hvar
Vrisnik
626
Zavala
Scendro
Rt Lovišće
787
Kučište
Orebić
Pakleni Otoci
Korčula
Račišće
Korčula
Otok

DALMATIA - DUBROVNIK

Pocrnje
Ljubinje
Tisac
1328
Bileća
Logor
Vidra
1016
Stražište
1246
584
1065
Vranjak
1073
1296
Mrkonjići
Motka
1396
Ravno
Dobromani
Zavala
Orahov Do
Slano
Majkovi
Trsteno
Komolac
Šipan
243
Sudurad
Orašac
Šipanska-
Luka
Lopud
Sv. Andrija
Dubrovnik
Lapad
Gruž
Srebreno
Mlini
Plat
Kupari
Cavtat
Mrkan
Obod
Cilipi

Panik
Mosko
Trebinje
Čičevo
Vilusi
Lastva
Vesac
677
Grab
1165
Štedar
Kupari
Gruda

Trubjela
Osječenica
Grahovo
Orjen
1722
1895
Orjen
1680
Subra
Crkvice
Vrbanje
Durici
Morinj
Baošić
Tivat
Meljine
Igalo
Zelenika
Herceg Novi
Obosnik
Oštri rt
Rt Platamuni

Drenovšćica
Vis
1475
Goli Vrh
1308
Dub
Risan
Perast
Stoliv
Dobrota
Lepetane
Kotor
Trojica
Lovćen
1749
Radovanići
Bečiči
Miločer
Sv. Stefan
Budva
Petrovac

Zeta
Frutak
Orja Luka
1436
Čevo
Malošin Do
Kučište
Célinac
1316
Hum
1414
1065
Gornji
Kokoti
Llino Brdo
702
Rijeka
Crnojevica
Cetinje
Virpazar
Velja

Bioč
Petrovici
Danilovgrad
Spuž DIOCLEA
Vr
Karkovina
Titograd
Gorica
Srpska
Morac
Kom
Sut
Golubovići
Sutomor

1:625 000

| 0 | 5 | 10 | 15 | 20 Miles |
| 0 | 5 | 10 | 15 | 20 | 25 | 30 Kilometres |

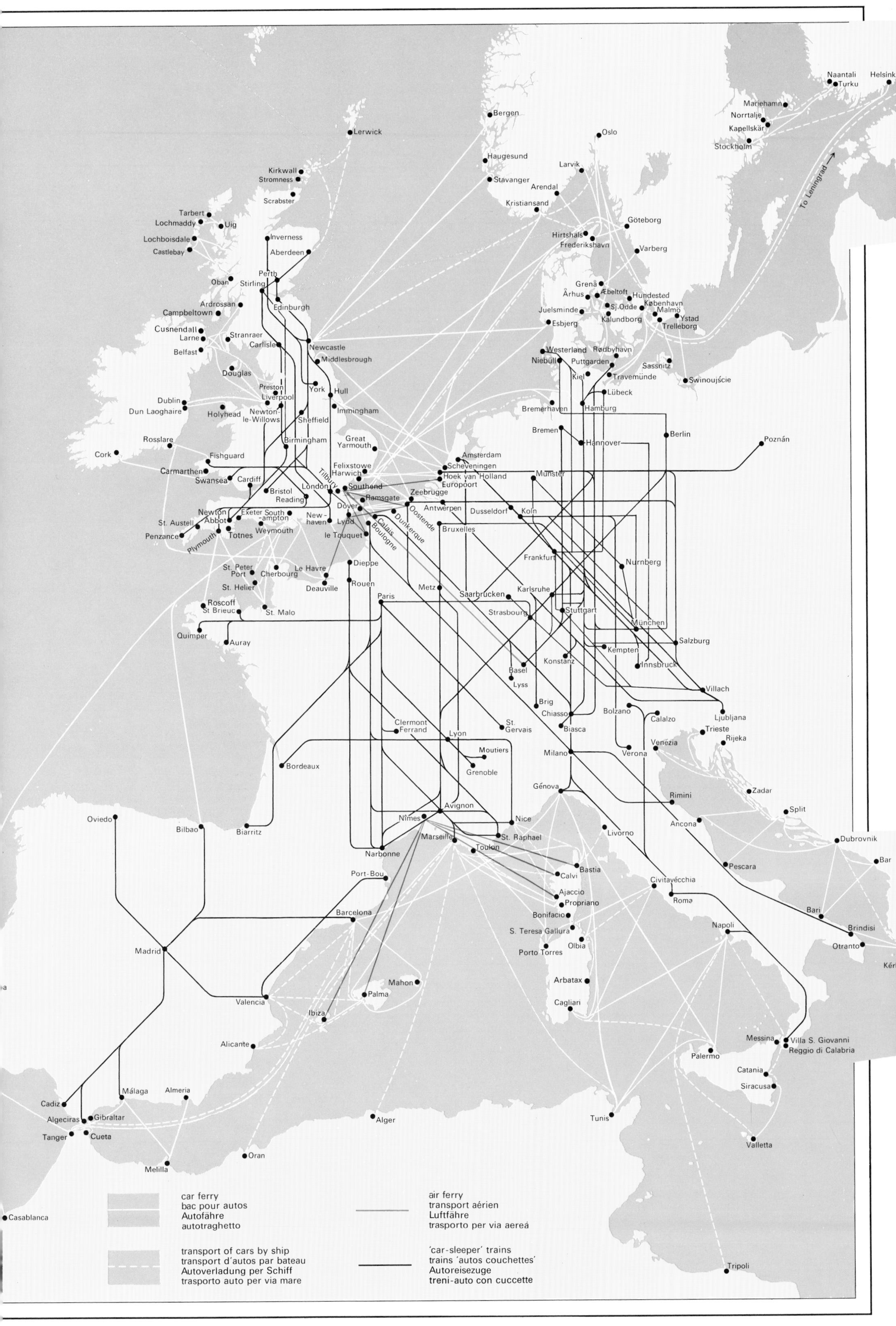

car ferry
bac pour autos
Autofähre
autotraghetto

air ferry
transport aérien
Luftfähre
trasporto per via aereá

transport of cars by ship
transport d'autos par bateau
Autoverladung per Schiff
trasporto auto per via mare

'car-sleeper' trains
trains 'autos couchettes'
Autoreisezuge
treni-auto con cuccette

1:12 500 000

| 0 | | 100 | | 200 | | 300 | | 400 | | 500 Miles |
| 0 | 100 | 200 | 300 | 400 | 500 | 600 | 700 | 800 Kilometres |

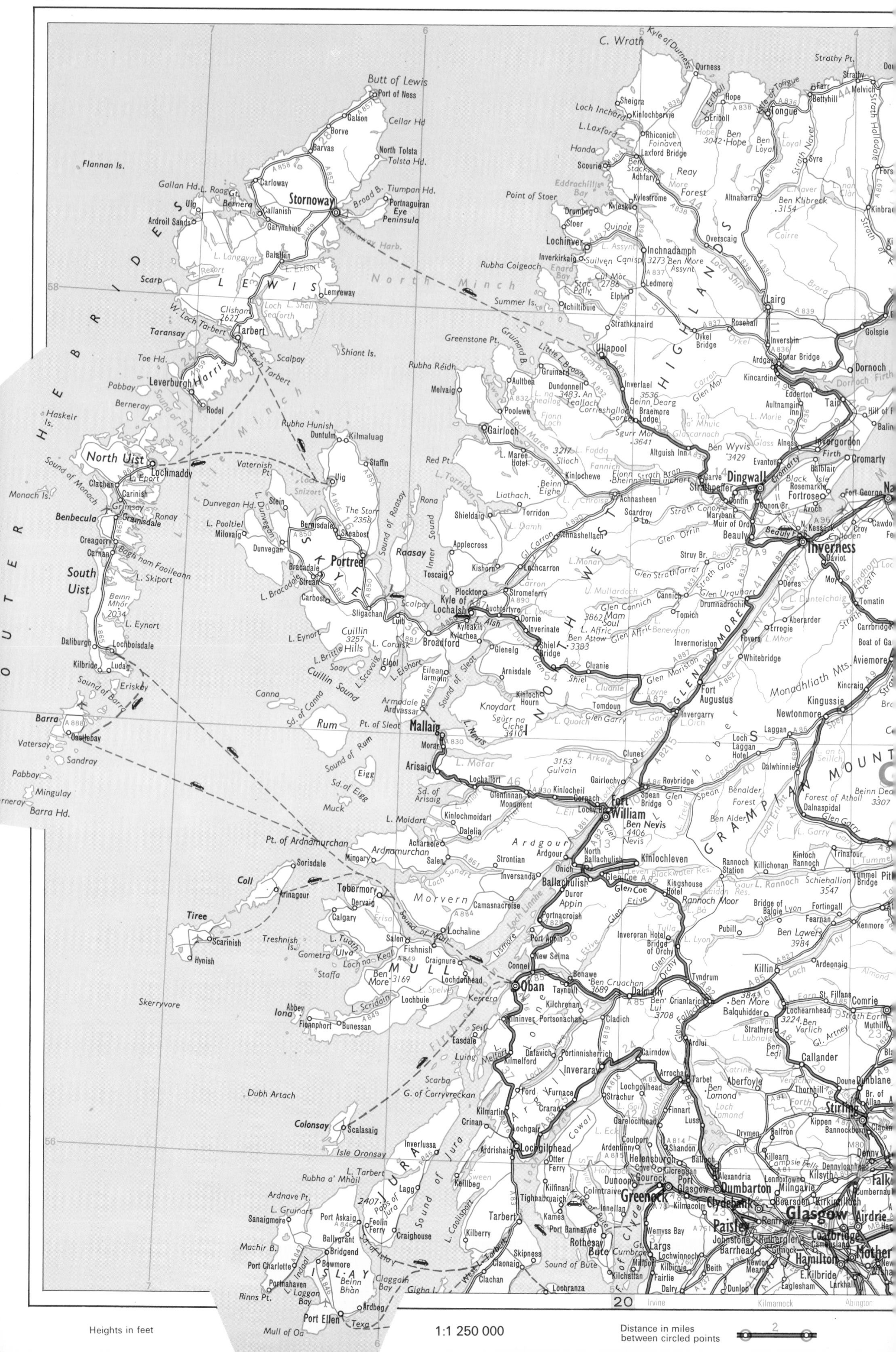

18

Heights in feet   1:1 250 000   Distance in miles
                                              between circled points

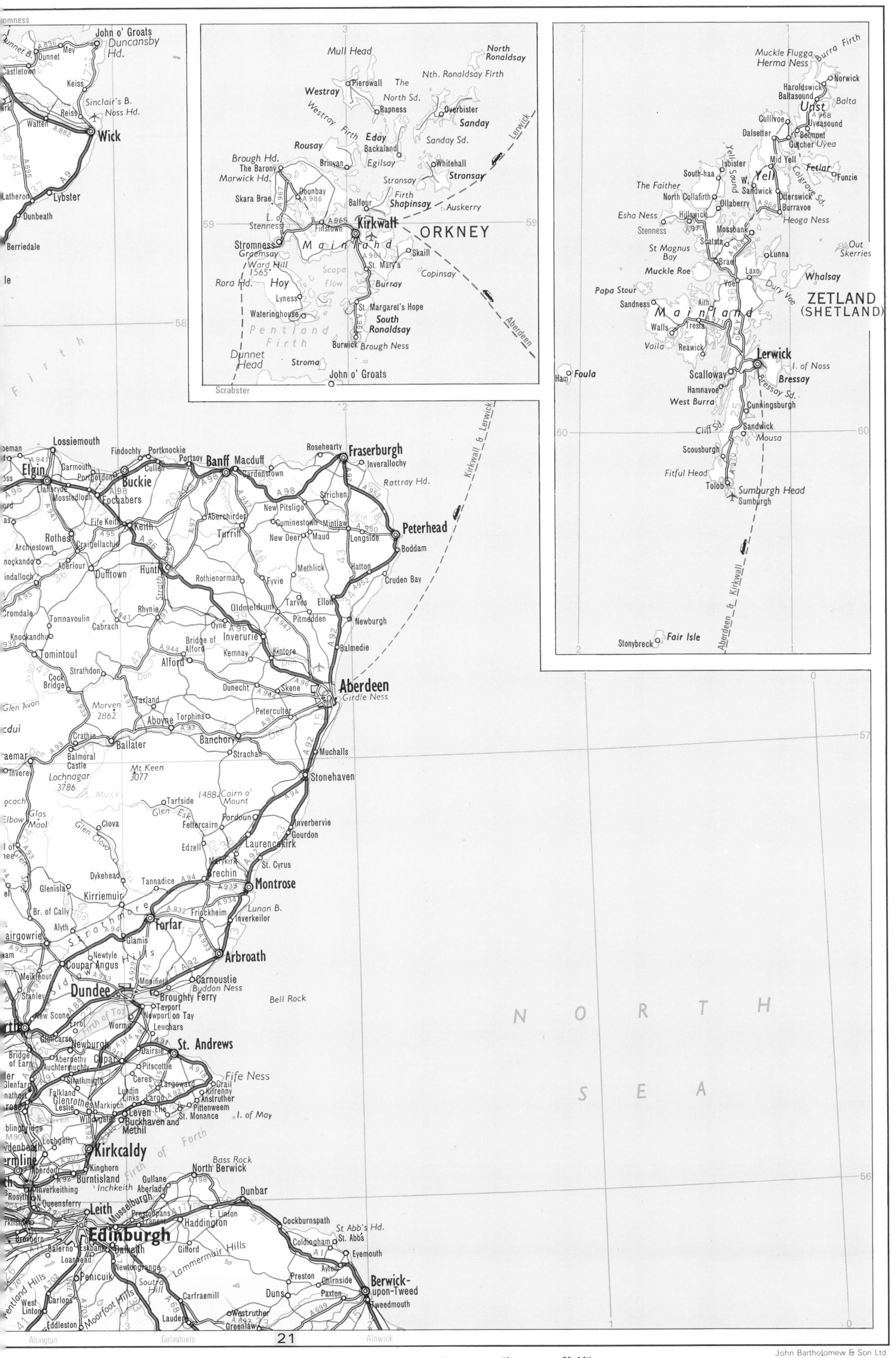

Caithness / John o' Groats area (top left inset)

Castletown
John o' Groats
Duncansby Hd.
Dunnet B.
Mey
Dunnet
A836
Keiss
Sinclair's B.
Reiss
Noss Hd.
Watten
A882
Wick
Latheron
A9
Lybster
A895
Dunbeath
Berriedale

ORKNEY (top centre inset)

Mull Head
North Ronaldsay
Westray
Pierowall
The North Sd.
Nth. Ronaldsay Firth
Rapness
Overbister
Sanday
Westray Firth
Rousay
Eday
Sanday Sd.
Brough Hd.
Brinyan
Backaland
The Barony
Egilsay
Whitehall
Stronsay
Marwick Hd.
Dounbay
A986
Stronsay
Skara Brae
A967
Balfour
Firth
Shapinsay
Auskerry
L. of Stenness
A965
Finstown
Kirkwall
ORKNEY
59
Stromness
Mainland
Skaill
Graemsay
Ward Hill 1565'
St. Mary's
Rora Hd.
Hoy
Scapa Flow
Burray
Copinsay
Lyness
Wateringhouse
St. Margaret's Hope
58
South Ronaldsay
Pentland Firth
Burwick
Brough Ness
Dunnet Head
Stroma
John o' Groats
Scrabster

ZETLAND (SHETLAND) (top right inset)

Muckle Flugga
Herma Ness
Burra Firth
Haroldswick
Norwick
Baltasound
Balta
Unst
Cullivoe
Uyeasound
Dalsetter
Belmont
Gutcher
Uyea
Yell Sound
Mid Yell
Fetlar
The Faither
South-haa
Yell
Funzie
North Collafirth
W. Sandwick
Uttsrwick
Esha Ness
Ollaberry
Burravoe
Hillswick
A966
Heoga Ness
Stenness
Scatsta
St Magnus Bay
Mossbank
Papa Stour
Brae
Lunna
Muckle Roe
Laxo
Out Skerries
Sandness
Voe
Aith
Dury Voe
Whalsay
Mainland
Walls
Trests
Vaila
Reawick
60
Lerwick
Ham
Foula
Scalloway
I. of Noss
Bressay
Hamnavoe
Cunningsburgh
West Burra
Bressay Sd.
Cliff
Sandwick
Mousa
Scousburgh
Fitful Head
Sumburgh Head
Tolob
Sumburgh
Stonybreck
Fair Isle
Aberdeen & Kirkwall
Kirkwall & Lerwick

Main map (north-east Scotland)

Lossiemouth
Findochty Portknockie
Rosehearty
Fraserburgh
Elgin
Garmouth
Cullen
Portsoy
Banff Macduff
Inverallochy
Buckie
Portgordon
Buckie
Gardenstown
A98
Rattray Hd.
Mosstodloch
Fochabers
A98
New Pitsligo
Strichen
Rothes
Fife Keith
Keith
Aberchirder
Cuminestown Mintlaw
A950
Peterhead
Craigellachie
Turriff
New Deer
Maud
Longside
Aberlour
Huntly
Rothienorman
Fyvie
Methlick
Hatton
Boddam
Dufftown
Rhynie
Oldmeldrum
Tarves
Ellon
Cruden Bay
Rothes
Oyne
Inverurie
Pitmedden
Newburgh
Cabrach
Alford
Kemnay
Kintore
Balmedie
Tomintoul
Strathdon
Cock Bridge
Dunecht
Skene
Aberdeen
Morven 2862
Tarland
Torphins
Peterculter
Girdle Ness
Crathie
Aboyne
Torphins
A93
Ballater
Banchory
Muchalls
Balmoral Castle
Strachan
Lochnagar 3786
Mt Keen 3077
Stonehaven
Tarfside
1488 Cairn o' Mount
Glas Maol
Glova
Tarfside
Inverbervie
Elbow
Glen Clova
Fettercairn
Pordoun
Gourdon
Edzell
Laurencekirk
Dykehead
Tannadice
St. Cyrus
Glenisla
Brechin
Kirriemuir
A94
Montrose
Br. of Cally
Friockheim
Inverkeilor
Alyth
Glamis
Newtyle
Forfar
Lunan B.
airgowrie
Coupar Angus
A92
Arbroath
Meikleour
Stanley
Monifieth
Carnoustie
New Scone
Dundee
Buddon Ness
Bell Rock
Broughty Ferry
Tayport
Wormit
Newport on Tay
Leuchars
Glencarse
Newburgh
St. Andrews
Bridge of Earn
Cupar
Dairsie
Glenfar
Abernethy
Strathmiglo
Ceres
Fife Ness
Falkland
Pitscottie
Largoward
Glenrothes
Leslie
Markinch
Crail
Drail
Lundin Links
Largo
Kilrenny
Anstruther
Leven
Elie
Pittenweem
Buckhaven and Methil
St. Monance
I. of May
blingbridge
Whitegate
Kirkcaldy
Kinghorn
Bass Rock
ydenbeath
Burntisland
North Berwick
rmline
Inverkeithing
Gullane
Aberlady
Inchkeith
A198
Rosyth
Queensferry
Leith
Musselburgh
Dunbar
Prestonpans
Edinburgh
Tranent
E. Linton
St Abb's Hd.
Dalkeith
Haddington
Cockburnspath
St. Abb's
Balerno
Gifford
Coldingham
Eyemouth
Loanhead
Newtongrange
Preston
Ayton
Penicuik
Lammermuir Hills
Chirnside
Paxton
Carlops
Soutra Hill
Duns
Berwick-upon-Tweed
West Linton
Carfraemill
Westruther
Tweedmouth
Eddleston
Moorfoot Hills
Lauder
Greenlaw
Abington
Galashiels
Alnwick

N O R T H   S E A

Firth of Forth

John Bartholomew & Son Ltd

0	10	20	30	40	50 Miles
0	10 20	30	40 50	60 70	80 Kilometres

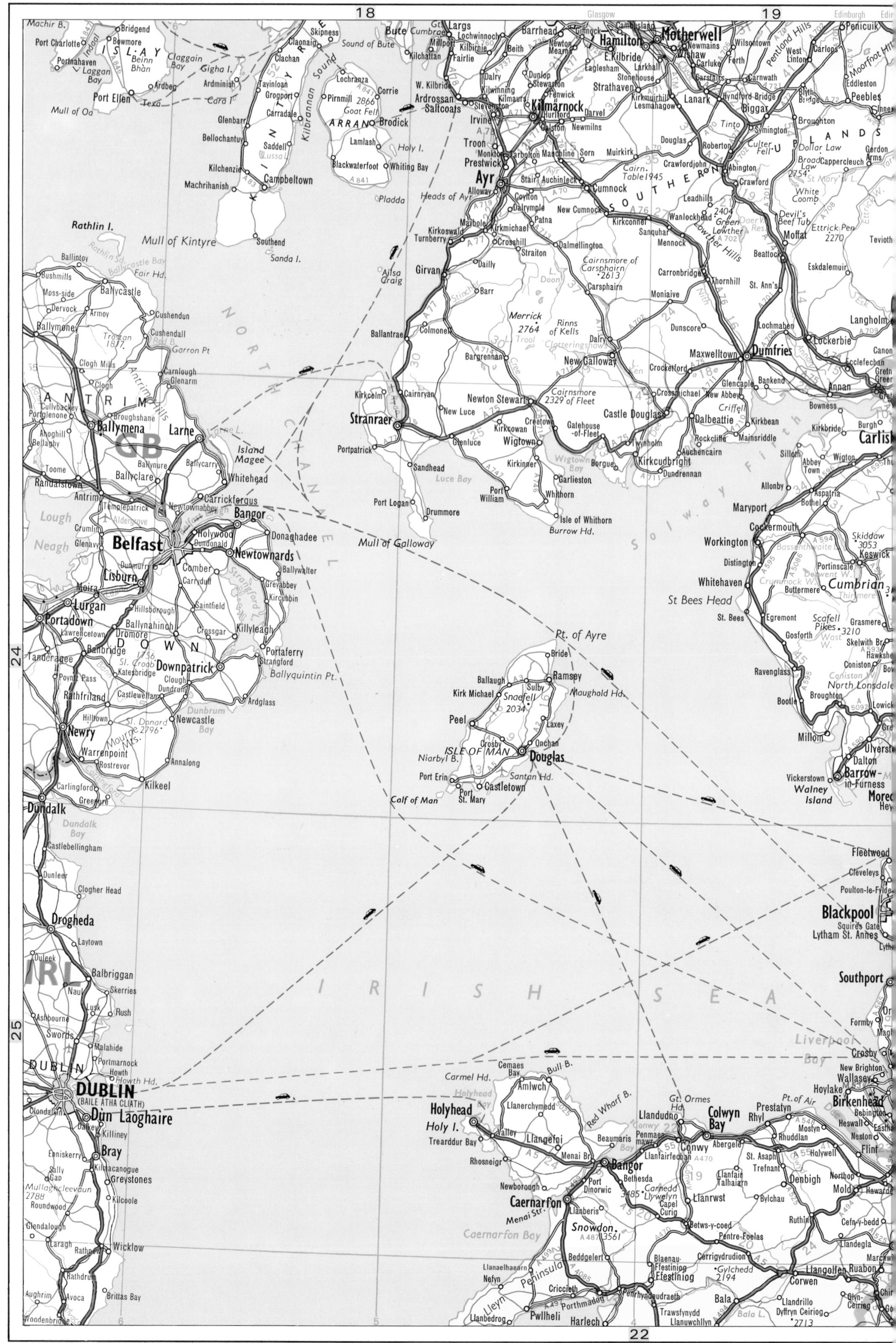

18

19

Glasgow

Edinburgh Edi

ISLAY
Machir B. · Bridgend
Port Charlotte · Bowmore
Beinn · Claggain
Portnahaven · Laggan Bay · Gigha I.
Port Ellen · Texa · Cara I.
Mull of Oa

KINTYRE
Skipness
Clachan
Tayinloan
Gropport
Carradale
Campbeltown
Machrihanish
Southend
Sanda I.

ARRAN
Pirnmill 2866
Goat Fell
Brodick
Lamlash
Holy I.
Blackwaterfoot · Whiting Bay
Pladda
Heads of Ayr

Largs
Bute Cumbrae
Millport
Kilchattan
Fairlie
W. Kilbride · Beith
Ardrossan · Dalry
Saltcoats · Kilwinning
Irvine · Stevenston
Troon · Monkton
Prestwick · Tarbolton
Ayr · Stair
Alloway · Coylton
Dalrymple
Maybole · Kirkmichael
Kirkoswald
Turnberry · Crosshill
Dailly · Straiton
Girvan
Barr
Colmonell
Ballantrae

Barrhead
Newton Mearns
Dunlop
Stewarton
Fenwick
Kilmarnock
Hurlford
Darvel
Newmilns
Sorn
Muirkirk
New Cumnock
Cumnock
Auchinleck
New Cumnock
Dalmellington
Carsphairn 2613
Carsphairn
Merrick 2764
Rinns of Kells
Loch Doon
Clatteringshaws
New Galloway
Dalry

Motherwell
Hamilton
Newmains
Wishaw
Carluke
E.Kilbride · Larkhall
Eaglesham · Stonehouse
Strathaven · Kirkmuirhill
Lesmahagow
Douglas
Roberton
Crawfordjohn
Crawford

SOUTHERN
Cairn Table 1945
LANDS
Tinto
Symington
Lanark
Carstairs
Carnwath
Biggar
Abington
Leadhills
Wanlockhead
Sanquhar
Mennock
Thornhill
Moniaive
Dunscore
Carronbridge
St. Ann's

Peebles
Broughton
Dollar Law
Culter Fell U
Green Lowther 2403
Lowther Hills
Beattock
Moffat
Ettrick Pen 2270
Eskdalemuir
Langholm
Lockerbie
Lochmaben
Dumfries
Maxwelltown
Crocketford
Glencaple
New Abbey
Bankend
Annan
Bowness
Ecclefechan
Canonbie
Gretna
Green

Cairnsmore of Fleet 2329
Creetown
Newton Stewart
New Luce
Glenluce
Wigtown
Kirkcowan
Bargrennan
Gatehouse-of-Fleet
Castle Douglas
Twynholm
Dalbeattie · Kirkbean
Kirkcudbright
Dundrennan
Rockcliffe · Mainsriddle
Auchencairn

Rathlin I.

North Castle Bay
Fair Hd.
Ballintoy
Bushmills
Moss-side
Dervock
Ballycastle
Armoy
Cushendun
Ballymoney
Trostan 1817
Cushendall
Clogh Mills
Garron Pt.
Carnlough
Glenarm

Kirkcolm
Cairnryan
Stranraer
Portpatrick
Sandhead
Port Logan
Drummore

Kirkinner
Garlieston
Whithorn
Isle of Whithorn
Burrow Hd.

Maxwelltown
Dumfries

Maryport
Cockermouth
Aspatria
Silloth
Abbey Town
Allonby
Wigton
Kirkbride
Burgh
Carlisle

CUMBRIAN
Workington
Distington
Whitehaven
St Bees Head
St. Bees
Egremont
Gosforth
Skiddaw 3053
Keswick
Crummock W.
Buttermere
Scafell Pikes 3210
Wast W.
Grasmere
Coniston

NORTH CHANNEL

ANTRIM
Cullybackey
Portglenone
Ahoghill
Ballymena
Ballymoney
Toome
Randalstown
Antrim
Templepatrick
Aldergrove
Crumlin
Glenavy
Lough Neagh
Moira
Lurgan
Lisburn
Portadown
Lawrencetown
Banbridge
Tandragee
Poyntz Pass

Larne
Island Magee
Ballycarry
Whitehead
Carrickfergus
Newtownabbey
Bangor
Holywood
Belfast
Dundonald · Donaghadee
Comber · Newtownards
Carryduff
Ballywalter
Saintfield
Greyabbey
Killyleagh
Kircubbin
Crossgar
Portaferry
Strangford
Ballynahinch
Uromore
Ballyquintin Pt.
Hillsborough
Hillsborough
St. Croob 1756
Clough
Dundrum
Castlewellan
Downpatrick
Ardglass
Rathfriland
Dunbrum Bay

Pt. of Ayre
Bride
Ramsey
Ballaugh
Sulby
Maughold Hd.
Kirk Michael
Snaefell 2034
Laxey
Peel
Crosby · Onchan
ISLE OF MAN
Niarbyl B.
Douglas
Port Erin
Santan Hd.
Castletown
Calf of Man · St. Mary
Port St. Mary

Vickerstown
Walney Island
Barrow-in-Furness
Ulverston
Dalton
Morecambe

IRISH SEA

Dundalk
Dundalk Bay
Castlebellingham
Dunleer
Clogher Head
Drogheda
Laytown

IRL
Oldleek
Balbriggan
Naul
Skerries
Lusk
Rush
Ashbourne
Swords

DUBLIN
Malahide
Portmarnock
Howth
Howth Hd.

DUBLIN
(BAILE ATHA CLIATH)
Clondalkin
Dún Laoghaire
Dalkey
Killiney
Enniskerry
Bray
Kilmacanogue
Greystones
Sally Gap
Mullaghcleevaun 2788
Kilcoole
Roundwood
Glendalough
Caragh
Rathnew
Wicklow
Rathdrum
Aughrim
Avoca
Brittas Bay
Woodenbridge

Fleetwood
Cleveleys
Poulton-le-Fylde
Blackpool
Squires Gate
Lytham St.Annes

Southport
Formby
Liverpool Bay
Crosby
New Brighton
Wallasey
Hoylake
Pt.of Air
Birkenhead
Bebington
Heswall
Neston

Cemaes Bay
Carmel Hd.
Bull B.
Amlwch
Holyhead
Holy I.
Valley
Treaddur Bay
Llangefni
Rhosneigr
Llanerchymedd
Gt. Ormes
Red Wharf B.
Llandudno
Colwyn Bay
Prestatyn
Rhyl
Beaumaris
Penmaenmawr
Abergele
Rhuddlan
St. Asaph
Holywell
Flint
Menai Br.
Bangor
Llanfairfechan
Conwy
Trefnant
Denbigh
Mold
Hawarde
Caernarfon
Menai Str.
Bethesda
Carnedd Llywelyn 3485
Llanfair Talhaiarn
Llanrwst
Bylchau
Ruthin
Cefn-y-bedd
Newborough
Port Dinorwic
Capel Curig
Betws-y-coed
Pentre-Foelas
Llandegla
Snowdon 3561
Llanberis
Blaenau Ffestiniog
Cerrigydrudion
Corwen
Llangollen
Ruabon
Caernarfon Bay
Beddgelert
Ffestiniog
Gylchedd 2194
Bala
Llandrillo
LLEYN Peninsula
Llanaelhaearn
Criccieth
Trawsfynydd
Bala L.
Dyffryn Ceiriog
Nefyn
Porthmadog
Penrhyndeudraeth
Bala L. 2713
Pwllheli
Llanbedrog
Harlech
Llanuwchllyn

24

25

NORTH CHANNEL

Newry
Warrenpoint
Rostrevor
Mourne Mts. 2796
Newcastle
Carlingford
Greenore
Annalong
Kilkeel

Distances in miles
between circled points   11

1:1 250 000

Heights in feet

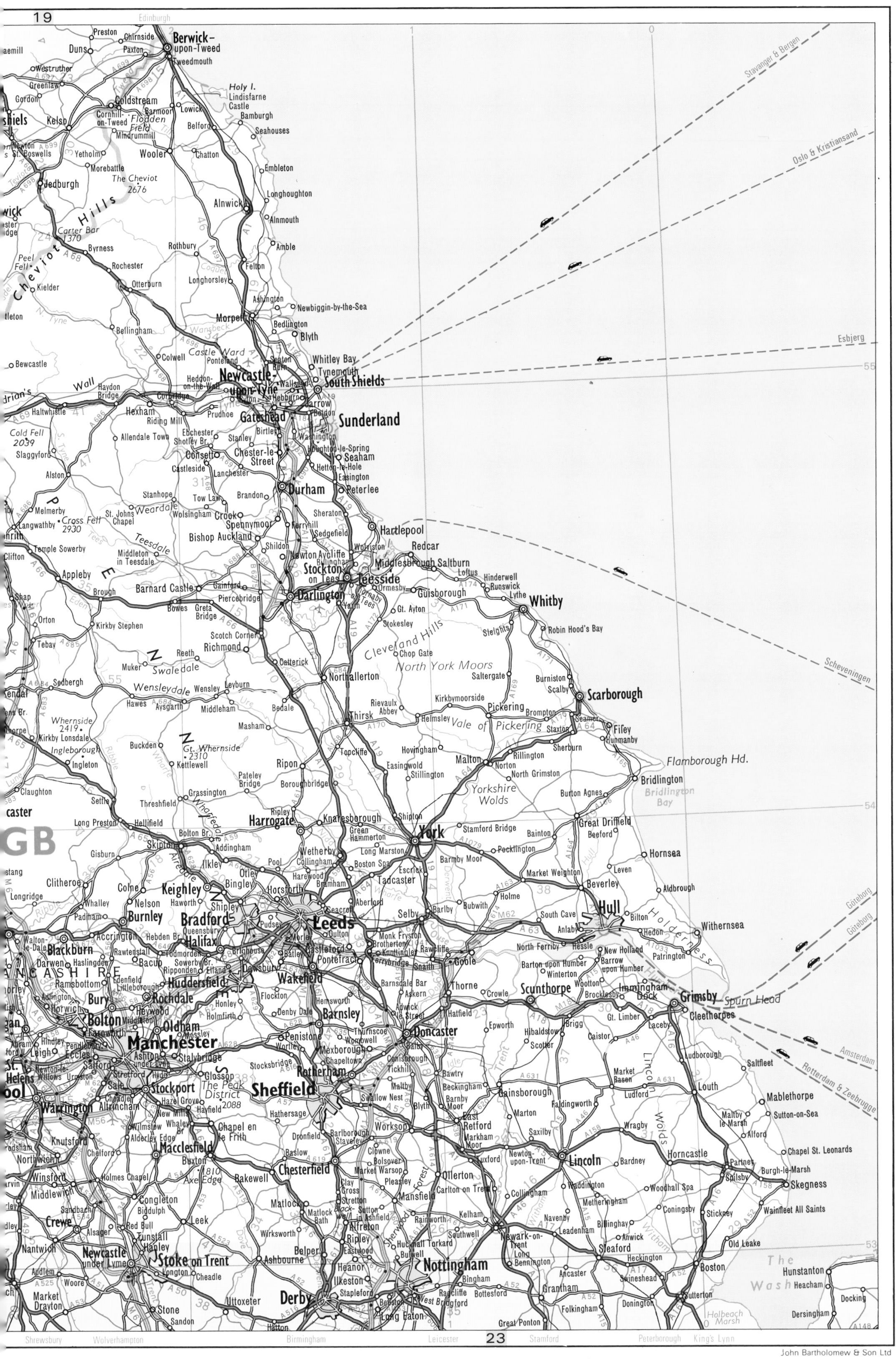

19

Berwick-upon-Tweed
Tweedmouth

GB

NCASHIRE

John Bartholomew & Son Ltd

0	10	20	30	40	50	Miles
0	10	20	30	40	50	60	70	80	Kilometres

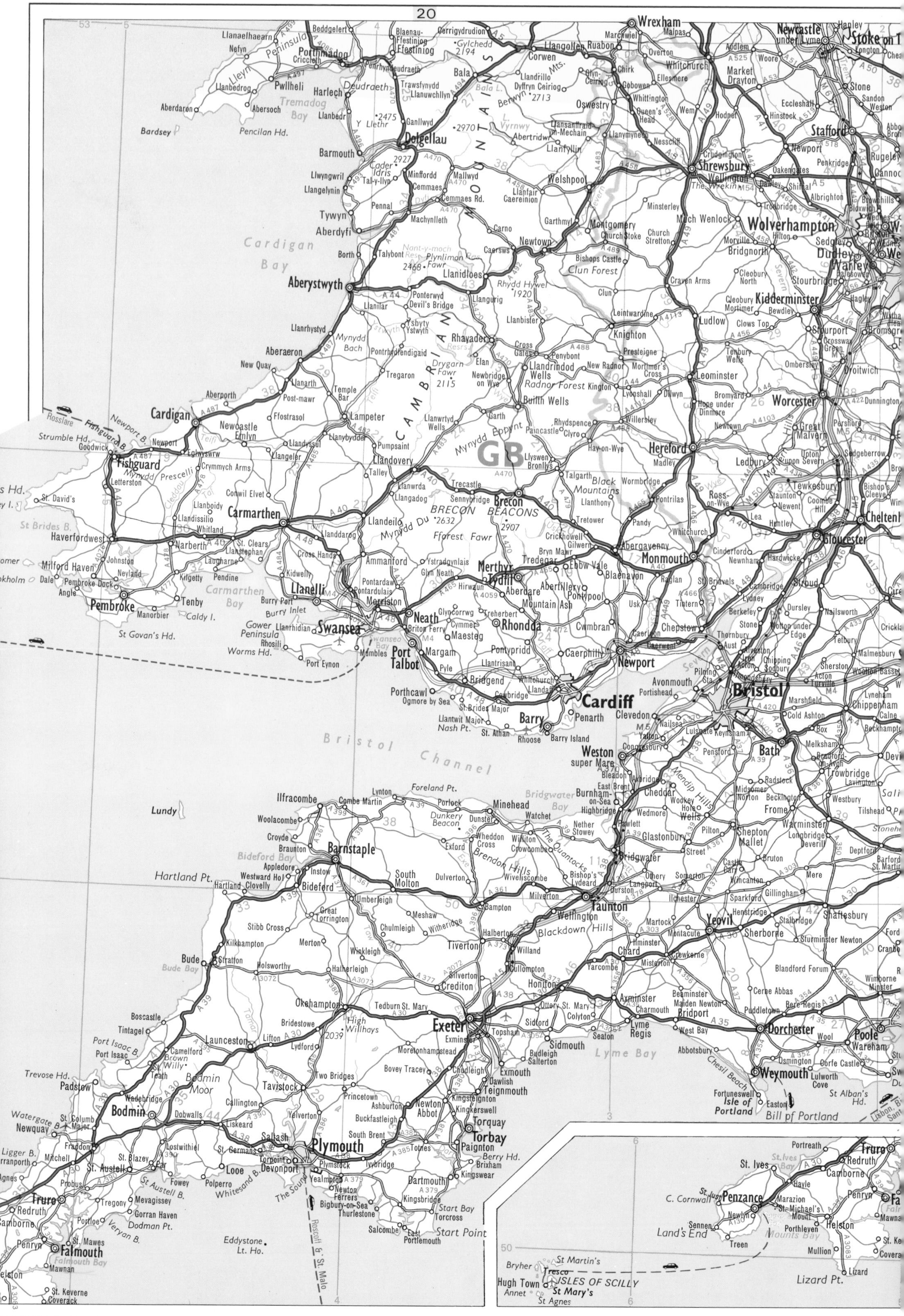

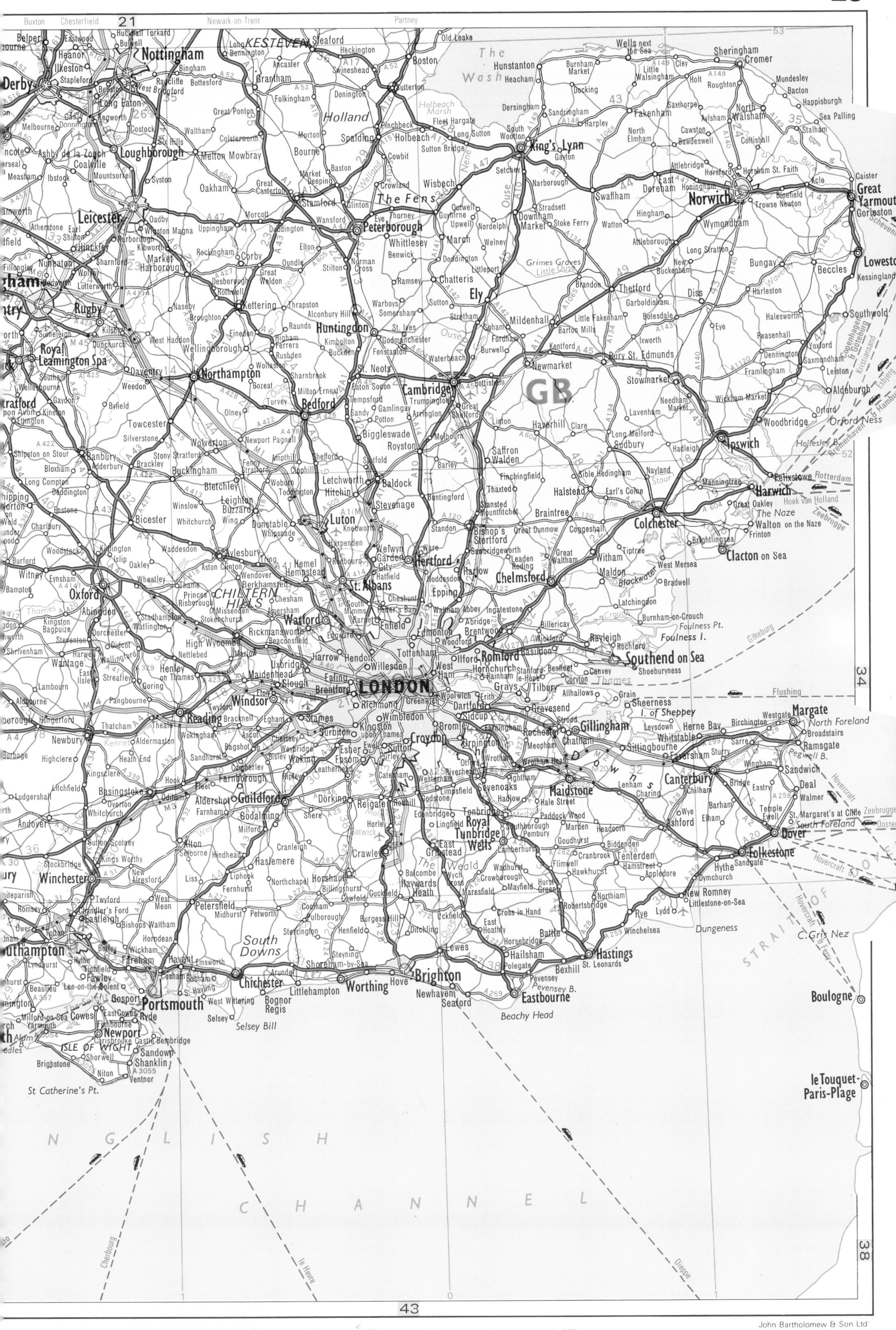

21
43
53
52
34
38
Buxton
Chesterfield
Newark-on-Trent
Partney
Nottingham
Derby
Leicester
Loughborough
Melton Mowbray
Grantham
Boston
Spalding
Stamford
Peterborough
The Wash
Hunstanton
King's Lynn
Wells next the Sea
Sheringham
Cromer
Norwich
Great Yarmouth
Lowestoft
Holland
The Fens
Wisbech
Ely
Thetford
Bury St. Edmunds
Newmarket
Cambridge
Huntingdon
Bedford
Northampton
Kettering
Rugby
Royal Leamington Spa
Coventry
Banbury
Oxford
Aylesbury
Luton
St. Albans
Hertford
Watford
Chelmsford
Colchester
Ipswich
Harwich
Felixstowe
Clacton on Sea
Southend on Sea
GB
CHILTERN HILLS
Reading
Windsor
Slough
LONDON
Croydon
Romford
Gravesend
Gillingham
Chatham
Rochester
Maidstone
Canterbury
Margate
Ramsgate
Deal
Dover
Folkestone
Ashford
Royal Tunbridge Wells
Hastings
Eastbourne
Brighton
Worthing
Chichester
Portsmouth
Southampton
Winchester
Guildford
Basingstoke
Newbury
Crawley
Horsham
South Downs
The Weald
Beachy Head
Selsey Bill
ISLE OF WIGHT
Newport
Ryde
St Catherine's Pt.
Dungeness
STRAIT OF DOVER
C. Gris Nez
Boulogne
le Touquet-Paris-Plage
ENGLISH CHANNEL
Cherbourg
le Havre
Dieppe
Hovercraft
Rotterdam
Hoek van Holland
Zeebrugge
Oostende
Flushing
Göteborg
Esbjerg
Bremerhaven & Hamburg
Kristiansand
Copenhagen & Goteborg
0 10 20 30 40 50 Miles
0 10 20 30 40 50 60 70 80 Kilometres

John Bartholomew & Son Ltd

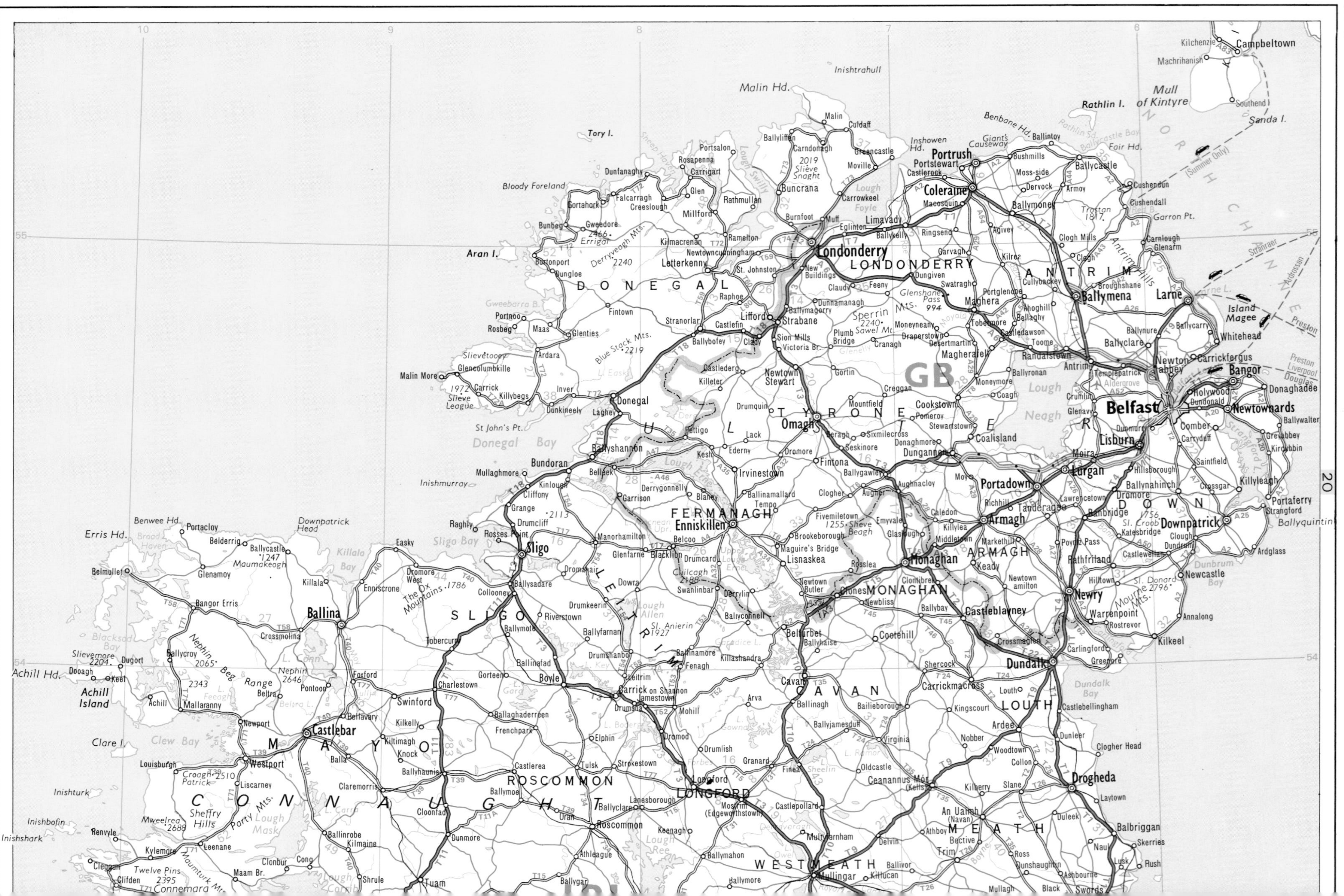

Distances in miles
between circled points

1:1 250 000

Heights in feet

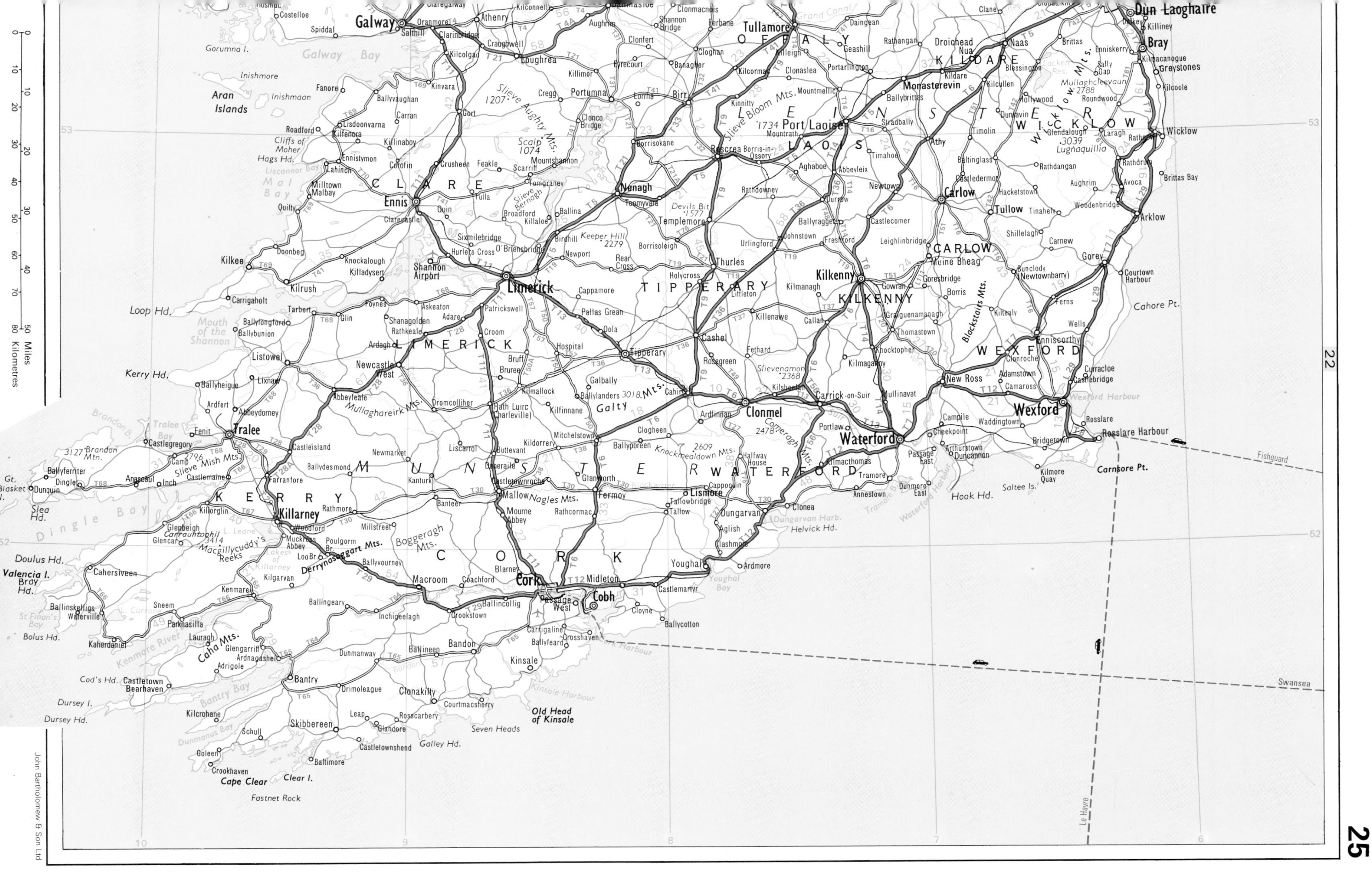

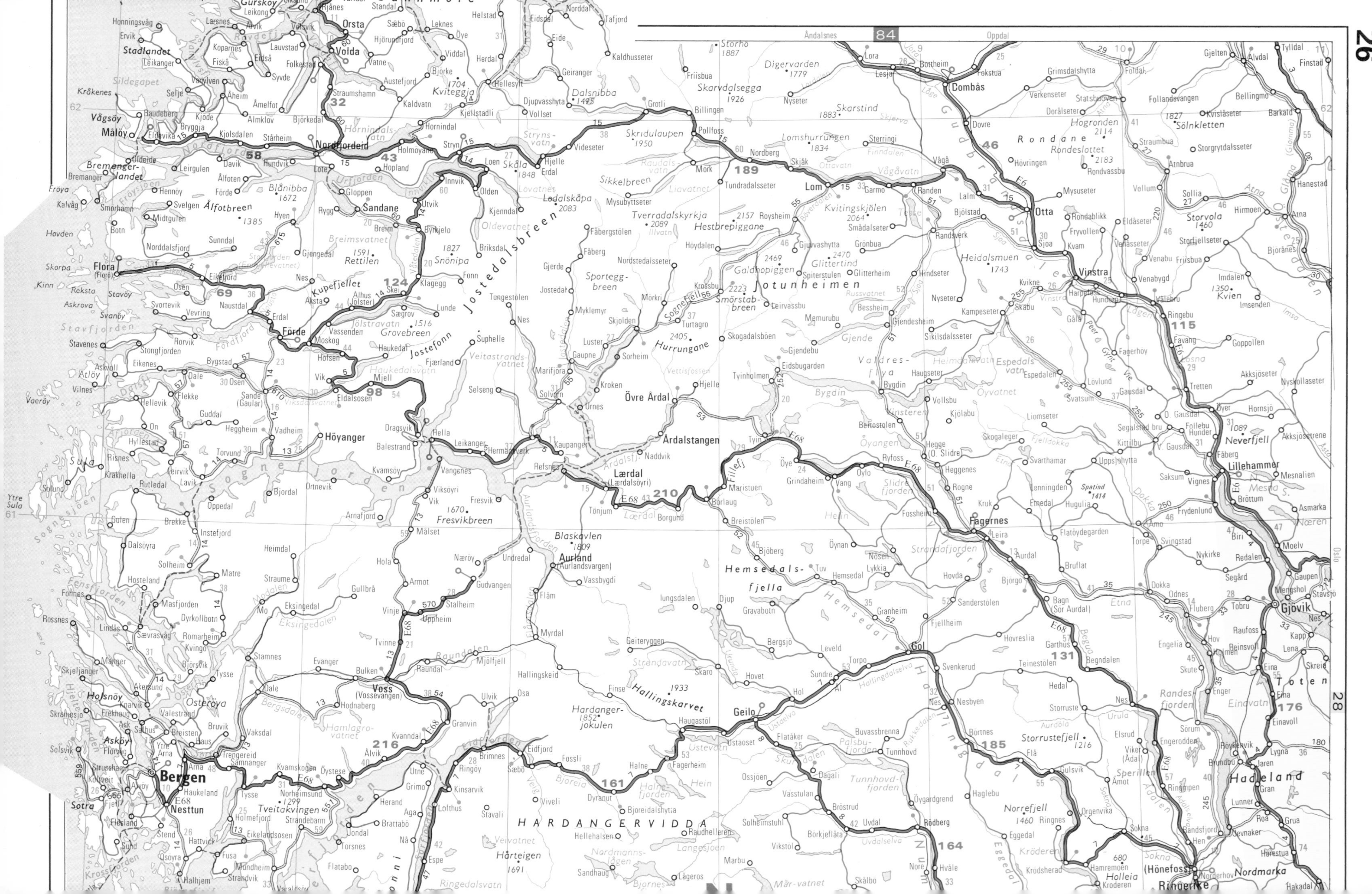

Oslo
28
84
Bergen
Nesttun
Voss (Vossevangen)
Geilo
Gol
Fagernes
Lillehammer
Gjövik
Otta
Vinstra
Lom
Dombås
Lærdal (Lærdalsöyri)
Aurland (Aurlandsvangen)
Höyanger
Förde
Florö (Florø)
Målöy
Nordfjordeid
Sandane
Volda
Örsta
Hönefoss
Ringerike
Hadeland
Jotunheimen
Rondane
Hallingskarvet
HARDANGERVIDDA
Jostedalsbreen
Sunnmøre
Sotra
Askøy

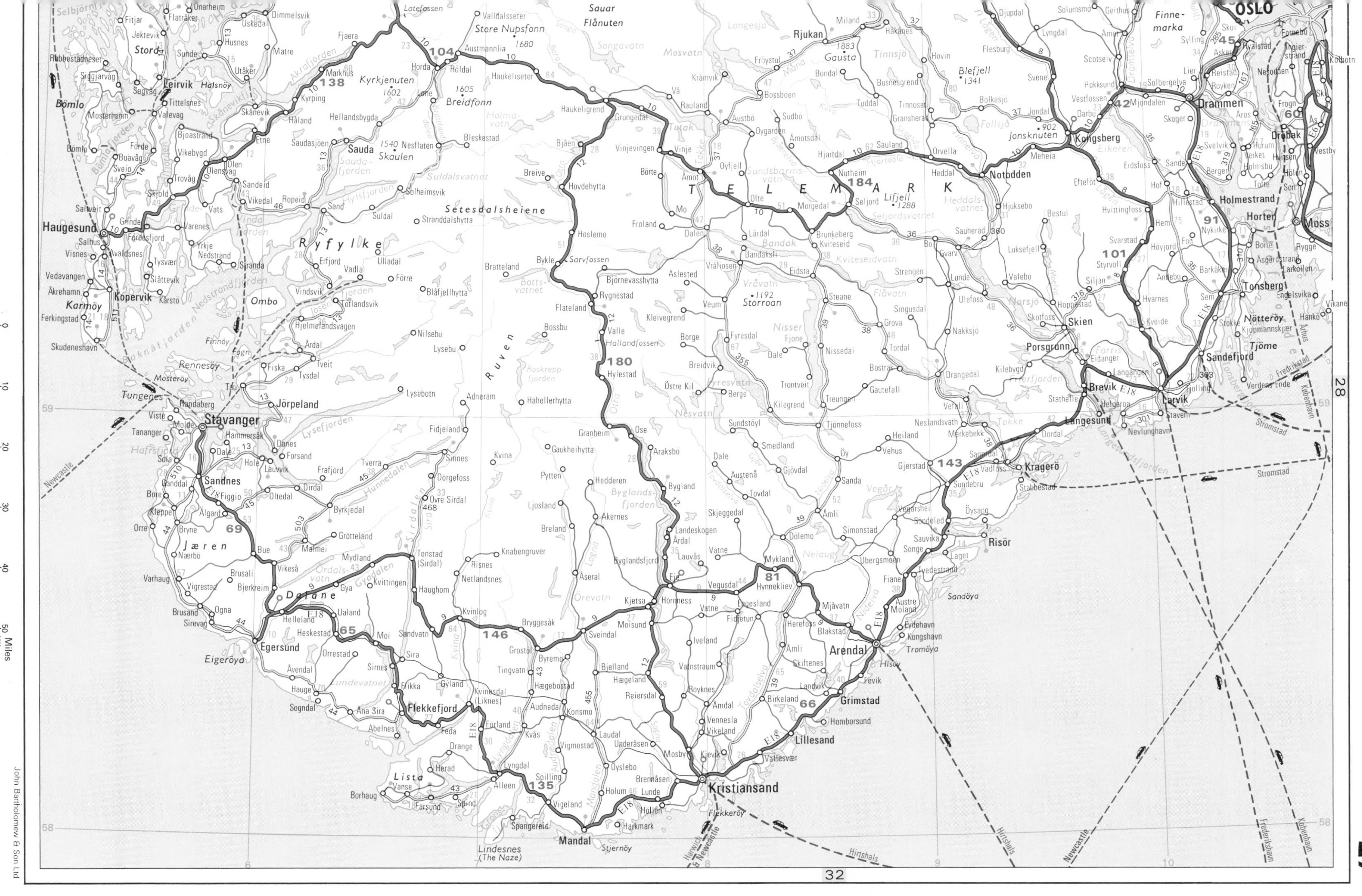

0	10	20	30	40	50	Miles
0	10	20	30	40	50	60	70	80	Kilometres

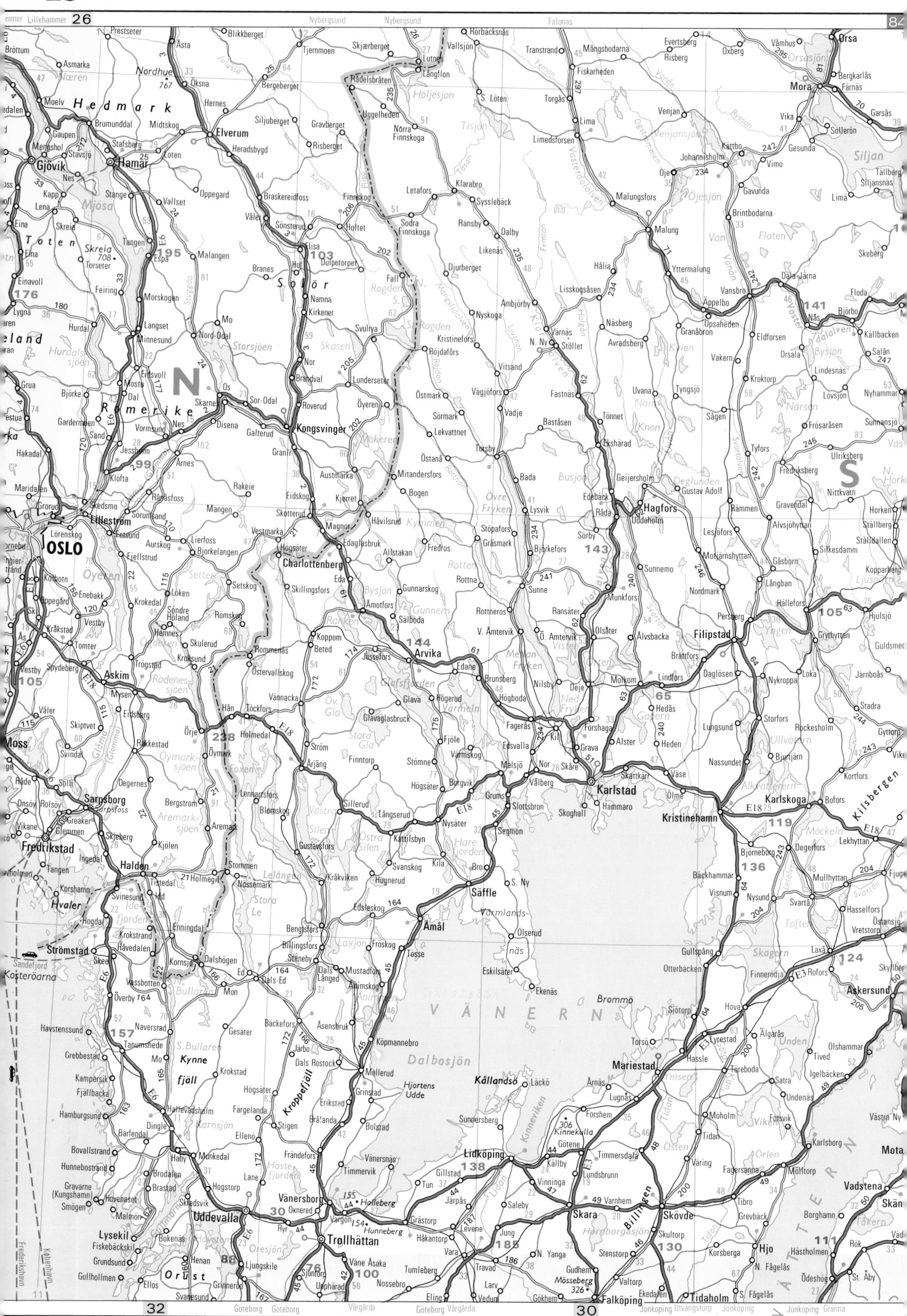
26
Lillehammer
Nybergsund
Falunas
84
Hedmark
Gjøvik
Hamar
Elverum
Brumunddal
Stange
Mjøsa
Toten
Solør
Kongsvinger
Romerike
Lillestrøm
OSLO
Askim
Moss
Sarpsborg
Fredrikstad
Halden
Strömstad
Hvaler
Kosteröarna
Charlottenberg
Arvika
Torsby
Hagfors
Filipstad
Karlstad
Kristinehamn
Karlskoga
Malung
Mora
Orsa
Siljan
Vansbro
Säffle
Åmål
VÄNERN
Dalbosjön
Vänersborg
Trollhättan
Uddevalla
Lysekil
Orust
Lidköping
Mariestad
Skara
Skövde
Falköping
Tidaholm
Hjo
Askersund
Vadstena
Degerfors
Töreboda
Kinnekulla
Kållandsö
Billingen
Kilsbergen
VÄTTERN
Mellerud
Bengtsfors
Kroppefjäll
Kynne fjäll
Tanumshede
Smögen
Ljungskile
Sunne
Kil
Grums
Årjäng
Töcksfors
Ed
Dals-Ed
Magnor
Eda
Munkfors
Storfors
Hällefors
Gullspång
Hova
Götene
Grästorp
Vara
32
30
Göteborg
Vårgårda
Jönköping
Gränna

1:1 250 000

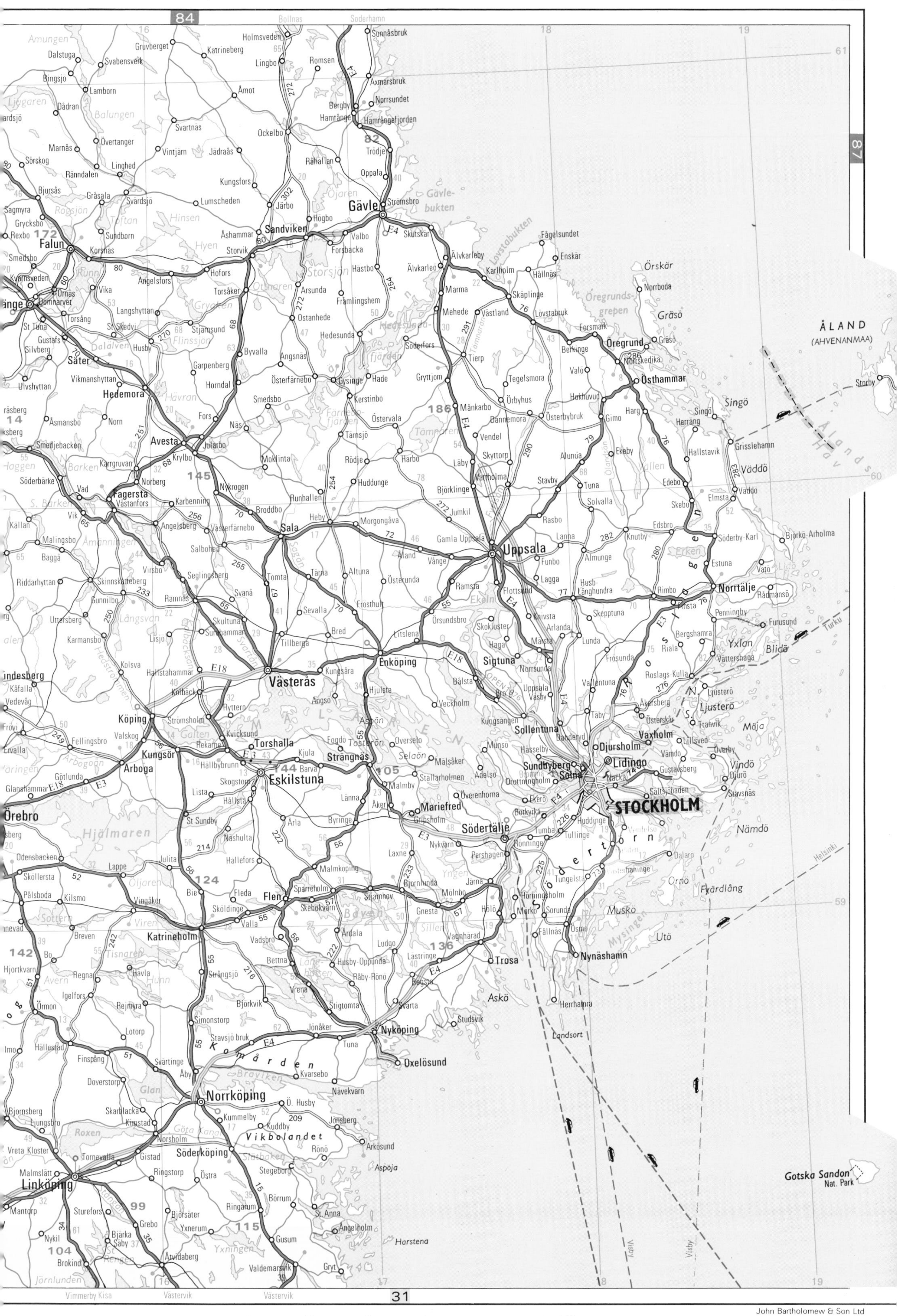

ÅLAND
(AHVENANMAA)

STOCKHOLM

Gävle

Uppsala

Falun

Västerås

Örebro

Eskilstuna

Södertälje

Nynäshamn

Norrköping

Linköping

Gotska Sandon
Nat. Park

John Bartholomew & Son Ltd

0	10	20	30	40	50 Miles
0	10	20	30 40	50 60	70 80 Kilometres

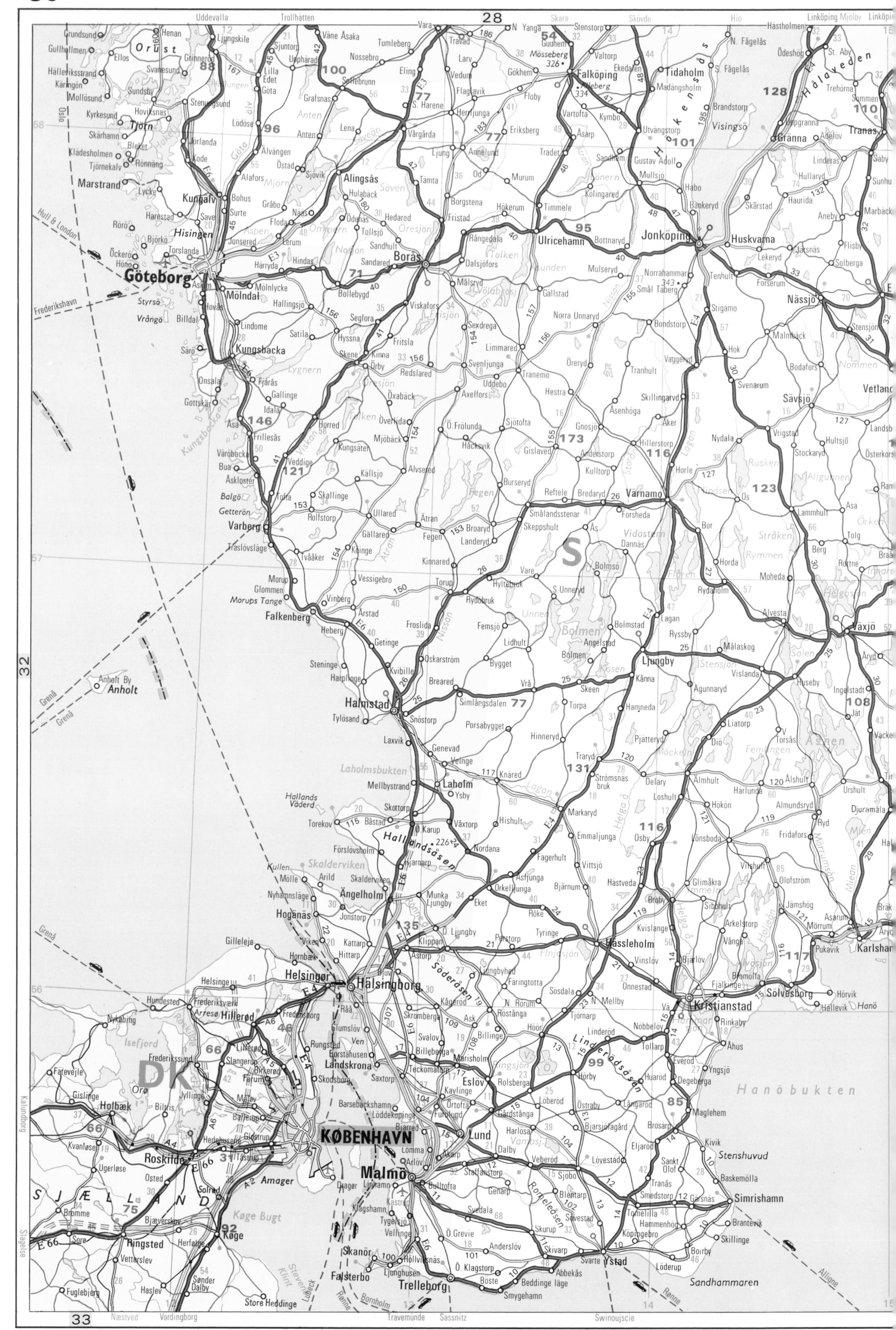
28
Uddevalla
Trollhättan
Skara
Skövde
Hio
Linköping Mjölby
Göteborg
Mölndal
Kungälv
Kungsbacka
Varberg
Falkenberg
Halmstad
Laholm
Ängelholm
Höganäs
Helsingborg
Helsingør
Landskrona
København
Malmö
Lund
Eslöv
Trelleborg
Ystad
Simrishamn
Kristianstad
Hässleholm
Sölvesborg
Karlshamn
Växjö
Ljungby
Värnamo
Jönköping
Huskvarna
Nässjö
Tranås
Gränna
Borås
Alingsås
Ulricehamn
Falköping
Tidaholm
Vara
Hillerød
Roskilde
Køge
Holbæk
Ringsted
Frederikssund
Anholt
Hisingen
Orust
Tjörn
Marstrand
Visingsö
Laholmsbukten
Hanöbukten
Køge Bugt
Isefjord
Amager
Hallandsåsen
Söderåsen
Linderödsåsen
Romeleåsen
Bolmen
Skanör
Falsterbo
S
DK
SJÆLLAND
Næstved
Vordingborg
Travemünde
Sassnitz
Swinoujscie
Rønne
Bornholm
Lübeck
Grenå
Frederikshavn
Hull & London
Allinge
33
32

1:1 250 000

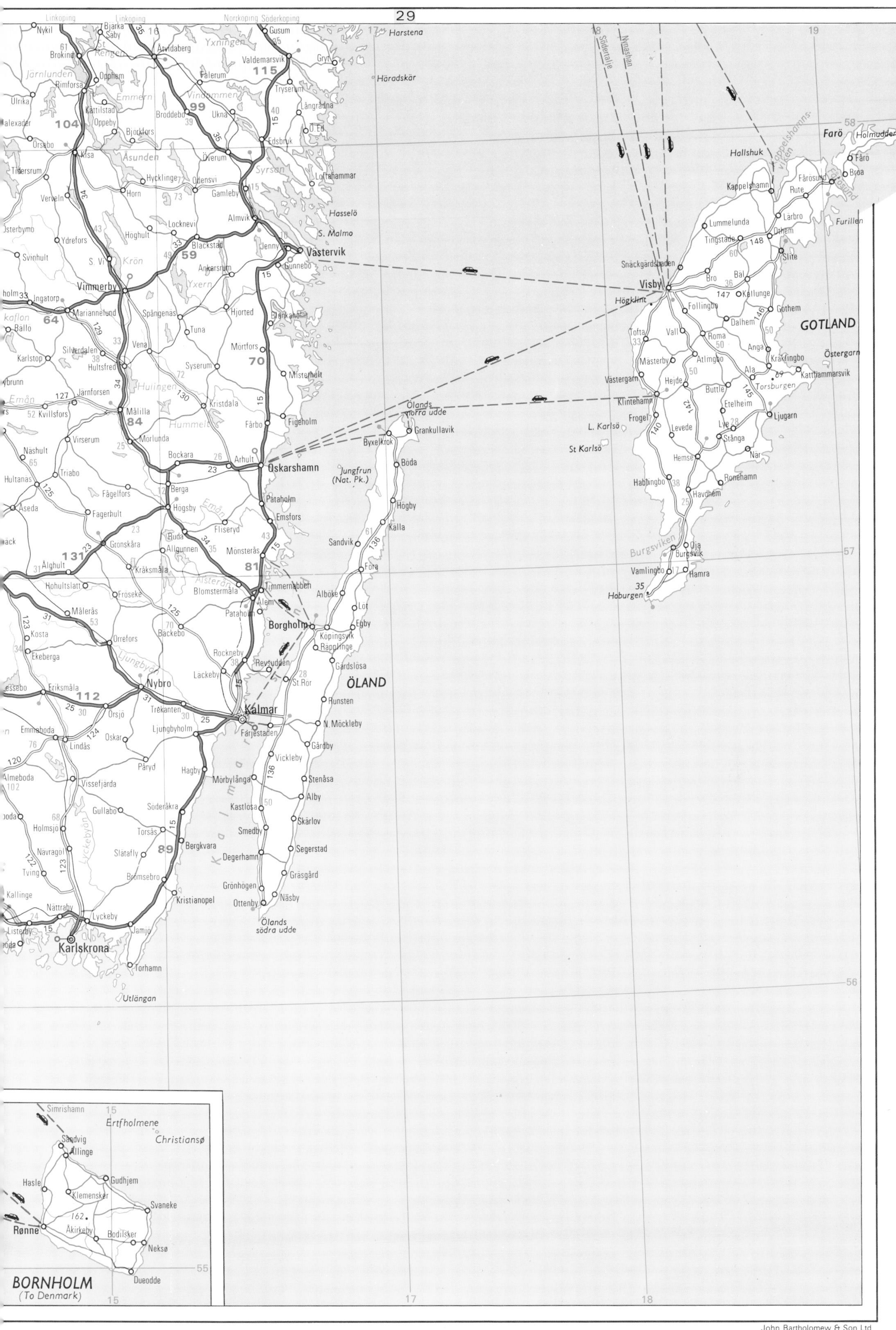

John Bartholomew & Son Ltd

| 0 | 10 | 20 | 30 | 40 | 50 | Miles |
| 0 | 10 20 | 30 | 40 50 | 60 | 70 80 | Kilometres |

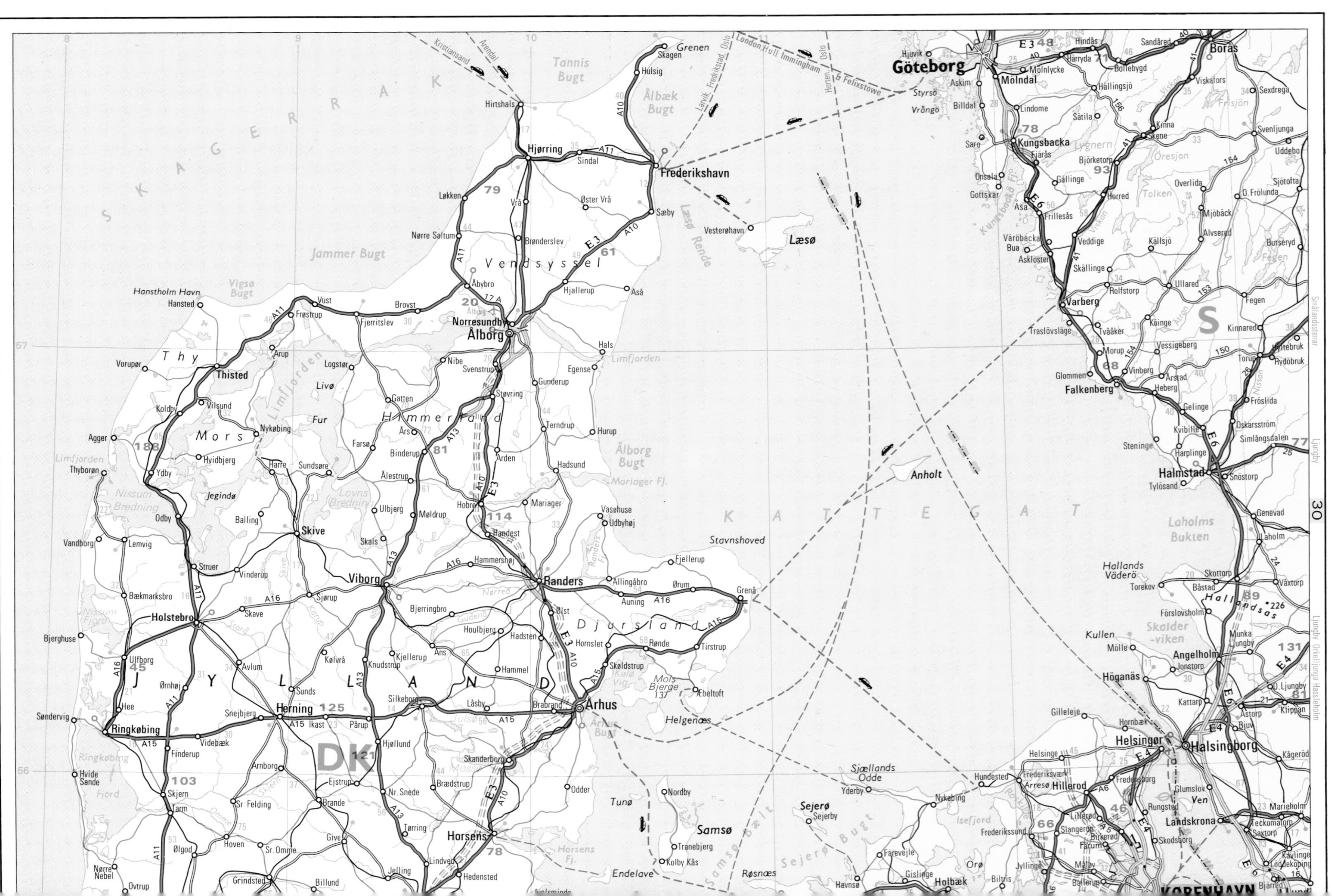

30
Göteborg
Borås
Mölndal
Kungsbacka
Varberg
Falkenberg
Halmstad
Ängelholm
Höganäs
Halsingborg
Helsingør
Landskrona
Hillerød
S
Laholms Bukten
Hallands Väderö
Kullen
K A T T E G A T
S K A G E R R A K
Anholt
Læsø
Læsø Rende
Frederikshavn
Skagen
Grenen
Ålbæk Bugt
Tannis Bugt
Hirtshals
Hjørring
Vendsyssel
Nørresundby
Ålborg
Ålborg Bugt
Limfjorden
Jammer Bugt
Thisted
Mors
Thy
Skive
Viborg
Randers
Djursland
Grenå
Æbeltoft
Århus
Samsø
Sejerø
Sjællands Odde
Horsens
Silkeborg
Herning
Holstebro
Ringkøbing
Lemvig
Thyborøn
J Y L L A N D
DK
Hanstholm Havn
Vigsø Bugt
Arendal
Kristiansand
Larvik, Fredrikstad, Oslo
London, Hull, Immingham
Horten, Oslo
Felixstowe

1:1 250 000

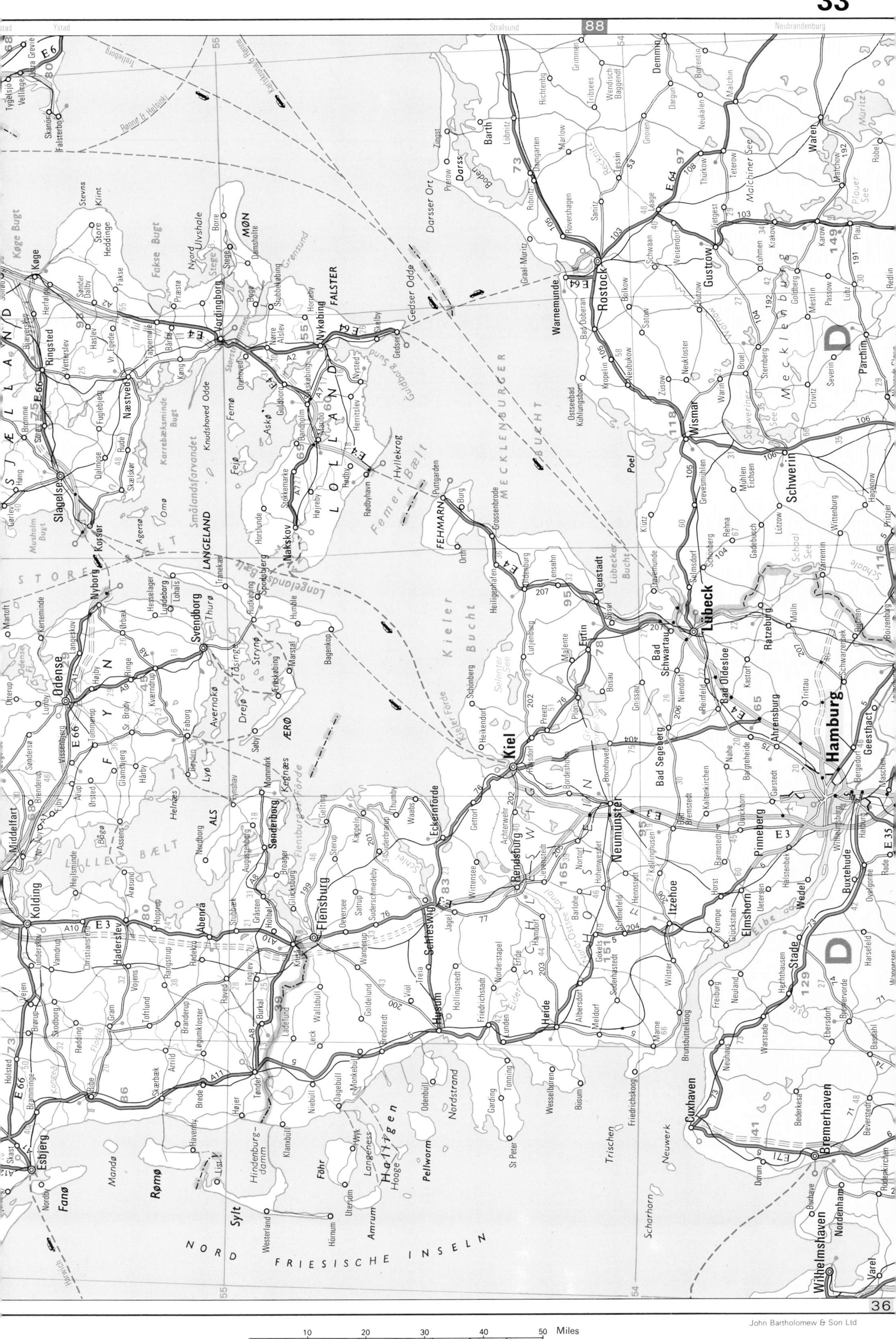

John Bartholomew & Son Ltd

| 0 | 10 | 20 | 30 | 40 | 50 | Miles |
| 0 | 10 | 20 | 30 | 40 | 50 | 60 | 70 | 80 | Kilometres |

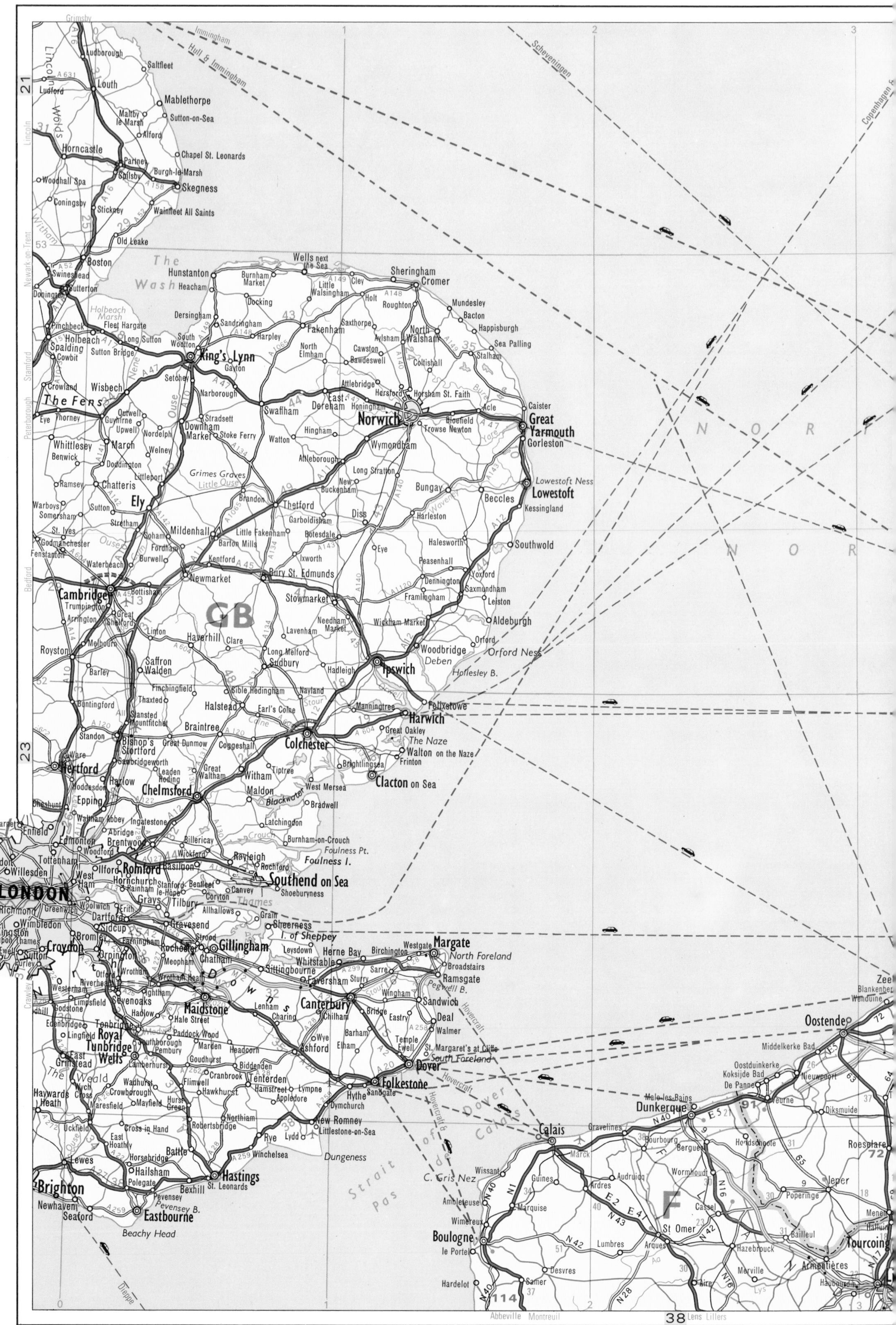

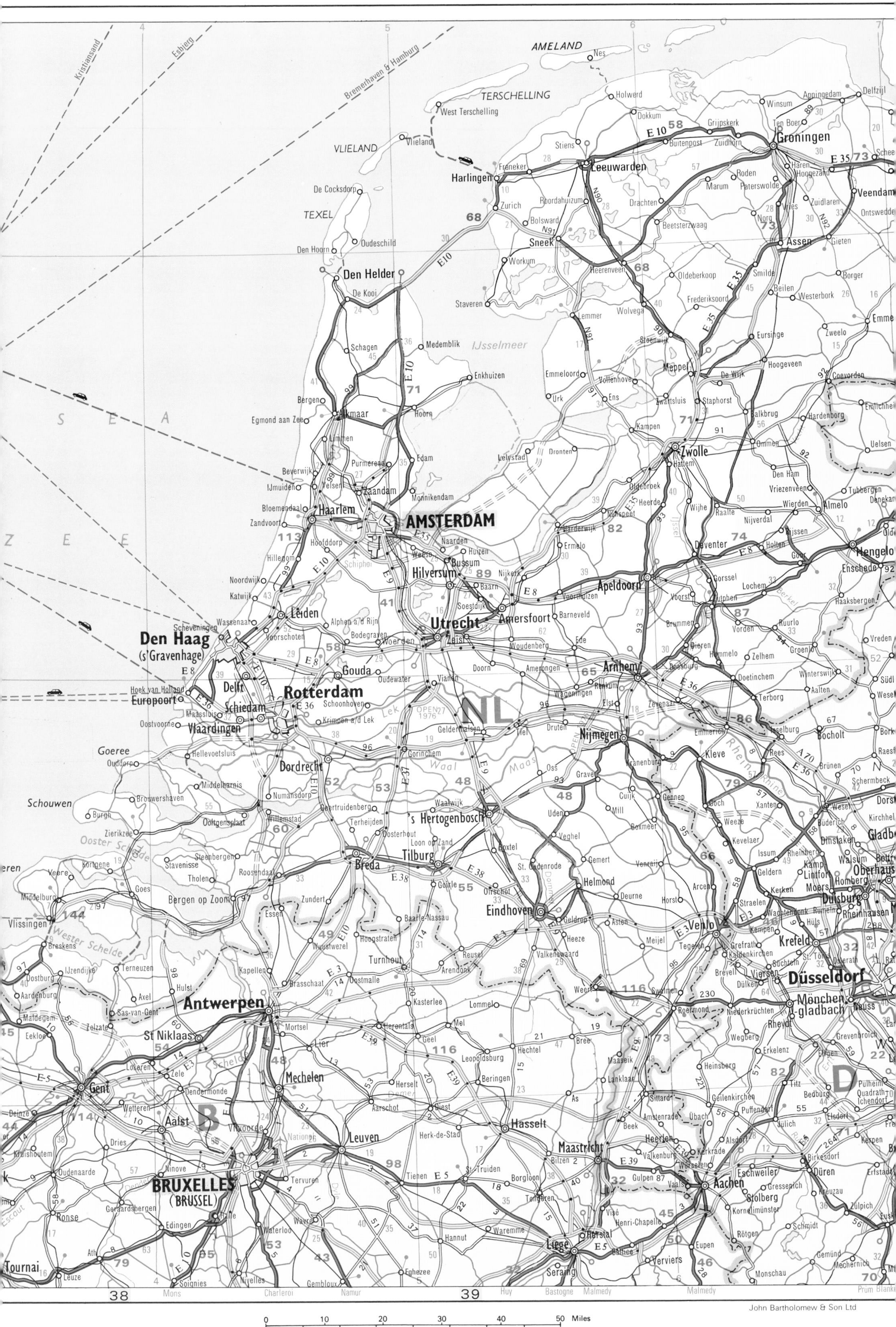

AMELAND
Nes
TERSCHELLING
West Terschelling
VLIELAND
Vlieland
Harlingen
Bremerhaven & Hamburg
Kristiansand
Esbjerg
TEXEL
De Cocksdorp
Zurich
Franeker
Leeuwarden
Stiens
Holwerd
Dokkum
Buitenpost
Grijpskerk
Winsum
Appingedam
Delfzijl
E 10 58
Zuidhorn
Groningen
E 35 73
Ten Boer
Roden
Haren
Hoogezand
Veendam
Drachten
Marum
Paterswolde
Norg
Vries
Zuidlaren
Ontswedde
Zuidhorn
Rhordahuizum
Bolsward
Beetsterzwaag
Assen
Gieten
73
Oudeschild
Den Hoorn
Sneek
Workum
Heerenveen
Oldeberkoop
Smilde
Borger
Emme
Den Helder
De Kooi
Staveren
Lemmer
Wolvega
Frederiksoord
Beilen
Westerbork
E 15
Schagen
Medemblik
IJsselmeer
Emmeloord
Vollenhove
Steenwijk
Eursinge
Zweelo
Bergen
Enkhuizen
Urk
Ens
Meppel
De Wijk
Hoogeveen
Coevorden
Entlichhel
Egmond aan Zee
Alkmaar
Hoorn
Dronten
Kampen
Staphorst
Zwartsluis
Hardenborg
Uelsen
Limmen
Lelystad
Balkbrug
Ommen
71
Beverwijk
Purmerend
Edam
Hattem
Zwolle
Den Ham
Tubbergen
IJmuiden
Velsen
Zaandam
Monnikendam
Oldebroek
Vriezenveen
Wierden
Almelo
Denekamp
Bloemendaal
Heerde
Wijhe
Raalte
Nijverdal
Haarlem
AMSTERDAM
Harderwijk
82
Deventer
74
Hengelo
Zandvoort
Hoofddorp
Naarden
Weesp
Huizen
Ermelo
Voorthuizen
Gorssel
Holten
Enschede
Hillegom
Schiphol
Bussum
Nijkerk
Apeldoorn
Lochem
Haaksbergen
113
Hilversum
89
Baarn
Voorst
Zutphen
87
Noordwijk
Katwijk
41
Soestdijk
E 8
Barneveld
Brummen
Vorden
Ruurlo
Vreden
Wassenaar
Leiden
Alphen a/d Rijn
Amersfoort
Ede
Hummelo
Zelhem
Groenlo
Süd
Scheveningen
Voorschoten
Bodegraven
Woerden
Zeist
Woudenberg
93
Doesburg
Winterswijk
Den Haag
('s Gravenhage)
58
E 8
Gouda
Oudewater
Vianen
Doorn
Amerongen
Arnhem
65
Doetinchem
Aalten
Wesel
Delft
Schoonhoven
Wageningen
Rhenen
Elst
Zevenaar
Terborg
86
Borken
Hoek van Holland
Europoort
Rotterdam
Schiedam
E 36
Krimpen a/d Lek
Lek
OPEN 27 1976
Geldermalsen
Tiel
Druten
Nijmegen
Kleve
Rees
E 36
Bocholt
Raesf
Vlaardingen
Oostvoorne
Helvoetsluis
Gorinchem
Waal
Oss
Grave
Cranenburg
Xanten
Brünen
Dorst
Goeree
Ouddorp
Dordrecht
52
53
E 37
48
93
Gennep
Boch
79
Weeze
Kevelaer
Issum
Dinslaken
Gladb
Middelharnis
Numansdorp
Waalwijk
48
Curjk
Mill
Boxmeer
Kamp
Linfort
Oberhaus
Brouwershaven
Geertruidenberg
's Hertogenbosch
Uden
Veghel
Venray
95
Geldern
Kerken
Moers
Duisburg
Schouwen
Burgh
Willemstad
60
Terheijden
Oosterhout
Loon op Zand
Boxtel
St. Oedenrode
Gemert
Horst
Arcen
Straelen
Homberg
Zierikzee
Steenbergen
Breda
Tilburg
E 38
Helmond
Deurne
58
Wachtendonk
Rumeln
Rheinhausen
Goes
Roosendaal
33
Goirle
55
Oirschot
33
Dormagen
Venlo
E 3
Grefrath
Kaldenkirchen
Krefeld
32
Bergen op Zoom
Zundert
Baarle-Nassau
Eindhoven
Geldrop
Asten
Meijel
Tegelen
Brevell
Viersen
Dülken
64
Düsseldorf
144
Essen
49
E 10
Hoogstraten
Reusel
14
31
Heeze
Valkenswaard
33
69
Weert
116
Swalmen
230
Niederkrüchten
Mönchen-gladbach
Neuss
Vlissingen
Breskens
IJzendijke
Terneuzen
Hulst
Wuustwezel
Turnhout
Arendonk
Roermond
Rheydt
Grevenbroich
W
Oostburg
Aardenburg
Axel
Sas-van-Gent
Brasschaat
Oostmalle
Kasterlee
Lommel
As
Maaseik
Lanklaar
Heinsberg
82
Titz
D
Maldegem
Zelzate
Antwerpen
Mortsel
Lier
Geel
Mol
Bree
Wegberg
Erkelenz
Birkesdorf
Eeklo
St Niklaas
54
48
Herentals
116
Leopoldsburg
Hechtel
Beringen
21
19
Sittard
Geilenkirchen
Titz
Bedburg
Quadrath
Ichendorf
Gent
114
Mechelen
13
Herselt
Aarschot
Diest
Hasselt
Herk-de-Stad
Maastricht
Amstenrade
Beek
Übach
Puffendorf
Jülich
Elsdorf
Kerpen
B
Aalst
Leuven
98
Tienen
E 5
St Truiden
Borgloon
Tongeren
40
32
Gulpen 87
Vaals
Aachen
Eschweiler
Stolberg
Düren
Erftstadt
Bruxelles
(Brussel)
Tervuren
Henri-Chapelle
45
Kornelimünster
Schmidt
Tournai
Leuze
79
Nivelles
Gembloux
Namur
Huy
Liège
46
Eupen
Monschau
Mechernich
70
Mons
Charleroi
Bastogne
Malmedy
Verviers
Seraing

John Bartholomew & Son Ltd

| 0 | 10 | 20 | 30 | 40 | 50 Miles |
| 0 | 10 | 20 | 30 | 40 | 50 | 60 | 70 | 80 Kilometres |

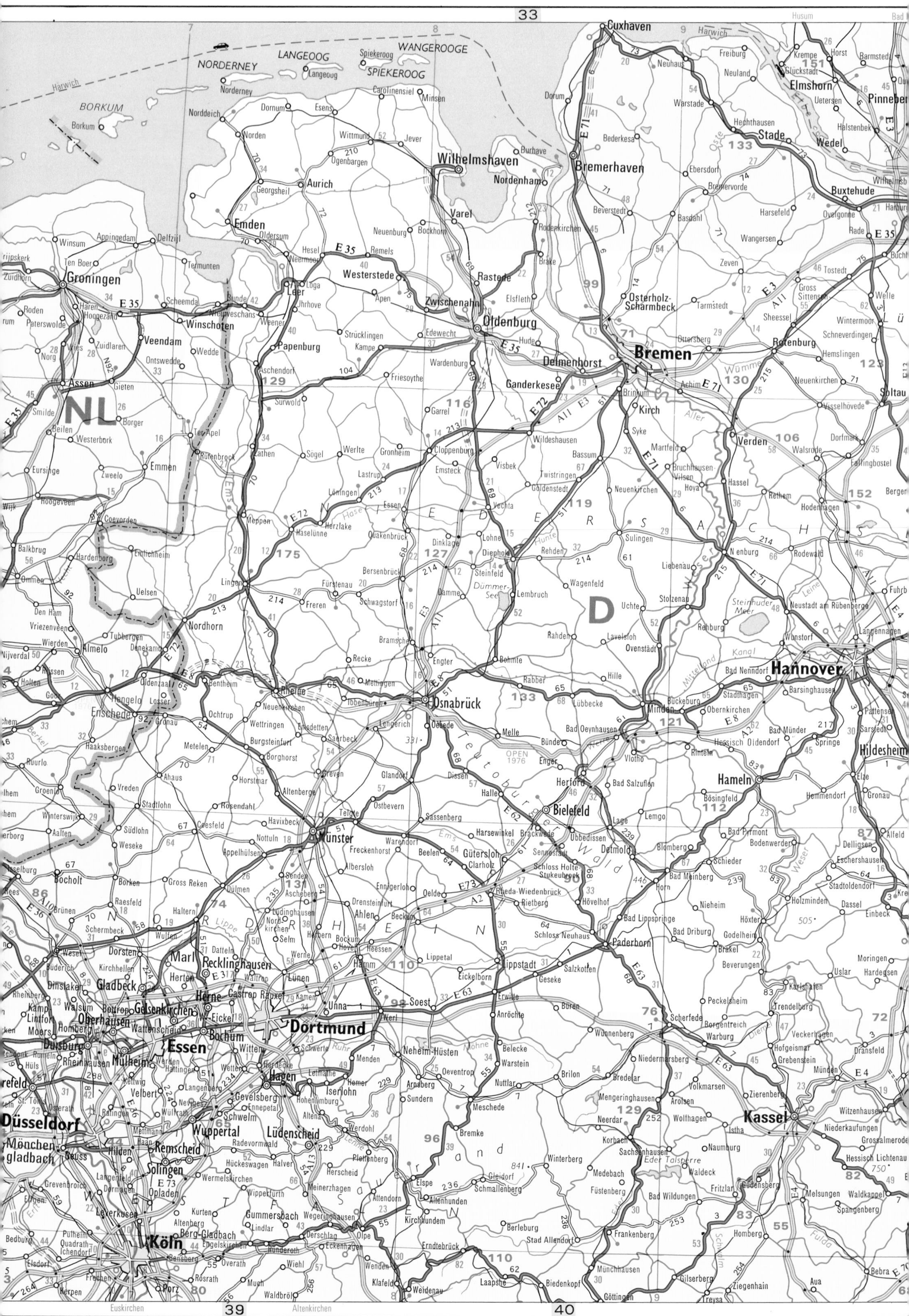
33
Cuxhaven
NORDERNEY
LANGEOOG
WANGEROOGE
SPIEKEROOG
BORKUM
Borkum
Norderney
Norden
Aurich
Emden
Wilhelmshaven
Nordenham
Bremerhaven
Stade
Elmshorn
Pinneberg
Wedel
Buxtehude
Groningen
Delfzijl
Winschoten
Veendam
Assen
Emmen
Leer
Papenburg
Westerstede
Oldenburg
Delmenhorst
Bremen
Rotenburg
Verden
Soltau
Cloppenburg
Vechta
Meppen
Lingen
Nordhorn
Almelo
Hengelo
Enschede
Rheine
Osnabrück
Minden
Hannover
Hildesheim
Hameln
Herford
Bielefeld
Detmold
Gütersloh
Paderborn
Münster
Hamm
Lippstadt
Soest
Unna
Dortmund
Bochum
Essen
Gelsenkirchen
Oberhausen
Duisburg
Mülheim
Recklinghausen
Marl
Gladbeck
Herne
Hagen
Iserlohn
Lüdenscheid
Wuppertal
Remscheid
Solingen
Düsseldorf
Mönchengladbach
Neuss
Leverkusen
Köln
Kassel
Göttingen
Biedenkopf
Warburg
NL
D
NIEDERSACHSEN
NORDRHEIN-WESTFALEN
SAUERLAND
Teutoburger Wald
Weser
Ems
Aller
Elbe
Steinhuder Meer
Dümmer-See
Eder Talsperre
39
40
Husum
Euskirchen
Altenkirchen

1:1 250 000

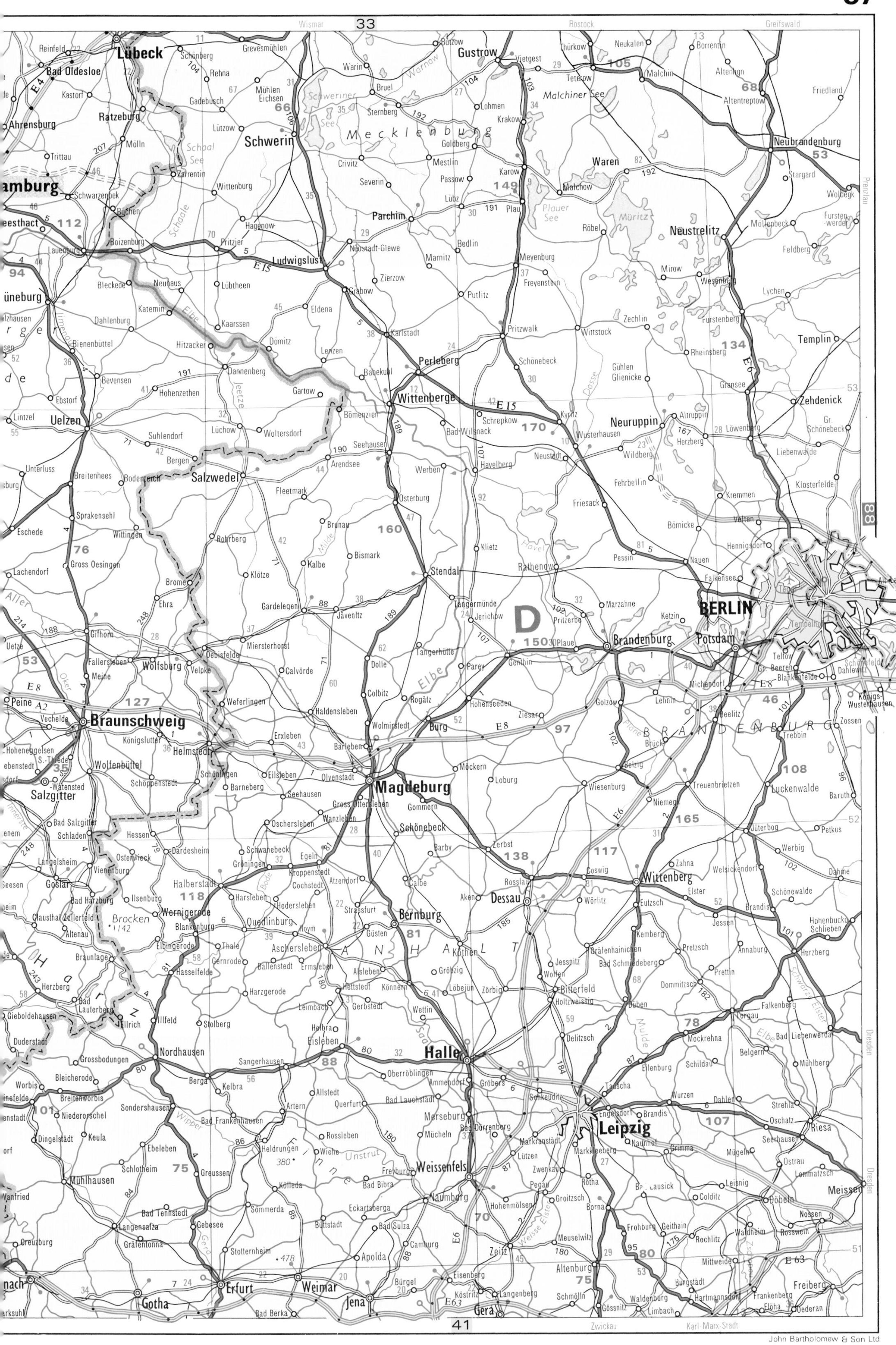

0	10	20	30	40	50 Miles	
0	10	20	30	40	50	60	70	80 Kilometres

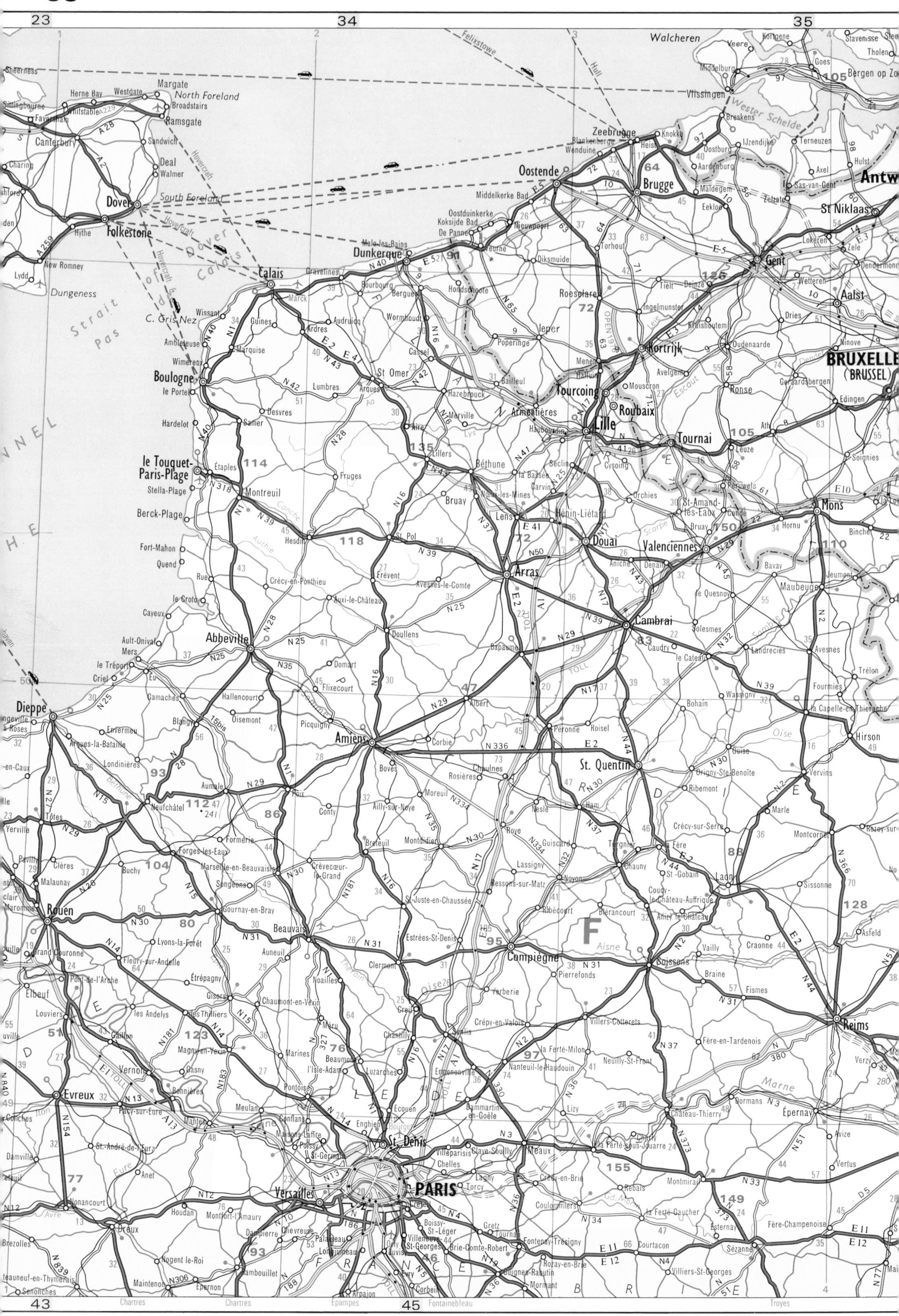
23
34
35
Walcheren
Vlissingen
Middelburg
Goes
Bergen op Zoom
Wester Schelde
Breskens
Terneuzen
Antwerpen
St Niklaas
Sheerness
Margate
North Foreland
Broadstairs
Ramsgate
Herne Bay
Westgate
Whitstable
Faversham
Sittingbourne
Canterbury
Sandwich
Deal
Walmer
Dover
South Foreland
Folkestone
Hythe
New Romney
Lydd
Dungeness
Felixstowe
Hull
Hovercraft
Strait of Dover
Pas de Calais
Zeebrugge
Blankenberge
Wenduine
Heist
Knokke
Oostende
Brugge
Middelkerke Bad
Oostduinkerke
Koksijde Bad
De Panne
Nieuwpoort
Veurne
Diksmuide
Torhout
Eeklo
Maldegem
Gent
Lokeren
Zele
Dendermonde
Wetteren
Aalst
Tielt
Deinze
Ingelmunster
Roeselare
Ieper
Poperinge
Kortrijk
Oudenaarde
Ninove
BRUXELLES
(BRUSSEL)
Geraardsbergen
Edingen
Ronse
Menen
Mouscron
Tourcoing
Roubaix
Lille
Tournai
Ath
Leuze
Soignies
Mons
Binche
Péruwelz
Condé
Hornu
Valenciennes
Denain
Maubeuge
Jeumont
Bavay
Le Quesnoy
Dunkerque
Malo-les-Bains
Gravelines
Bourbourg
Bergues
Hondschoote
Wormhoudt
Cassel
Calais
Marck
Ardres
Audruicq
Guînes
Wissant
C. Gris Nez
Ambleteuse
Marquise
Wimereux
Boulogne
le Portel
Desvres
Lumbres
St Omer
Arques
Hazebrouck
Bailleul
Armentières
Merville
Aire
Lillers
Béthune
Bruay
Noeux-les-Mines
Lens
Hénin-Liétard
Carvin
Seclin
Cysoing
Orchies
St-Amand-les-Eaux
Douai
Arras
Cambrai
Hardelot
Samer
le Touquet-Paris-Plage
Étaples
Montreuil
Stella-Plage
Berck-Plage
Fort-Mahon
Quend
Rue
Le Crotoy
Cayeux
Hesdin
Fruges
St. Pol
Frévent
Avesnes-le-Comte
Auxi-le-Château
Doullens
Crécy-en-Ponthieu
Abbeville
Ault-Onival
Mers
le Tréport
Eu
Criel
Dieppe
Gamaches
Hallencourt
Oisement
Blangy
Domart
Flixecourt
Picquigny
Amiens
Corbie
Albert
Bapaume
Péronne
Roisel
St. Quentin
Caudry
le Cateau
Solesmes
Landrecies
Avesnes
Trélon
Fourmies
Bohain
Wassigny
La Capelle-en-Thiérache
Hirson
Guise
Origny-Ste-Benoite
Ribemont
Vervins
Marle
Crécy-sur-Serre
Montcornet
Rozoy-sur-Serre
La Fère
Tergnier
Chauny
St-Gobain
Laon
Sissonne
Coucy-le-Château-Auffrique
Anizy-le-Château
Blérancourt
Noyon
Ham
Nesle
Roye
Guiscard
Lassigny
Ressons-sur-Matz
Ribécourt
Chaulnes
Rosières
Moreuil
Boves
Ailly-sur-Noye
Conty
Breteuil
Montdidier
St-Juste-en-Chaussée
Crèvecœur-le-Grand
Poix
Aumale
Neufchâtel
Londinières
Envermeu
Arques-la-Bataille
Tôtes
Yerville
Pavilly
Clères
Malaunay
Maromme
Rouen
Buchy
Forges-les-Eaux
Formerie
Marseille-en-Beauvaisis
Songeons
Gournay-en-Bray
Beauvais
Lyons-la-Forêt
Fleury-sur-Andelle
Étrépagny
Gisors
Grand Couronne
Pont-de-l'Arche
Elbeuf
Louviers
les Andelys
Gaillon
Vernon
Évreux
Pacy-sur-Eure
Bonnières
Mantes
Magny-en-Vexin
Chaumont-en-Vexin
Auneuil
Noailles
Méru
Marines
Pontoise
Beaumont
l'Isle-Adam
Luzarches
Chantilly
Creil
Clermont
Estrées-St-Denis
Compiègne
Pierrefonds
Verberie
Senlis
Crépy-en-Valois
Villers-Cotterets
La Ferté-Milon
Nanteuil-le-Haudouin
Ermenonville
Dammartin-en-Goële
Neuilly-St-Front
Lizy
Soissons
Vailly
Braine
Fismes
Craonne
Asfeld
Reims
Fère-en-Tardenois
Château-Thierry
Dormans
Épernay
Avize
Vertus
Verzy
Meulan
Conflans
Maisons-Laffite
Poissy
St-Germain
Ecouen
Enghien
Bourget
St. Denis
Villeparisis
Claye-Souilly
Meaux
Chelles
Lagny
Torcy
PARIS
Versailles
Créteil
Crécy-en-Brie
Coulommiers
La Ferté-sous-Jouarre
Charly
Rebais
Montmirail
La Ferté-Gaucher
Esternay
Sézanne
Fère-Champenoise
Boissy-St-Léger
Gretz
Tournan
Fontenay-Trésigny
Brie-Comte-Robert
Villeneuve-St-Georges
Rozay-en-Brie
Courtacon
Villiers-St-Georges
Nangis
Mormant
Guignes-Rabutin
Corbeil
Évry
Juvisy
Longjumeau
Palaiseau
Chevreuse
Dampierre
Rambouillet
Épernon
Maintenon
Montfort-l'Amaury
Houdan
Dreux
Nonancourt
Anet
St.-André-de-l'Eure
Damville
Conches
Nogent le Roi
Brezolles
Châteauneuf-en-Thymerais
Senonches
Arpajon
Chartres
Étampes
Fontainebleau
Troyes
Somme
Oise
Aisne
Marne
Seine
Eure
Avre
Bresle
Béthune
Authie
Canche
Scarpe
Escaut
Sambre
Lys
Leie
Thérain
Epte
F
43
45

1:1 250 000

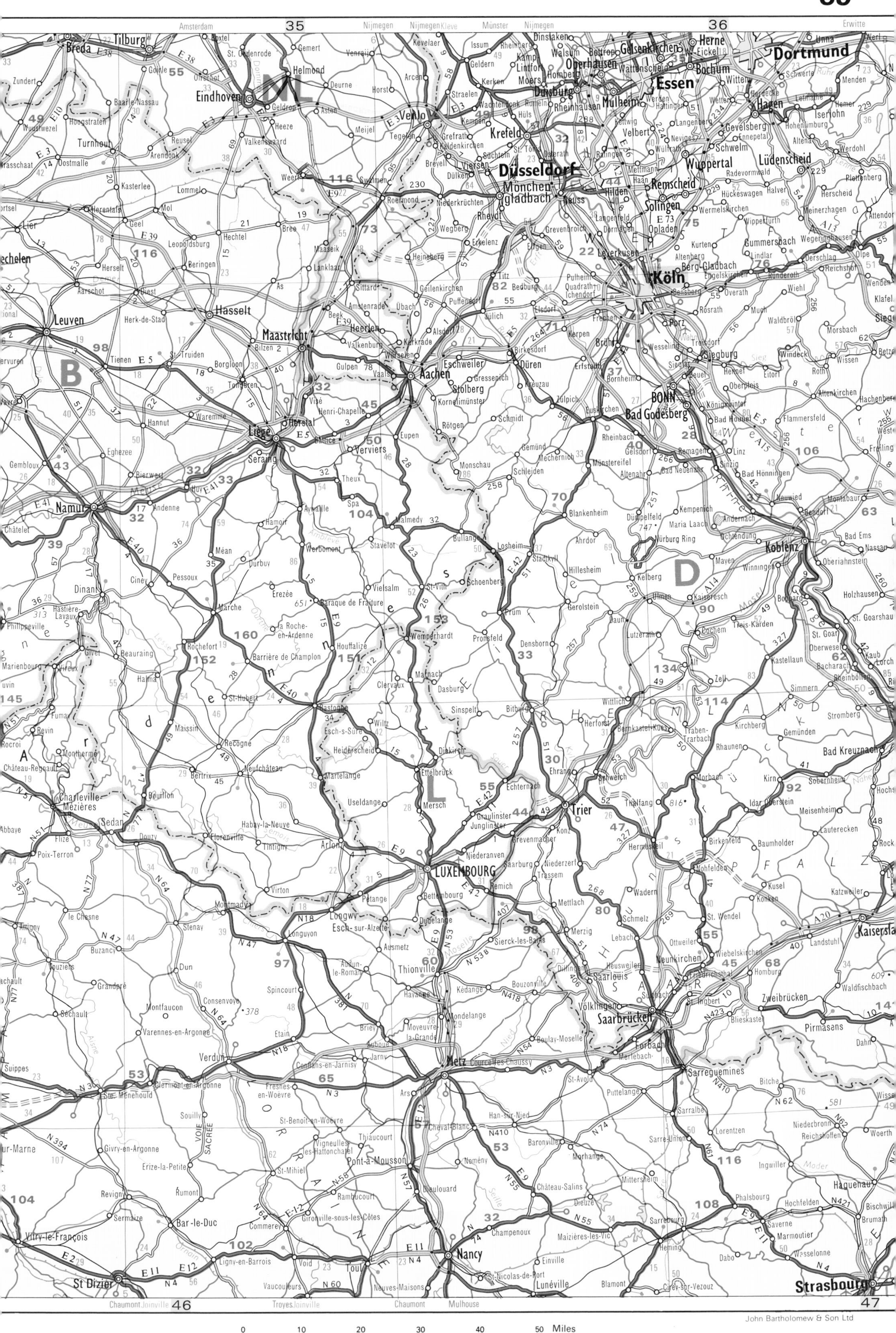

0   10   20   30   40   50 Miles
0   10   20   30   40   50   60   70   80 Kilometres

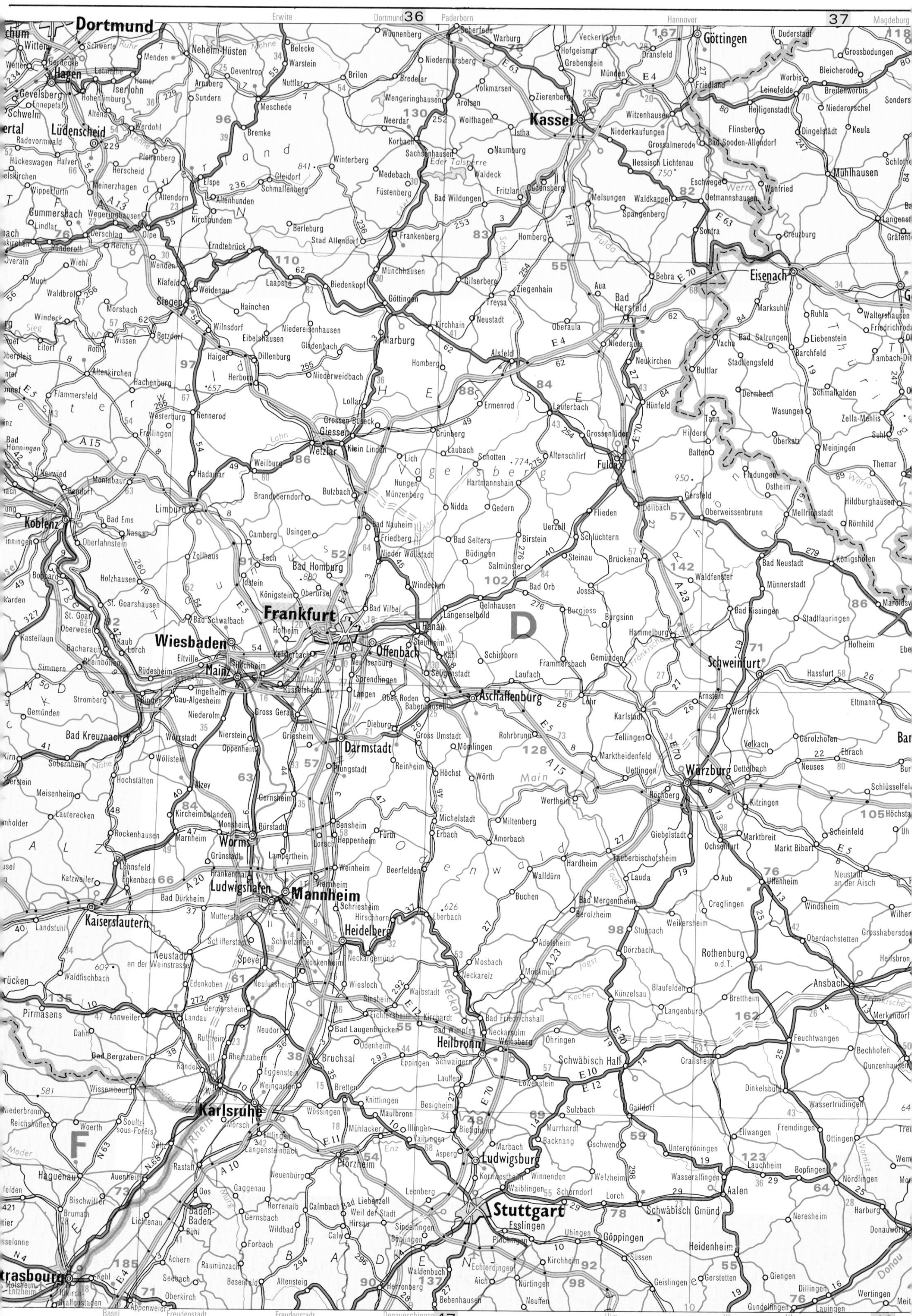

36
37
47
48
Erwite
Dortmund
Paderborn
Hannover
Magdeburg
Basel
Freudenstadt
Donaueschingen
Ulm
Dortmund
Hagen
Lüdenscheid
Iserlohn
Schwerte
Menden
Neheim-Hüsten
Arnsberg
Sundern
Meschede
Warstein
Belecke
Brilon
Wünnenberg
Scherfede
Warburg
Göttingen
Duderstadt
Grossbodungen
Bleicherode
Worbis
Leinefelde
Heiligenstadt
Dingelstädt
Keula
Mühlhausen
Kassel
Hofgeismar
Grebenstein
Münden
Dransfeld
Witzenhausen
Niederkaufungen
Grossalmerode
Hessisch Lichtenau
Bad Sooden-Allendorf
Eschwege
Wanfried
Oetmannshausen
Waldkappel
Spangenberg
Melsungen
Sontra
Creuzburg
Eisenach
Gräfenroda
Langensalza
Arolsen
Volkmarsen
Wolfhagen
Zierenberg
Istha
Naumburg
Waldeck
Fritzlar
Gudensberg
Bad Wildungen
Homberg
Korbach
Sachsenhausen
Mengeringhausen
Niedermarsberg
Bredelar
Medebach
Winterberg
Fürstenberg
Eder Talsperre
Frankenberg
Berleburg
Stad Allendorf
Schmallenberg
Gleidorf
Elspe
Altenhundern
Attendorn
Kirchhundem
Olpe
Plettenberg
Herscheid
Meinerzhagen
Werdohl
Altena
Hohenlimburg
Letmathe
Hemer
Wetter
Herdecke
Gevelsberg
Ennepetal
Schwelm
Witten
Radevormwald
Hückeswagen
Halver
Wipperfürth
Gummersbach
Wegeringhausen
Derschlag
Lindlar
Runderoth
Reichs
Overath
Wiehl
Much
Waldbröl
Morsbach
Windeck
Sieg
Eitorf
Roth
Wissen
Betzdorf
Siegen
Klafeld
Weidenau
Wenden
Erndtebrück
Laasphe
Biedenkopf
Münchhausen
Göttingen
Marburg
Kirchhain
Neustadt
Treysa
Ziegenhain
Gilserberg
Alsfeld
Oberaula
Bad Hersfeld
Niederaula
Bebra
Aua
Neukirchen
Hünfeld
Vacha
Buttlar
Dermbach
Marksuhl
Ruhla
Bad Salzungen
Liebenstein
Barchfeld
Stadtlengsfeld
Schmalkalden
Wasungen
Zella-Mehlis
Suhl
Meiningen
Themar
Hildburghausen
Römhild
Mellrichstadt
Oberweissenbrunn
Gersfeld
Ostheim
Fladungen
Batten
Hilders
Tann
Oberkatz
Waltershausen
Friedrichroda
Tambach-Dietharz
Rotenburg
Fulda
Lauterbach
Grossenlüder
Altenschlirf
Schotten
Hartmannshain
Vogelsberg
Laubach
Grünberg
Lich
Hungen
Münzenberg
Nidda
Gedern
Ermenrod
Homberg
Lollar
Grossen Buseck
Giessen
Klein Linden
Wetzlar
Weilburg
Hadamar
Limburg
Rennerod
Westerburg
Hachenburg
Altenkirchen
Flammersfeld
Fröllingen
Herborn
Dillenburg
Haiger
Wilnsdorf
Eibelshausen
Niedereisenhausen
Gladenbach
Niederweidbach
Hainchen
Butzbach
Brandoberndorf
Bad Nauheim
Friedberg
Nieder Wöllstadt
Bad Selters
Büdingen
Birstein
Uerzell
Schlüchtern
Flieden
Döllbach
Steinau
Salmünster
Brückenau
Waldfenster
Bad Neustadt
Münnerstadt
Königshofen
Maroldsweisach
Bad Kissingen
Stadtlauringen
Hofheim
Schweinfurt
Hassfurt
Eltmann
Hammelburg
Burgsinn
Jossa
Burgjoss
Bad Orb
Gelnhausen
Langenselbold
Windecken
Bad Vilbel
Hanau
Steinheim
Kahl
Seligenstadt
Schimborn
Frammersbach
Laufach
Gemünden
Lohr
Aschaffenburg
Babenhausen
Ober Roden
Dieburg
Gross Umstadt
Mömlingen
Rohrbrunn
Karlstadt
Arnstein
Werneck
Zellingen
Marktheidenfeld
Uettingen
Würzburg
Dettelbach
Volkach
Gerolzhofen
Ebrach
Neuses
Schlüsselfeld
Kitzingen
Höchberg
Wertheim
Miltenberg
Amorbach
Höchst
Wörth
Michelstadt
Erbach
Reinheim
Darmstadt
Pfungstadt
Griesheim
Langen
Sprendlingen
Neu Isenburg
Offenbach
Frankfurt
Hofheim
Kelsterbach
Rüsselsheim
Gross Gerau
Mainz
Wiesbaden
Hochheim
Eltville
Rüdesheim
Bingen
Ingelheim
Gau-Algesheim
Niederolm
Nierstein
Oppenheim
Wörrstadt
Wöllstein
Alzey
Bad Kreuznach
Sobernheim
Kirn
Meisenheim
Lauterecken
Rockenhausen
Kirchheimbolanden
Monsheim
Marnheim
Worms
Bürstadt
Lampertheim
Gernsheim
Bensheim
Heppenheim
Lorsch
Fürth
Weinheim
Beerfelden
Viernheim
Mannheim
Ludwigshafen
Frankenthal
Grünstadt
Eisenberg
Enkenbach
Göllheim
Katzweiler
Kaiserslautern
Landstuhl
Bad Dürkheim
Mutterstadt
Schifferstadt
Neustadt an der Weinstrasse
Speyer
Schwetzingen
Hockenheim
Heidelberg
Neckargemünd
Schriesheim
Hirschhorn
Eberbach
Mosbach
Neckarelz
Möckmühl
Walldürn
Buchen
Hardheim
Tauberbischofsheim
Lauda
Bad Mergentheim
Berolzheim
Weikersheim
Stuppach
Dörzbach
Künzelsau
Langenburg
Blaufelden
Rothenburg o.d.T.
Brettheim
Ansbach
Gerabronn
Crailsheim
Feuchtwangen
Dinkelsbühl
Wassertrüdingen
Oettingen
Nördlingen
Harburg
Donauwörth
Bopfingen
Lauchheim
Aalen
Wasseralfingen
Ellwangen
Untergröningen
Gaildorf
Schwäbisch Hall
Öhringen
Neckarsulm
Bad Friedrichshall
Bad Wimpfen
Heilbronn
Weinsberg
Löwenstein
Sulzbach
Murrhardt
Backnang
Gschwend
Welzheim
Lorch
Schwäbisch Gmünd
Göppingen
Süssen
Geislingen
Heidenheim
Giengen
Gerstetten
Dillingen
Lauingen
Gundelfingen
Wertingen
Neresheim
Ochsenfurt
Giebelstadt
Aub
Creglingen
Marktbreit
Markt Bibart
Scheinfeld
Uffenheim
Windsheim
Neustadt an der Aisch
Oberdachstetten
Merkendorf
Bechhofen
Gunzenhausen
Fremdingen
Ludwigsburg
Bietigheim
Marbach
Winnenden
Schorndorf
Waiblingen
Kornwestheim
Stuttgart
Esslingen
Plochingen
Uhingen
Kirchheim
Nürtingen
Neuffen
Aich
Echterdingen
Waldenbuch
Bebenhausen
Herrenberg
Böblingen
Sindelfingen
Leonberg
Calw
Hirsau
Weil der Stadt
Bad Liebenzell
Calmbach
Wildbad
Herrenalb
Neuenbürg
Pforzheim
Mühlacker
Maulbronn
Vaihingen
Illingen
Asperg
Besigheim
Lauffen
Schwaigern
Eppingen
Knittlingen
Bretten
Bruchsal
Odenheim
Bad Langenbrücken
Sinsheim
Kirchardt
Waibstadt
Wiesloch
Eichtersheim
Neulussheim
Germersheim
Rülzheim
Rheinzabern
Kandel
Landau
Edenkoben
Annweiler
Dahn
Bad Bergzabern
Pirmasens
Waldfischbach
Wissembourg
Karlsruhe
Eggenstein
Weingarten
Wössingen
Ettlingen
Langensteinbach
Mörsch
Rastatt
Gaggenau
Oos
Baden-Baden
Bühl
Gernsbach
Forbach
Achern
Raumünzach
Seebach
Oberkirch
Kehl
Appenweier
Besenfeld
Altensteig
Hagenau
Bischwiller
Brumath
Soultz-sous-Forêts
Woerth
Reichshoffen
Niederbronn
Seltz
Auenheim
Lichtenau
Strasbourg
Molsheim
Illkirch
Grafenstaden
Koblenz
Bendorf
Neuwied
Montabaur
Bad Ems
Nassau
Oberlahnstein
Boppard
St. Goarshausen
St. Goar
Oberwesel
Kaub
Lorch
Bacharach
Rheinböllen
Simmern
Kastellaun
Stromberg
Gemünden
Holzhausen
Zollhaus
Bad Schwalbach
Idstein
Königstein
Oberursel
Bad Homburg
Camberg
Usingen
Esch
Bad Honningen
Karden
Hochstätten
Nahe
Rhein
Main
Neckar
Donau
Lahn
Fulda
Werra
Lenne
Ruhr
Möhne
Eder
Jagst
Kocher
Tauber
Enz
Murg
Moder
Nidda
Wörnitz
Fränkische Saale
Vogelsberg
Odenwald
Spessart
Rhön
Westerwald
Taunus
Sauerland
Hessen
Thüringen
Baden
Pfalz
D
F
E 4
E 5
E 63
E 70
E 10
E 12
E 11
A 15
A 13
A 20
A 23
A 10
N 63
N 68
N 4

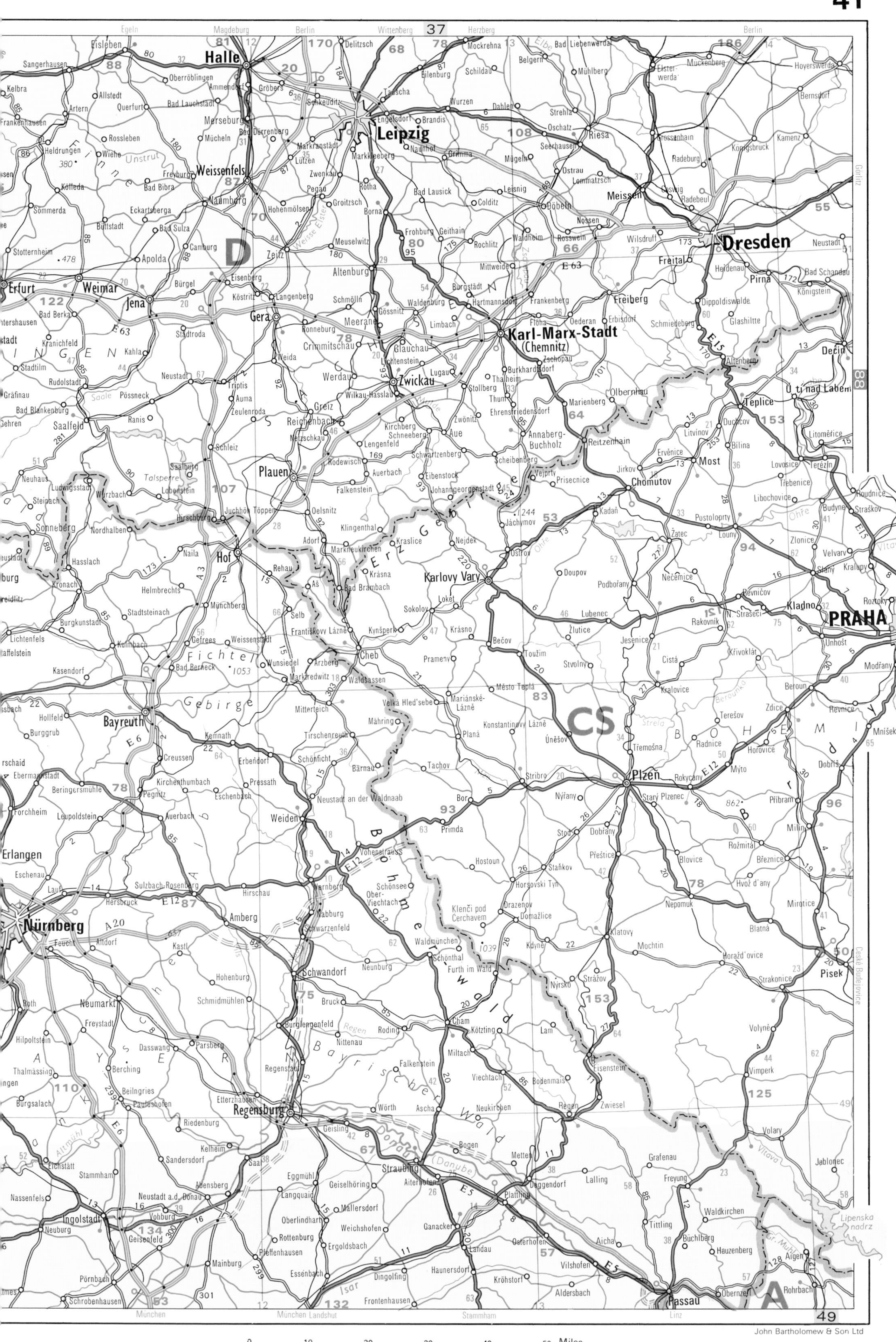

John Bartholomew & Son Ltd

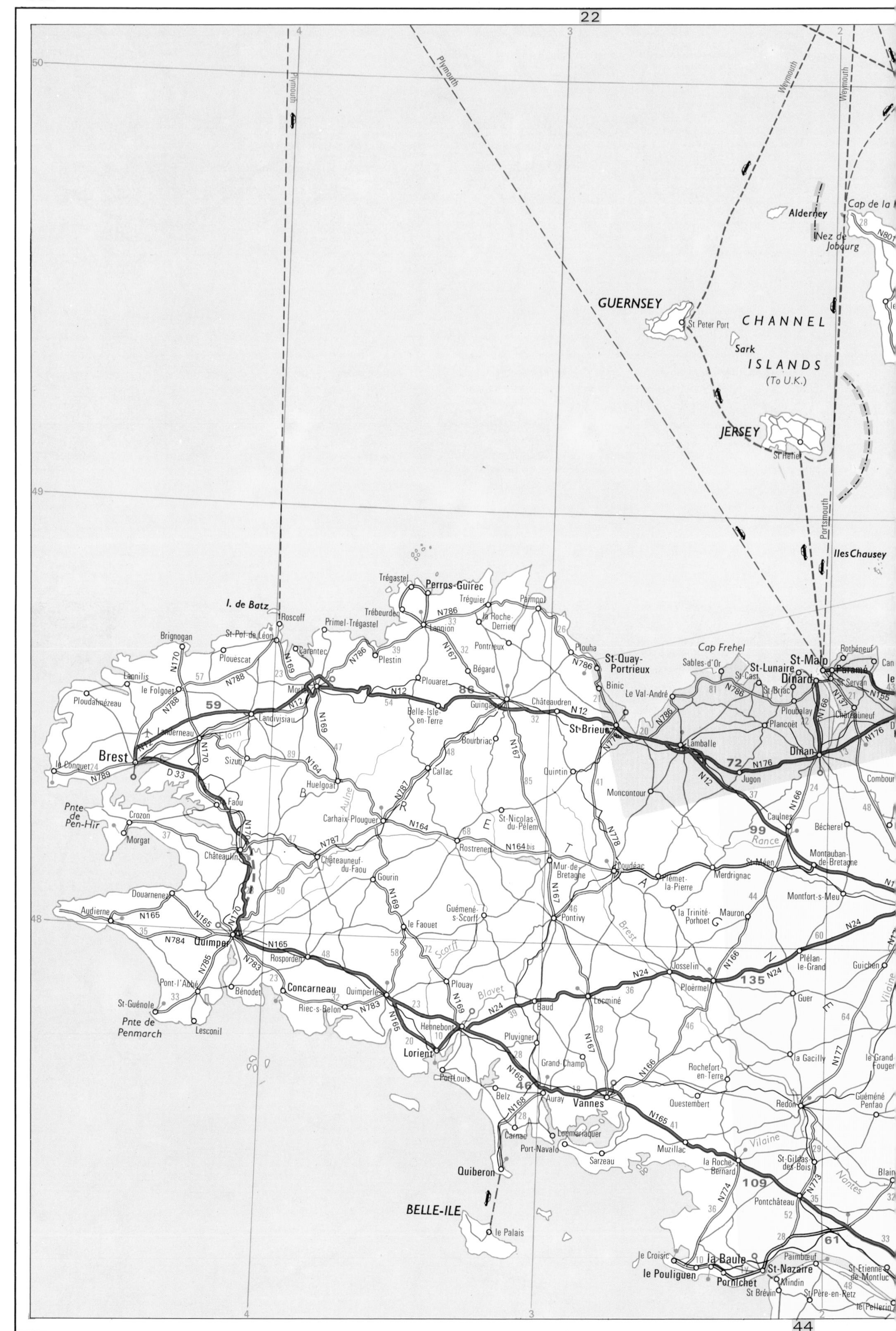

1:1 250 000

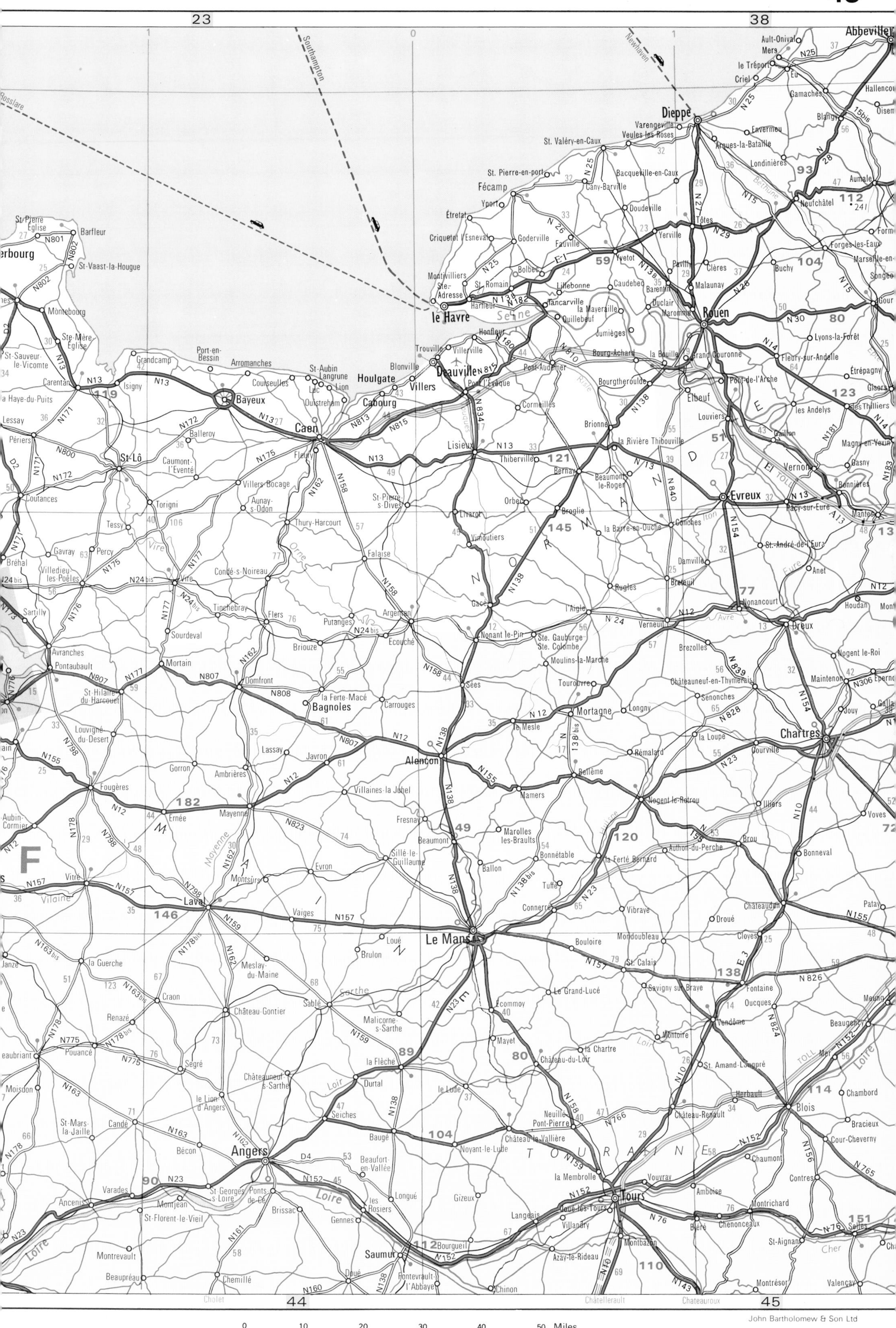

23
38
Southampton
Newhaven
Rosslare
Abbeville
Ault-Onival
Mers
le Tréport
Criel
Eu
Gamaches
Hallencourt
Dieppe
Varengeville
Veules-les-Roses
St. Valéry-en-Caux
Envermeu
Arques-la-Bataille
Londinières
Blangy
Aumale
Neufchâtel
St. Pierre-en-port
Fécamp
Yport
Étretat
Cany-Barville
Bacqueville-en-Caux
Doudeville
Tôtes
Yerville
Criquetot l'Esneval
Goderville
Fauville
Yvetot
Pavilly
Clères
Buchy
Forges-les-Eaux
Montivilliers
Ste-Adresse
St. Romain
Bolbec
Lillebonne
Caudebec
Barentin
Malaunay
le Havre
Harfleur
Tancarville
la Mayeraille
Quillebeuf
Duclair
Maromme
Rouen
Seine
Jumièges
Lyons-la-Forêt
Honfleur
Trouville
Villerville
Deauville
Bourg-Achard
la Bouille
Grand Couronne
Fleury-sur-Andelle
Étrépagny
Gisors
Pont-Audemer
Pont-l'Évêque
Bourgtheroulde
Elbeuf
Pont-de-l'Arche
les Andelys
Louviers
Gaillon
Vernon
Magny-en-Vexin
Barfleur
St-Pierre-Église
erbourg
St-Vaast-la-Hougue
Montebourg
Ste-Mère-Église
St-Sauveur-le-Vicomte
Carentan
Isigny
Grandcamp
Port-en-Bessin
Arromanches
Courseulles
St-Aubin
Langrune
Luc
Lion
Ouistreham
Houlgate
Cabourg
Blonville
Villers
Bayeux
Caen
la Haye-du-Puits
Lessay
Périers
Balleroy
Cormeilles
Lisieux
Brionne
la Rivière Thibouville
Thiberville
Bernay
Beaumont-le-Roger
Evreux
Pacy-sur-Eure
Mantes
St-Lô
Caumont-l'Eventé
Fleury
Villers-Bocage
Aunay-s-Odon
St-Pierre-s-Dives
Orbec
Livarot
Broglie
la Barre-en-Ouche
Conches
Coutances
Torigni
Tessy
Thury-Harcourt
Falaise
Vimoutiers
St.-André-de-l'Eure
Anet
Damville
Bréhal
Gavray
Percy
Villedieu-les-Poêles
Vire
Condé-s-Noireau
Flers
Tinchebray
Argentan
Putanges
Gacé
Rugles
Breteuil
Nonancourt
Houdan
Sartilly
Sourdeval
Briouze
Écouché
Nonant le-Pin
Ste. Gauburge-Ste. Colombe
l'Aigle
Verneuil
Dreux
Avranches
Pontaubault
Mortain
Domfront
Moulins-la-Marche
Brezolles
Nogent le-Roi
St-Hilaire-du-Harcouet
la Ferte-Macé
Bagnoles
Carrouges
Sées
Tourouvre
Châteauneuf-en-Thymerais
Senonches
Maintenon
Épernon
Louvigné-du-Desert
Lassay
Javron
Alençon
le Mesle
Mortagne
Longny
Jouy
Chartres
la Loupe
Courville
Rémalard
Gorron
Ambrières
Fougères
Villaines-la-Juhel
Mamers
Bellême
Nogent le-Rotrou
Illiers
Voves
Ernée
Mayenne
Fresnay
Marolles-les-Braults
Beaumont
Bonnétable
la Ferté Bernard
Authon-du-Perche
Brou
Bonneval
Vitré
Montsûrs
Evron
Sillé-le-Guillaume
Ballon
Tuffé
Laval
Vaiges
Le Mans
Connerré
Vibraye
Châteaudun
Droué
Patay
Cloyes
Mondoubleau
Bouloire
St. Calais
la Guerche
Loué
Brulon
Meslay-du-Maine
Craon
Château-Gontier
Sablé
Malicorne-s-Sarthe
le Grand-Lucé
Écommoy
Savigny sur Braye
Fontaine
Oucques
Vendôme
Montoire
Beaugency
Renazé
Pouancé
Segré
Châteauneuf-s-Sarthe
la Flèche
Durtal
Mayet
Château-du-Loir
la Chartre
St. Amand-Longpré
le Lion-d'Angers
le Lude
Herbault
Blois
Chambord
Moisdon
Candé
St-Mars-la-Jaille
Seiches
Baugé
Château-Renault
Neuillé-Pont-Pierre
Bécon
Angers
Noyant-le-Lude
Château-la-Vallière
Bracieux
Cour-Cheverny
Beaufort-en-Vallée
la Membrolle
Vouvray
Amboise
Chaumont
Contres
Ancenis
Varades
Montjean
St-Georges-s-Loire
Ponts-de-Cé
les Rosiers
Longué
Gizeux
Tours
Joué-les-Tours
Montrichard
St-Florent-le-Vieil
Brissac
Gennes
Langeais
Villandry
Bléré
Chenonceaux
St-Aignan
Selles
Montrevault
Saumur
Bourgueil
Azay-le-Rideau
Montbazon
Cher
Beaupréau
Chemillé
Doué
Fontevrault-l'Abbaye
Chinon
Montrésor
Valençay
Chalet
Châtellerault
Chateauroux
44
45
Loire
Sarthe
Mayenne
Vilaine
Orne
Vire
Eure
Avre
Iton
Risle
Huisne
Loir
NORMANDIE
MAINE
TOURAINE
F
John Bartholomew & Son Ltd
0 10 20 30 40 50 Miles
0 10 20 30 40 50 60 70 80 Kilometres

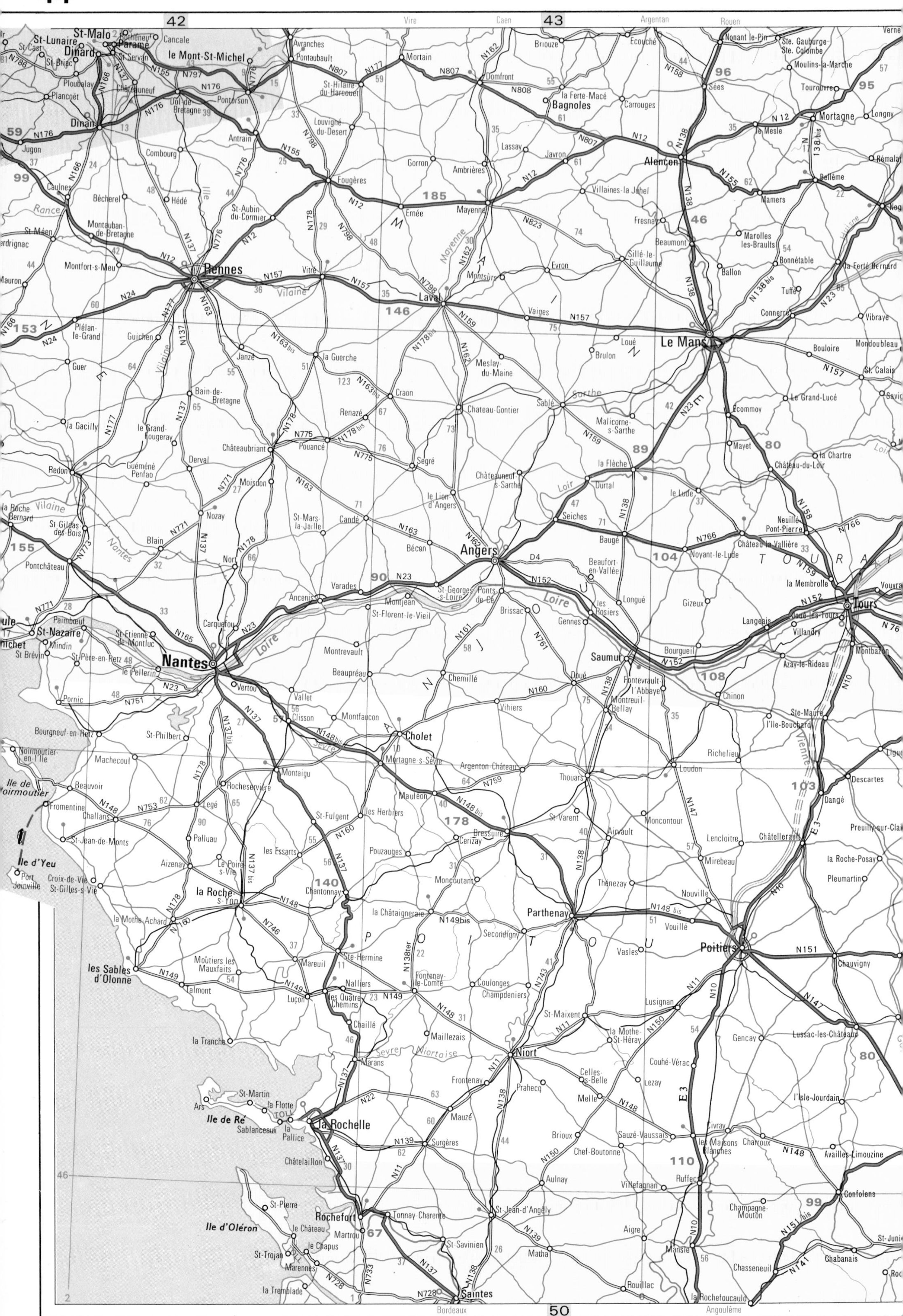

1:1 250 000

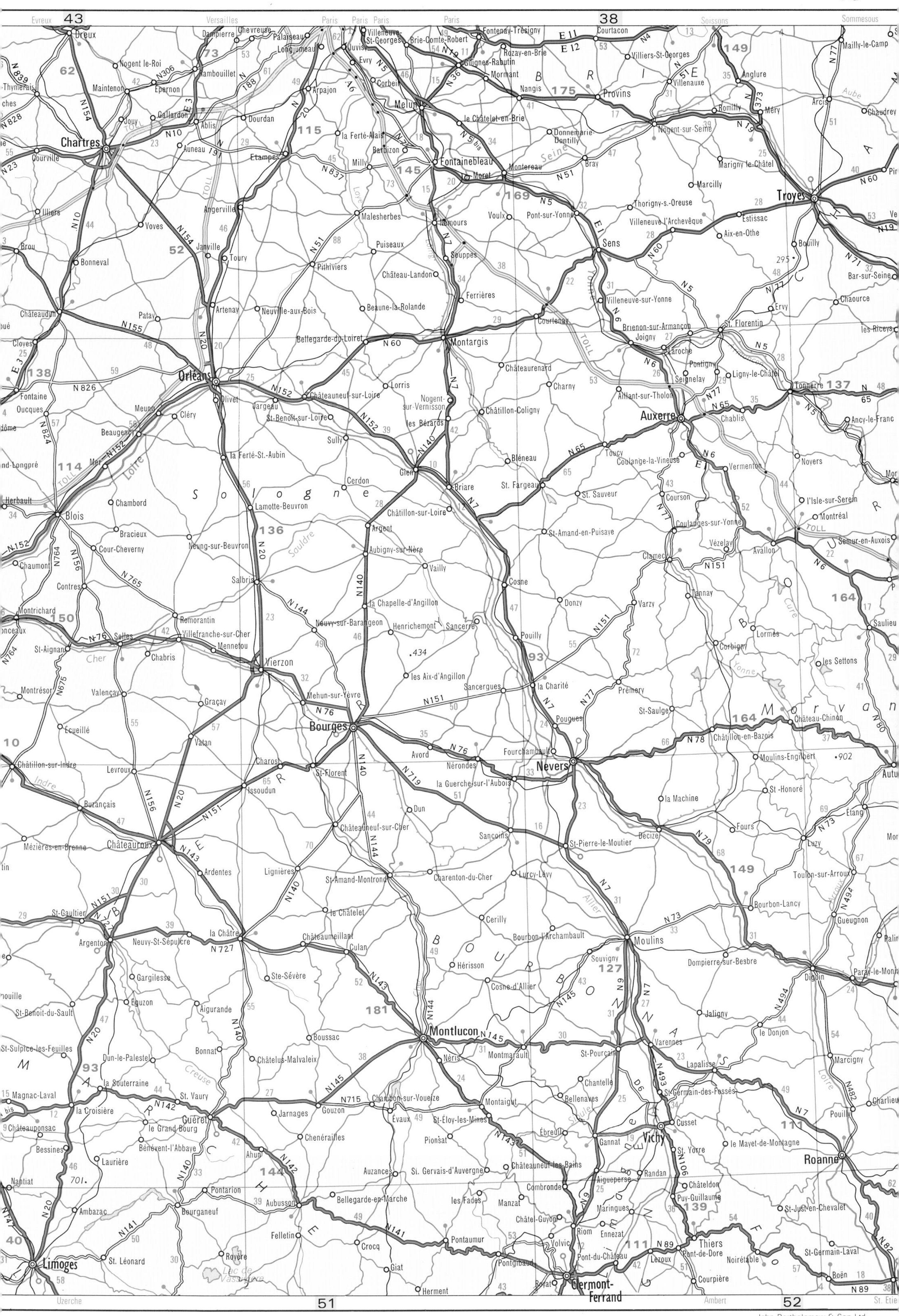

0   10   20   30   40   50   Miles
0   10   20   30   40   50   60   70   80   Kilometres

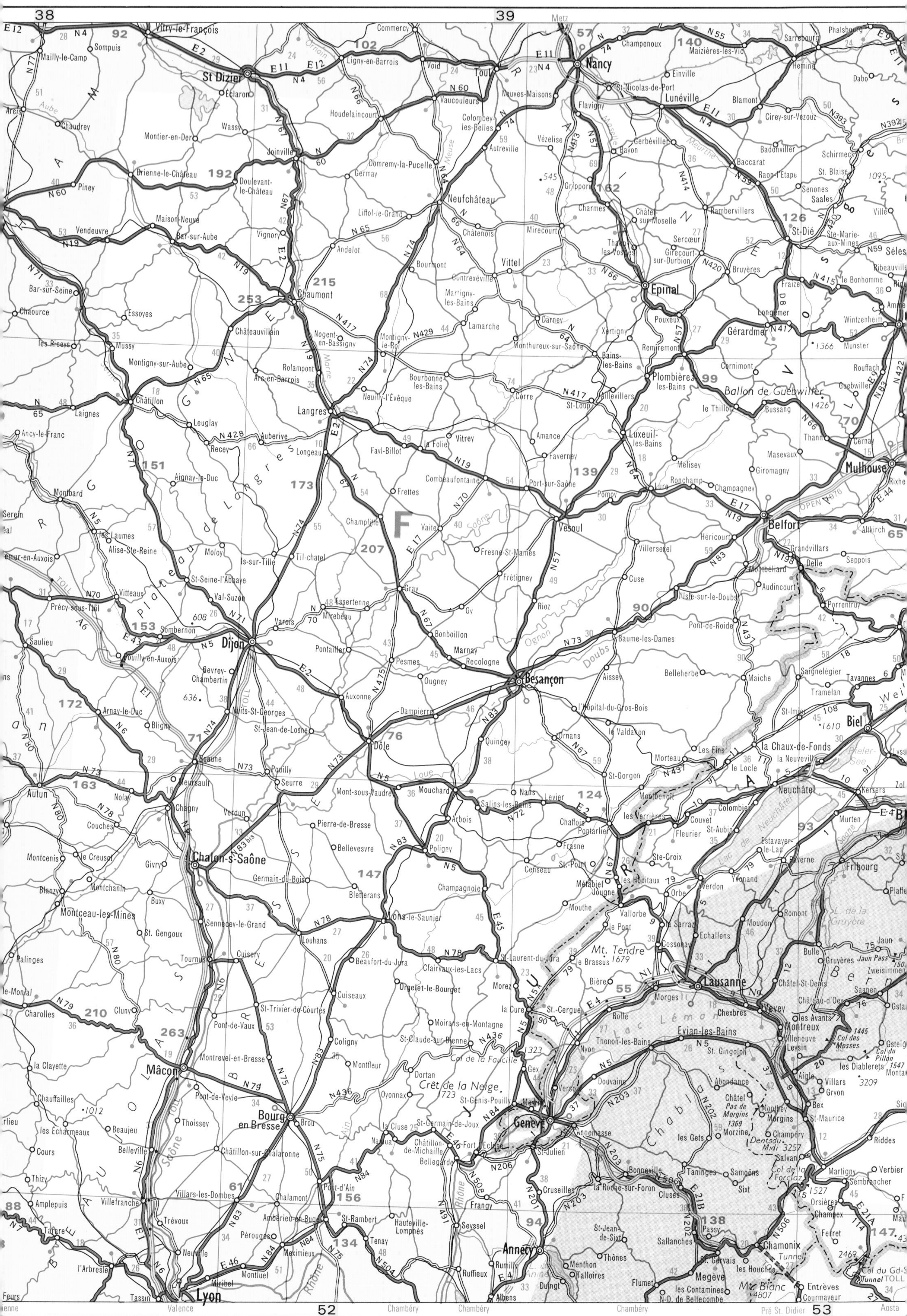

38
39
Metz
Vitry-le-François
Mailly-le-Camp
Sompuis
St Dizier
Eclaron
Commercy
Ligny-en-Barrois
Void
Toul
Nancy
Champenoux
Maizières-les-Vic
Sarrebourg
Phalsbourg
Heming
Dabo
Einville
St-Nicolas-de-Port
Lunéville
Blamont
Cirey-sur-Vezouz
Neuves-Maisons
Vaucouleurs
Houdelaincourt
Colombey-les-Belles
Flavigny
Chaudrey
Wassy
Montier-en-Der
Joinville
Brienne-le-Château
Doulevant-le-Château
Piney
Domremy-la-Pucelle
Germay
Autreville
Vézelise
Gerbéviller
Bayon
Badonviller
Schirmeck
Baccarat
Raon-l'Etape
St. Blaise
Senones
Saales
Neufchâteau
Liffol-le-Grand
Châtenois
Mirecourt
Charmes
Châtel-sur-Moselle
Rambervillers
St-Dié
Ste-Marie-aux-Mines
Sélestat
Ribeauvillé
Maison-Neuve
Bar-sur-Aube
Vendeuvre
Vignory
Andelot
Bourmont
Contrexéville
Vittel
Thaon-les-Vosges
Epinal
Sercœur
Girecourt-sur-Durbion
Bruyères
Fraize
le Bonhomme
Bar-sur-Seine
Chaource
Essoyes
Chaumont
Martigny-les-Bains
Lamarche
Darney
Xertigny
Pouxeux
Gérardmer
Longemer
Wintzenheim
Munster
Les Riceys
Mussy
Châteauvillain
Nogent-en-Bassigny
Montigny-le-Roi
Monthureux-sur-Saône
Bains-les-Bains
Remiremont
Cornimont
Rouffach
Guebwiller
Montigny-sur-Aube
Rolampont
Arc-en-Barrois
Bourbonne-les-Bains
Neuilly-l'Évêque
Corre
Plombières-les-Bains
le Thillot
Ballon de Guebwiller
Bussang
Châtillon
Laignes
Langres
Fillevillers
St-Loup
Ancy-le-Franc
Leuglay
Recey
Auberive
Longeau
Fayl-Billot
la Folie
Vitrey
Amance
Faverney
Luxeuil-les-Bains
Thann
Cernay
Mulhouse
Masevaux
Giromagny
Mélisey
Aignay-le-Duc
Plateau de Langres
Combeaufontaine
Port-sur-Saône
Ronchamp
Champagney
Rixheim
Montbard
Frettes
Pomoy
Lure
Belfort
Champlitte
F
Vaite
Vesoul
Héricourt
Altkirch
Les Laumes
Alise-Ste-Reine
Semur-en-Auxois
Moloy
Is-sur-Tille
Til-chatel
Fresne-St-Mamès
Villersexel
Grandvillars
Delle
Seppois
Montbéliard
St-Seine-l'Abbaye
Val-Suzon
Essertenne
Mirebeau
Gray
Frétigney
Cuse
Audincourt
Porrentruy
Précy-sous-Thil
Vitteaux
Sombernon
Dijon
Varois
Gy
Rioz
l'Isle-sur-le-Doubs
Saulieu
Pouilly-en-Auxois
Gevrey-Chambertin
Pontailler
Pesmes
Bonboillon
Marnay
Recologne
Baume-les-Dames
Pont-de-Roide
Auxonne
Ougney
Besançon
Aissey
Belleherbe
Maiche
Saignelégier
Tavannes
Tramelan
Arnay-le-Duc
Nuits-St-Georges
Bligny
St-Jean-de-Losne
Dampierre
l'Hôpital-du-Gros-Bois
le Valdahon
Ornans
St-Imier
Biel
Autun
Beaune
Meursault
Nolay
Pouilly
Seurre
Dôle
Quingey
Morteau
Les Fins
la Chaux-de-Fonds
le Locle
la Neuveville
Couches
Chagny
Verdun
Mont-sous-Vaudrey
Mouchard
Nans
Levier
Salins-les-Bains
Arbois
Chaffois
Pontarlier
les Verrières
Couvet
Fleurier
Neuchâtel
Kerzers
Murten
Montcenis
le Creusot
Givry
Chalon-s-Saône
Pierre-de-Bresse
Bellevesvre
Poligny
Frasne
St-Point
Ste-Croix
Colombier
St-Aubin
Estavayer-le-Lac
Payerne
Fribourg
Blanzy
Montchanin
Germain-du-Bois
Bletterans
Champagnole
Censeau
Métabief
Jougne
les Hôpitaux
Orbe
Yvonand
Montceau-les-Mines
Buxy
Sennecey-le-Grand
Louhans
Lons-le-Saunier
Mouthe
Vallorbe
le Pont
La Sarraz
Echallens
Moudon
Romont
St. Gengoux
Tournus
Cuisery
Beaufort-du-Jura
Clairvaux-les-Lacs
St-Laurent-du-Jura
Mt. Tendre
le Brassus
Cossonay
Lausanne
Bulle
Gruyères
Jaun Pass
Palinges
Orgelet-le-Bourget
Morez
Bière
Morges
Châtel-St-Denis
Vevey
Saanen
Charolles
Cluny
Pont-de-Vaux
St-Trivier-de-Courtes
Cuiseaux
Moirans-en-Montagne
la Cure
St.-Cergue
Rolle
Lac Léman
Chexbres
Montreux
Château-d'Oex
Les Avants
Col des Mosses
Gsteig
la Clayette
Mâcon
Montrevel-en-Bresse
Coligny
St-Claude-sur-Bienne
Col de la Faucille
Gex
Nyon
Thonon-les-Bains
Evian-les-Bains
St. Gingolph
Villeneuve
Leysin
Col du Pillon
les Diablerets
Aigle
Villars
Gryon
Bex
St-Maurice
Montfleur
Dortan
Crêt de la Neige
Oyonnax
Versoix
Douvaine
Abondance
Châtel
Pas de Morgins
Monthey
Morgins
Chauffailles
Pont-de-Veyle
Thoissey
Bourg en Bresse
Brou
St-Genis-Pouilly
Meyrin
Genève
Annemasse
Chablais
Champéry
Dents du Midi
Riddes
Beaujeu
les Echarmeaux
Belleville
Châtillon-sur-Chalaronne
Nantua
la Cluse
St-Germain-de-Joux
Châtillon-de-Michaille
Bellegarde
Fort l'Ecluse
St-Julien
les Gets
Morzine
Salvan
Cours
Thizy
Villars-les-Dombes
Chalamont
Pont-d'Ain
Cruseilles
la Roche-sur-Foron
Bonneville
Cluses
Taninges
Samoëns
Sixt
Col de la Forclaz
Martigny
Sembrancher
Amplepuis
Villefranche
Trévoux
Ambérieu-en-Bugey
St-Rambert
Hauteville-Lompnès
Frangy
Seyssel
Orsières
Champex
Tarare
Pérouges
Meximieux
Tenay
Annecy
St-Jean-de-Sixt
Passy
Sallanches
Chamonix
Ferret
l'Arbresle
Neuville
Montluel
Miribel
Rumilly
Ruffieux
Menthon
Talloires
Thônes
St. Gervais
les Houches
Col du Gd-St-Bernard
Tunnel
Lyon
Feurs
Tassin
Albens
Duingt
Flumet
Megève
les Contamines
N.-D. de Bellecombe
Mt. Blanc
Entrèves
Courmayeur
Valence
Chambéry
Pré St. Didier
Aosta
52
53

1:1 250 000

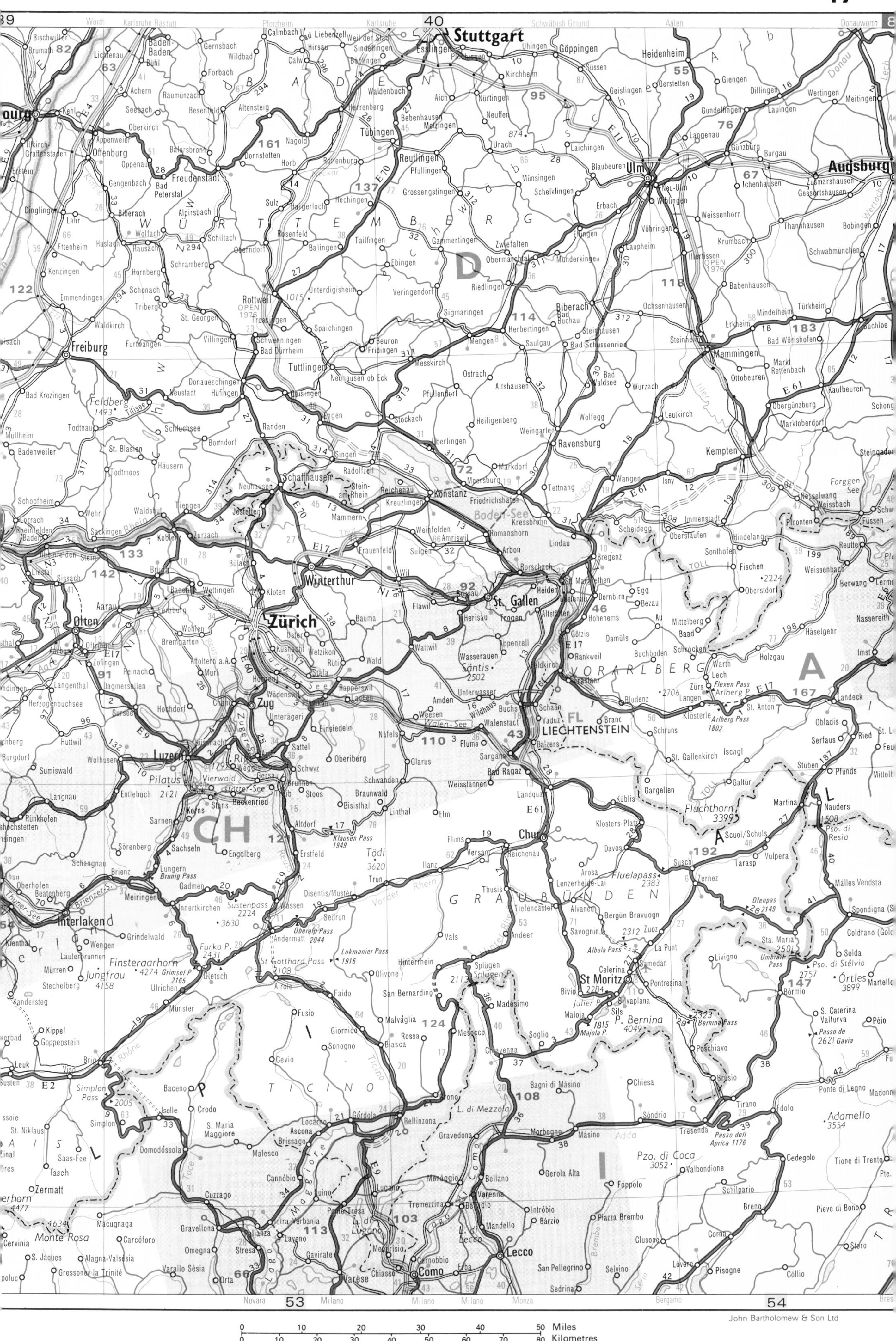
40
Stuttgart
Esslingen
Göppingen
Heidenheim
Aalen
Schwäbisch Gmünd
Donauwörth
Pforzheim
Karlsruhe
Baden-Baden
Calw
Böblingen
Herrenberg
Tübingen
Reutlingen
Nürtingen
Kirchheim
Geislingen
Ulm
Neu-Ulm
Günzburg
Augsburg
Dillingen
Lauingen
Gundelfingen
Offenburg
Freudenstadt
Nagold
Horb
Hechingen
Balingen
Rottweil
Tuttlingen
Sigmaringen
Riedlingen
Biberach
Memmingen
Mindelheim
Kaufbeuren
Kempten
Freiburg
Feldberg
Titisee
Donaueschingen
Villingen
Schwenningen
Singen
Radolfzell
Stockach
Überlingen
Ravensburg
Weingarten
Friedrichshafen
Konstanz
Boden-See
Lindau
Bregenz
Dornbirn
Feldkirch
Bludenz
Schaffhausen
Winterthur
Zürich
Baden
Aarau
Olten
Zug
Luzern
Pilatus
Rigi
Schwyz
Altdorf
St. Gallen
Rorschach
Appenzell
Säntis
Wil
Frauenfeld
Kreuzlingen
Glarus
Walen-See
Zürich-See
Zuger-See
Vierwald-stätter-See
Brienzer-See
Thuner-See
Interlaken
Meiringen
Grindelwald
Wengen
Mürren
Lauterbrunnen
Jungfrau
4158
Finsteraarhorn
4274
Grimsel P.
2165
Furka P.
2431
Sustenpass
2224
Oberalp Pass
2044
St Gotthard Pass
2108
Lukmanier Pass
1916
Klausen Pass
1949
Brunig Pass
Andermatt
Airolo
Faido
Biasca
Bellinzona
Locarno
Ascona
Lugano
L. di Lugano
Lago Maggiore
Lago di Como
L. di Lecco
Como
Lecco
Varese
Chiasso
Stresa
Pallanza
Domodóssola
Simplon Pass
2005
Brig
Visp
Zermatt
Monte Rosa
4634
Chur
Vaduz
LIECHTENSTEIN
FL
Bad Ragaz
Landquart
Davos
Klosters-Platz
Arosa
Lenzerheide-Lai
Thusis
Tiefencastel
Flims
Ilanz
Disentis/Mustér
Splügen
Splügen P.
San Bernardino
Chiavenna
Sondrio
Tirano
St Moritz
Maloja
Majola P.
1815
Julier P.
2284
Albula Pass
2312
Fluelapass
2383
Ofenpass
2149
P. Bernina
4049
Bernina Pass
2323
Poschiavo
Scuol/Schuls
Zernez
Sta. Maria
Umbrail Pass
2501
Pso. di Stélvio
2757
Ortles
3899
Bórmio
Passo di Gavia
2621
Adamello
3554
Pzo. di Coca
3052
Passo dell Aprica 1176
Landeck
Arlberg Pass
1802
Flexen Pass
St. Anton
Nauders
Pso. di Resia
1508
Fluchthorn
3399
Füssen
Reutte
Oberstdorf
Sonthofen
Immenstadt
Tödi
3620
Feldberg
1493
G R A U B Ü N D E N
V O R A R L B E R G
T I C I N O
W Ü R T T E M B E R G
B A D E N
D
CH
A
I
FL
53
54
Novara
Milano
Monza
Bergamo
Brescia
John Bartholomew & Son Ltd
0 10 20 30 40 50 Miles
0 10 20 30 40 50 60 70 80 Kilometres

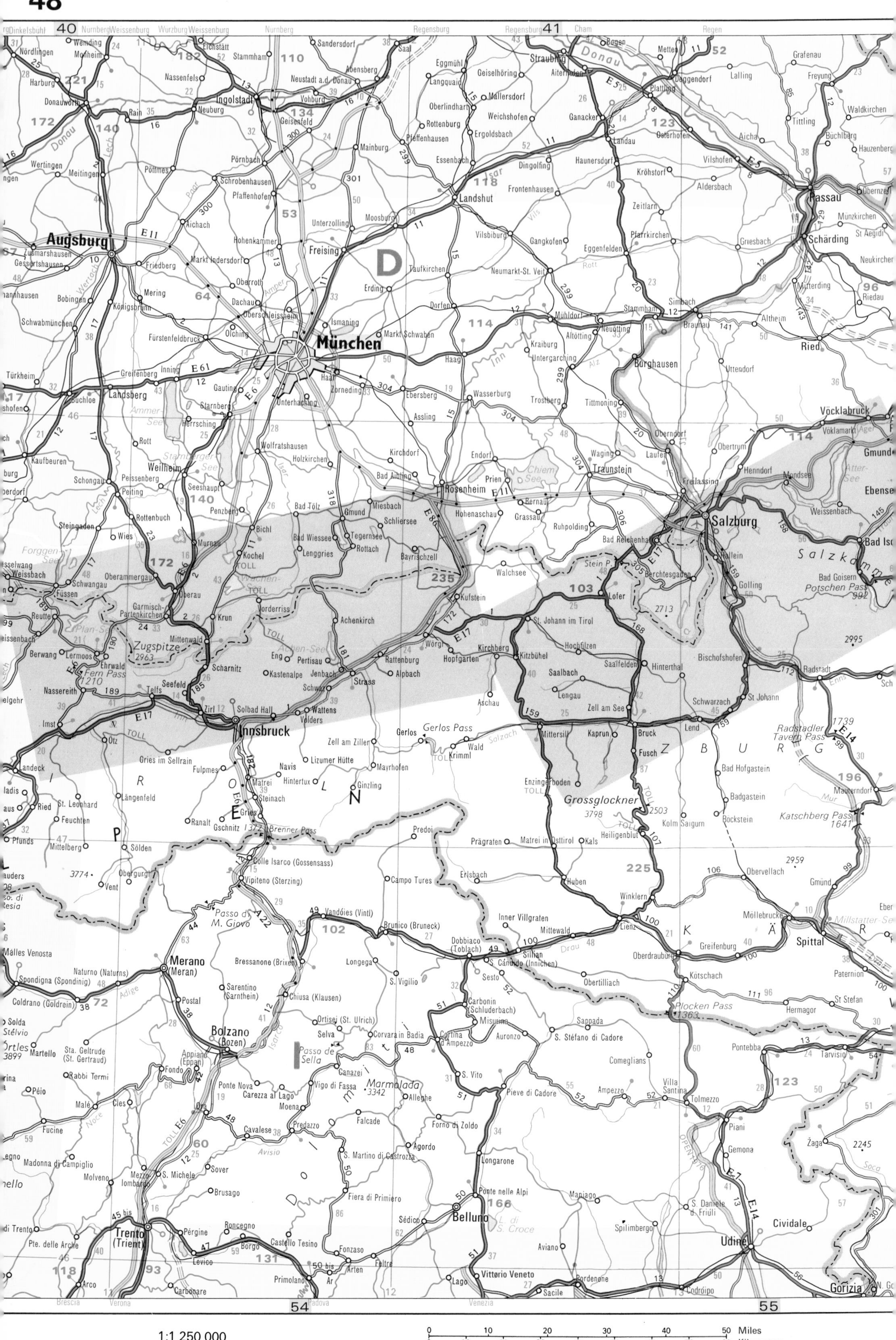

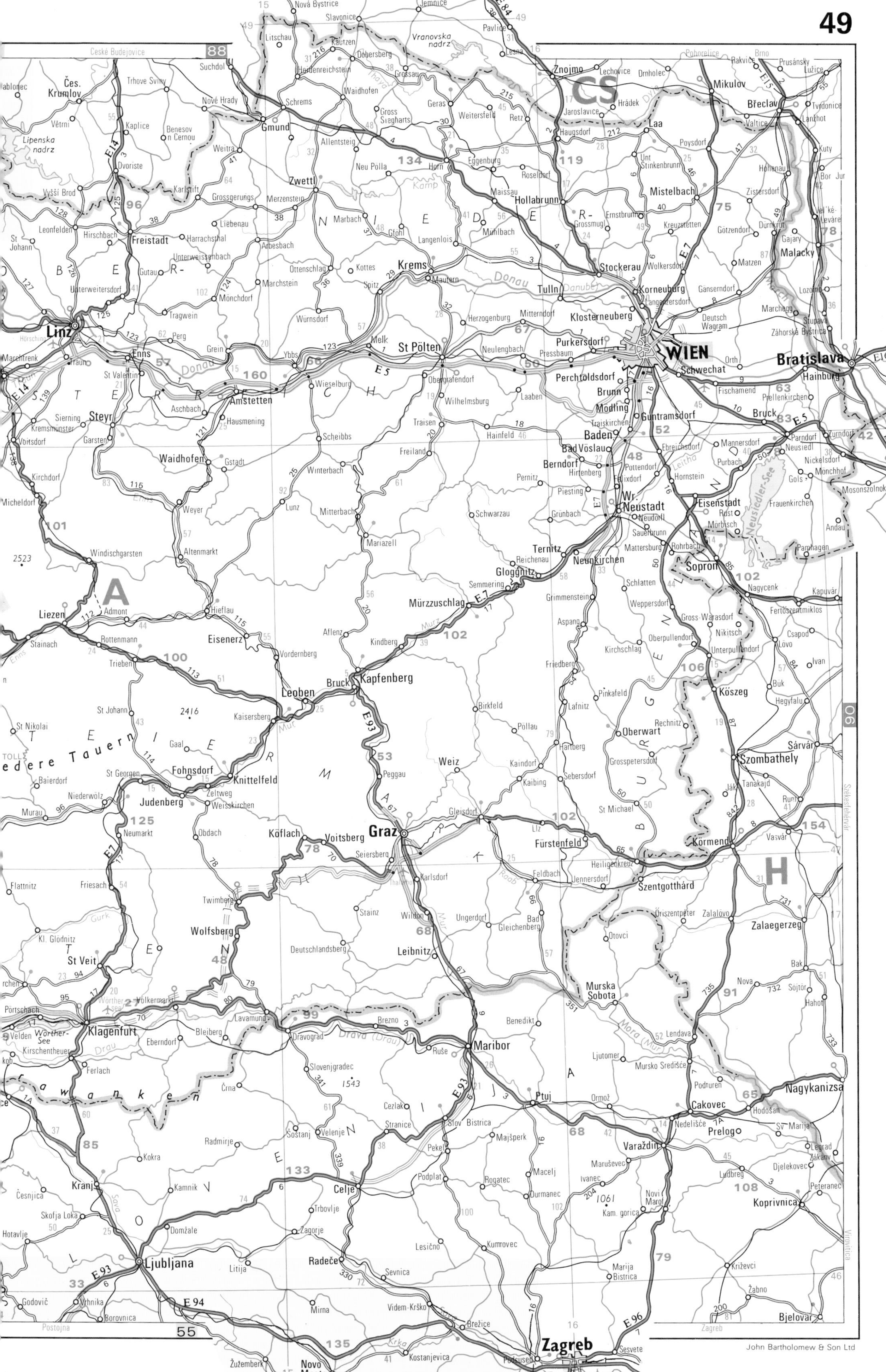

Ceské Budejovice 88

Ces. Krumlov
Jablonec
Vetrni
Lipenska nadrz
Trhove Sviny
Nové Hrady
Kaplice
Benesov n Cernou
Vyssí Brod
Dvoriste
St Johann
Leonfelden
Hirschbach
Unterweitersdorf
Freistadt
Harrachsthal
Gutau
Unterweissenbach
Linz
Hörsching
Marchtrenk
Traun
Enns
St Valentin
Steyr
Sierning
Kremsmünster
Voitsdorf
Garsten
Kirchdorf
Micheldorf
Windischgarsten
Liezen
Admont
Stainach
Rottenmann
Trieben
Eisenerz
Vordernberg
St Nikolai
St Johann
Kaisersberg
Leoben
Bruck
Kapfenberg
Knittelfeld
Zeltweg
Weisskirchen
Judenberg
Obdach
Köflach
Voitsberg
Graz
Seiersberg
Karlsdorf
Twimberg
Wolfsberg
Deutschlandsberg
Leibnitz
Wildon
Stainz

Nová Bystrice
Slavonice
Litschau
Kautzen
Schrems
Waidhofen
Gmünd
Weitra
Allentsteig
Zwettl
Grossgerungs
Merzenstein
Liebenau
Karlstift
Mönchdorf
Tragwein
Perg
Grein
Ybbs
Amstetten
Aschbach
Hausmening
Waidhofen
Gstadt
Scheibbs
Winterbach
Lunz
Mitterbach
Mariazell
Altenmarkt
Weyer
Hieflau
Aflenz
Kindberg
Mürzzuschlag
Semmering
Glognitz
Reichenau
Neunkirchen
Gloggnitz
Aspang
Friedberg
Pinkafeld
Lafnitz
Oberwart
Rechnitz
Hartberg
Kaindorf
Sebersdorf
St Michael
Gleisdorf
Feldbach
Bad Gleichenberg
Fürstenfeld
Jennersdorf
Heiligenkreuz
Szentgotthárd
Öriszentpéter
Zalalovo
Otovci
Murska Sobota
Lendava
Ljutomer
Mursko Sredisce
Ormoz
Ptuj
Maribor
Ruse
Slovenjgradec
Crna
Cezlak
Slov. Bistrica
Stranice
Velenje
Sostanj
Soštanj
Majsperk
Pekel
Rogatec
Celje
Trbovlje
Zagorje
Lesicno
Kumrovec
Radmirje
Kamnik
Litija
Radece
Sevnica
Videm-Krsko
Brezice
Mirna
Krka
Novo Mesto
Kostanjevica
Zagreb
Sesvete
Ljubljana
Vrhnika
Godovic
Borovnica
Postojna

Jemnice
Dobersberg
Grossau
Lesná
Pavlice
Vranovska nadrz
Znojmo
Jaroslavice
Hrádek
Laa
Poysdorf
Mikulov
Breclav
Valtice
Lanzhot
Tvrdonice
Hohenau
Bor Jur
Gajary
Malacky
Zistersdorf
Dürnkrut
Marchegg
Záhorská Bystrica
Stupava
Devínska
Bratislava
Hainburg
Parndorf
Neusiedl
Nickelsdorf
Gols
Mönchhof
Mosonszolnok
Andau
Pamhagen
Frauenkirchen
Neusiedler-See
Eisenstadt
Rust
Mörbisch
Sopron
Nagycenk
Fertöszentmiklos
Kapuvár
Sárvár
Szombathely
Ják
Tanakajd
Rum
Vasvár
Körmend
Zalaegerszeg
Bak
Nova
Sojtor
Hahon
Lendava
Podturen
Cakovec
Nedelisce
Prelog
Nagykanizsa
Varazdin
Hodosan
Marija
Ludbreg
Djelekovec
Koprivnica
Peteranec
Legrad
Krizevci
Bjelovar
Virovitica

Haugsdorf
Retz
Hollabrunn
Stockerau
Korneuburg
Klosterneuburg
Tulln
Langenzersdorf
Deutsch Wagram
Gänserndorf
WIEN
Schwechat
Purkersdorf
Pressbaum
Perchtoldsdorf
Brunn
Mödling
Guntramsdorf
Baden
Bad Vöslau
Berndorf
Hirtenberg
Pottenstein
Piesting
Wr. Neustadt
Neudörfl
Sauerbrunn
Mattersburg
Rohrbach
Pöttsching
Wöllersdorf
Ternitz
Grimmenstein
Schlatten
Wepersdorf
Oberpullendorf
Unterpullendorf
Kirchschlag
Nikitsch
Lövo
Csapod
Bük
Köszeg
Hegyfalu

St Veit
Friesach
Flattnitz
Kl. Glödnitz
St Nikolai
Baierdorf
Niederwölz
Murau
Neumarkt
Fohnsdorf
St Georgen
Gaal
Klagenfurt
Wörther-See
Velden
Pörtschach
Völkermarkt
Ferlach
Kirschentheuer
Bleiberg
Ebendorf
Lavamund
Drávograd
Brezno
Benedikt

John Bartholomew & Son Ltd

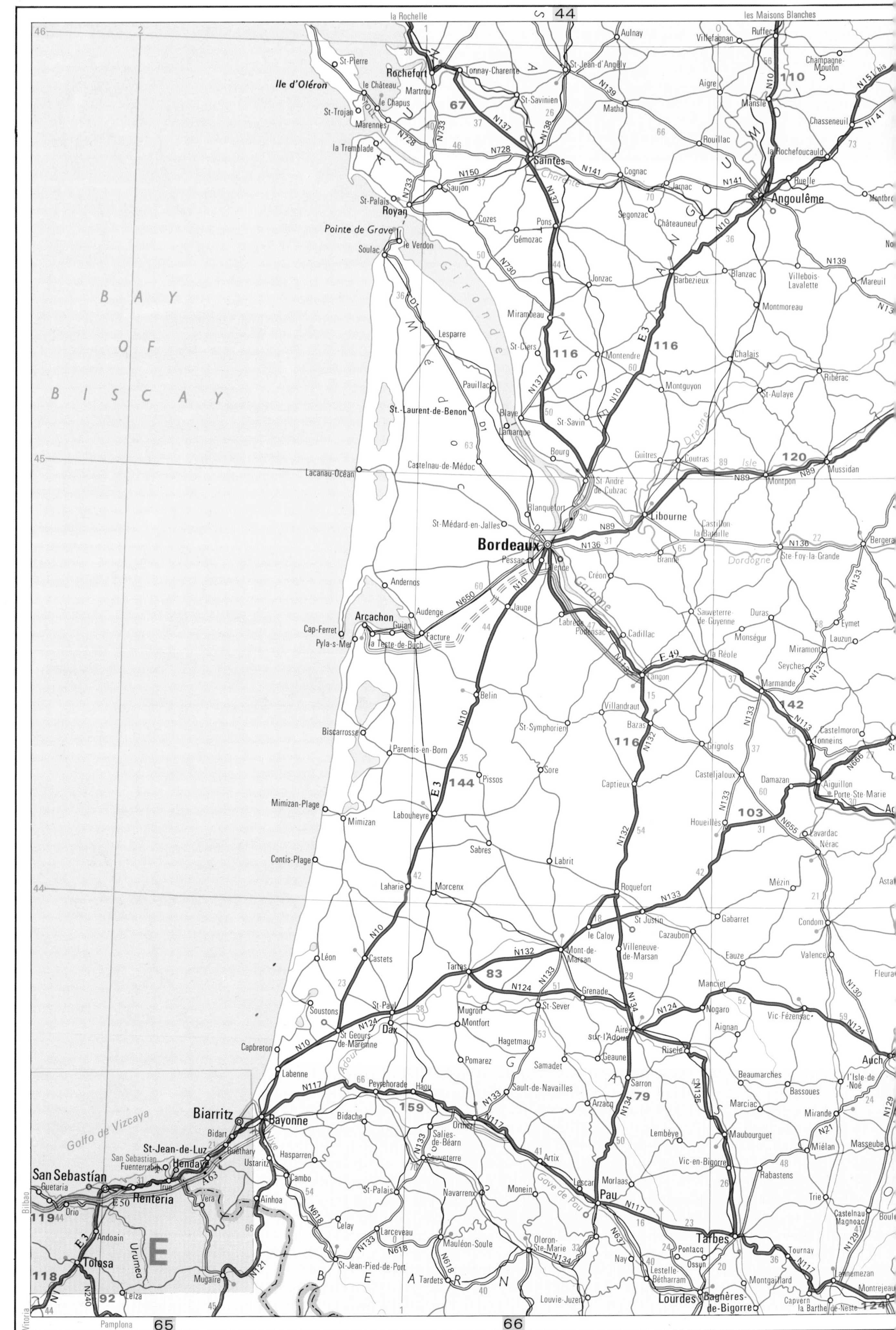
la Rochelle
44
les Maisons Blanches
Ile d'Oléron
St-Pierre
Rochefort
Tonnay-Charente
Aulnay
Villefagnan
Ruffec
Champagne-Mouton
le Château
le Chapus
St-Trojan
Marennes
Martrou
St-Jean-d'Angély
St-Savinien
Aigre
Mansle
Matha
la Tremblade
Saintes
Rouillac
Chasseneuil
la Rochefoucauld
Cognac
Jarnac
Ruelle
Angoulême
Saujon
St-Palais
Royan
Pointe de Grave
Soulac
le Verdon
Cozes
Gémozac
Pons
Segonzac
Châteauneuf
Charente
Gironde
Jonzac
Barbezieux
Blanzac
Villebois-Lavalette
Mareuil
Montmoreau
BAY
OF
BISCAY
Lesparre
Mirambeau
St-Ciers
Montendre
Chalais
Ribérac
Pauillac
St.-Laurent-de-Benon
Blaye
Lamarque
Montguyon
St-Aulaye
St-Savin
Bourg
Guîtres
Coutras
Dronne
Isle
Mussidan
Montpon
Lacanau-Océan
Castelnau-de-Médoc
St-André-de-Cubzac
Blanquefort
Libourne
St-Médard-en-Jalles
Castillon-la-Bataille
Bordeaux
Pessac
Talence
Branne
Dordogne
Ste-Foy-la-Grande
Bergerac
Créon
Garonne
Andernos
Arcachon
Audenge
Gujan
Facture
Cap-Ferret
Pyla-s-Mer
la Teste-de-Buch
Jauge
Labrède
Podensac
Cadillac
Sauveterre-de-Guyenne
Duras
Monségur
Eymet
Lauzun
Miramont
Seyches
la Réole
Langon
Marmande
Belin
Villandraut
Bazas
St-Symphorien
Castelmoron
Tonneins
Biscarrosse
Parentis-en-Born
Grignols
Sore
Pissos
Captieux
Casteljaloux
Damazan
Aiguillon
Porte-Ste-Marie
Mimizan-Plage
Mimizan
Labouheyre
Houeillès
Lavardac
Nérac
Sabres
Contis-Plage
Labrit
Mézin
Laharie
Morcenx
Roquefort
St Justin
Gabarret
le Caloy
Cazaubon
Condom
Léon
Castets
Mont-de-Marsan
Villeneuve-de-Marsan
Eauze
Valence
Tartas
Grenade
Manciet
Soustons
St-Paul
Mugron
Montfort
St-Sever
Nogaro
Vic-Fézensac
Capbreton
St Geours-de-Maremne
Dax
Hagetmau
Aire-sur-l'Adour
Aignan
Riscle
Auch
Pomarez
Samadet
Geaune
Labenne
Peyrehorade
Hagetmau
Sarron
Beaumarchès
Bassoues
l'Isle-de-Noé
Sault-de-Navailles
Biarritz
Bayonne
Bidache
Orthez
Arzacq
Marciac
Mirande
Bidart
St-Jean-de-Luz
Guéthary
Salies-de-Béarn
Sauveterre
Lembèye
Maubourguet
San Sebastián
Fuenterrabía
Hendaya
Ustaritz
Hasparren
Artix
Miélan
Masseube
San Sebastián
Guetaria
Renteria
Irun
Cambo
St-Palais
Lescar
Vic-en-Bigorre
Rabastens
Orio
Vera
Ainhoa
Navarrenx
Monein
Pau
Morlaas
Trie
Castelnau-Magnoac
Andoain
Celay
Larceveau
Mauléon-Soule
Oloron-Ste-Marie
Nay
Tarbes
Pontacq
Ossun
Tournay
Tolosa
Urumea
E
St-Jean-Pied-de-Port
Lestelle-Bétharram
Mugaire
Leiza
Tardets
Lannemezan
Montgaillard
Montréjeau
Lourdes
Bagnères-de-Bigorre
Capvern
la Barthe-de-Neste
Louvie-Juzon
Golfo de Vizcaya
Gave de Pau
Adour
BÉARN
Pamplona
65
66
Bilbao
Vitoria

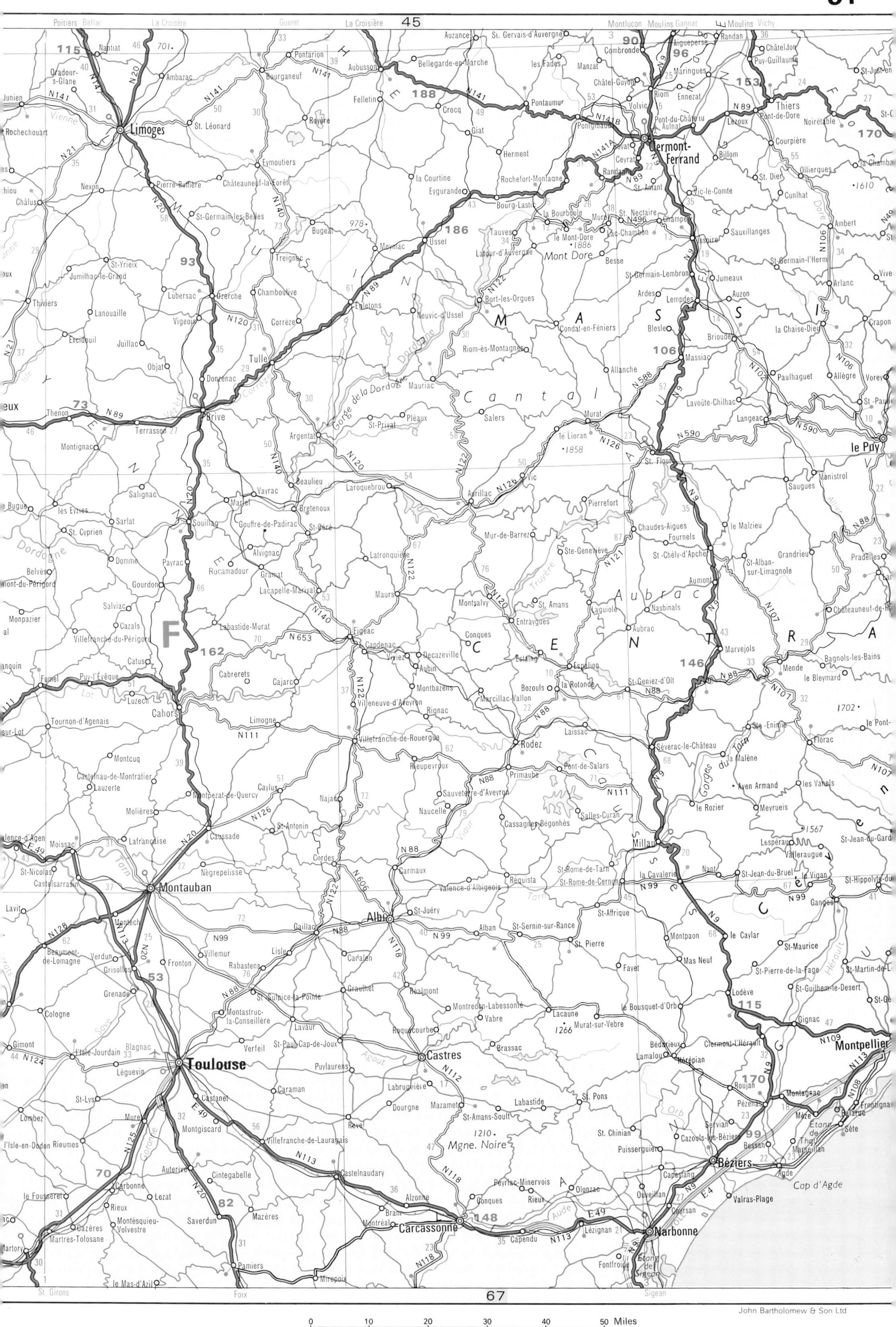

John Bartholomew & Son Ltd

0   10   20   30   40   50 Miles
0   10   20   30   40   50   60   70   80 Kilometres

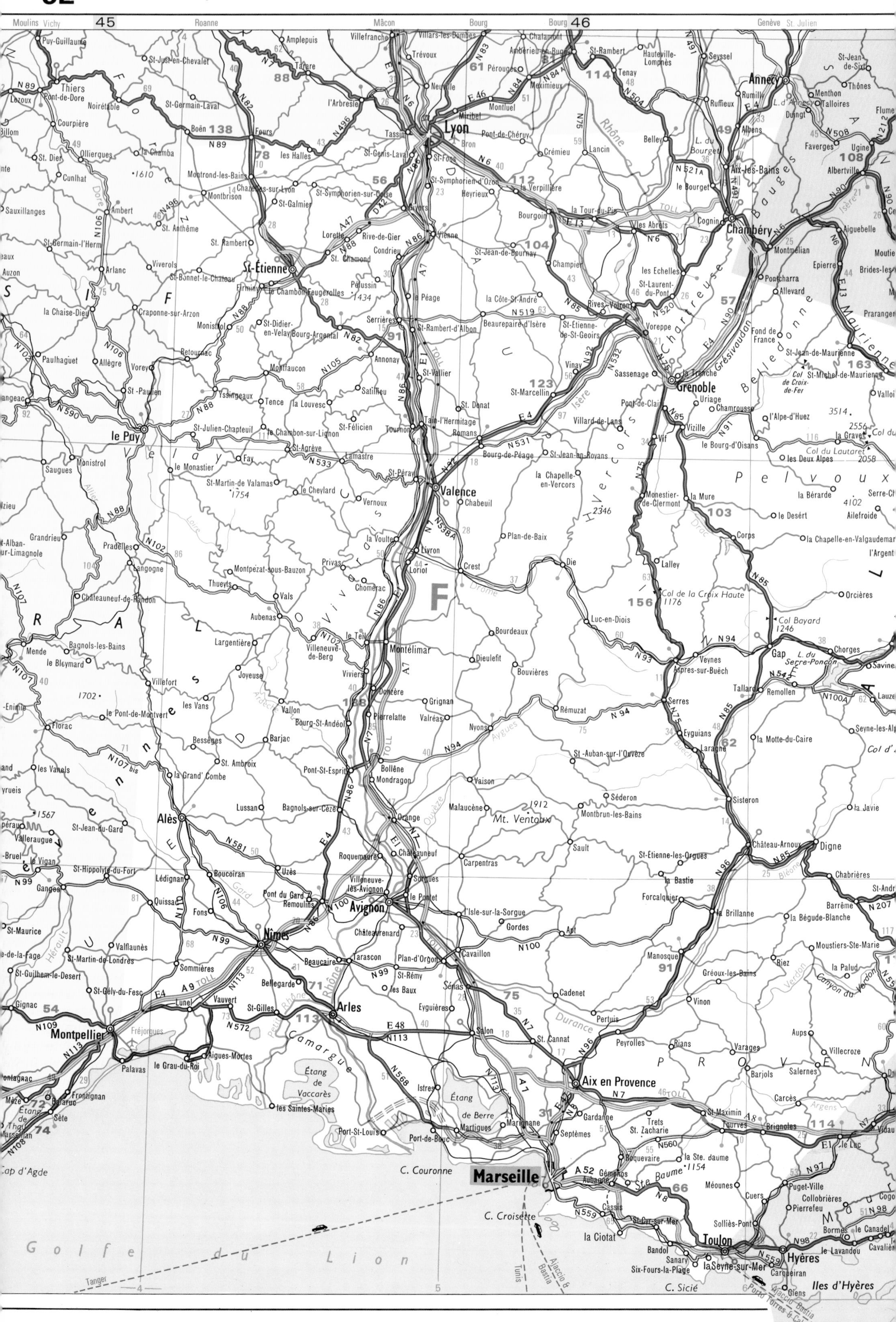
Moulins Vichy
45
Roanne
Mâcon
Bourg
Bourg 46
Genève St. Julien
Puy-Guillaume
Amplepuis
Villefranche
Villars-les-Dombes
Chalamont
Trévoux
Ambérieu-en-Bugey
St-Rambert
Hauteville-Lompnès
Seyssel
St-Jean-de-Sixt
St-Just-en-Chevalet
Tarare
Pérouges
Tenay
Annecy
Thônes
Thiers
Lezoux
Pont-de-Dore
Noirétable
Neuville
Meximieux
Rumilly
Menthon
Talloires
Billom
Courpière
St-Germain-Laval
l'Arbresle
Miribel
Montluel
Ruffieux
L. d'Annecy
Duingt
Flumet
Boën
Feurs
Lyon
Tassin
Bron
Pont-de-Chéruy
Belley
L. du Bourget
Albens
Faverges
Ugine
St. Dier
Olliergues
la Chamba
les Halles
St-Genis-Lavol
St-Fons
Crémieu
Lancin
Rhône
Cunlhat
Montrond-les-Bains
Chazelles-sur-Lyon
St-Symphorien-sur-Coise
St-Symphorien-d'Ozon
Heyrieux
la Verpillière
Aix-les-Bains
le Bourget
Albertville
Sauxillanges
Ambert
Montbrison
St-Galmier
Givors
la Tour-du-Pin
Bourgoin
les Abrets
Cognin
Chambéry
Bauges
Aiguebelle
St. Anthème
St-Germain-l'Herm
St. Rambert
Lorette
Rive-de-Gier
Vienne
Condrieu
St. Chamond
St-Jean-de-Bournay
Montmélian
Moutie
Arlanc
Viverols
St-Étienne
St-Bonnet-le-Château
Firminy
le Chambon-Feugerolles
Pélussin
Champier
les Echelles
St-Laurent-du-Pont
Pontcharra
Allevard
Epierre
Brides-les-Bains
la Chaise-Dieu
Craponne-sur-Arzon
Monistrol
St-Didier-en-Velay
Bourg-Argental
le Péage
la Côte-St-André
Beaurepaire-d'Isère
Rives
Voiron
Voreppe
Chartreuse
St-Étienne-de-St-Geoirs
Fond de France
Prarangei
Paulhaguet
Allègre
Vorey
Retournac
Montfaucon
Annonay
Serrières
St-Rambert-d'Albon
Sassenage
la Tronche
Grenoble
St-Jean-de-Maurienne
St-Michel-de-Maurienne
Maurienne
Belledonne
Grésivaudan
Col de Croix-de-Fer
St-Paulien
Yssingeaux
Tence
la Louvesc
Satillieu
St-Vallier
Vinay
St-Marcellin
Isère
Pont-de-Claix
Uriage
Chamrousse
l'Alpe-d'Huez
le Puy
Velay
St-Julien-Chapteuil
le Chambon-sur-Lignon
St-Félicien
Tournon
Tain-l'Hermitage
St. Donat
Romans
Villard-de-Lans
Vif
Vizille
le Bourg-d'Oisans
la Grave
Col du Lautaret
les Deux Alpes
Saugues
Monistrol
le Monastier
Fay
St-Agrève
Lamastre
St-Péray
Bourg-de-Péage
St-Jean-en-Royans
Vercors
St-Martin-de-Valamas
le Cheylard
Valence
Chabeuil
la Chapelle-en-Vercors
Monestier-de-Clermont
la Mure
Pelvoux
la Bérarde
Serre-Ch
Vernoux
le Désert
Ailefroide
Grandrieu
Pradelles
Langogne
Privas
la Voulte
Livron
Loriol
Plan-de-Baix
Crest
Die
Corps
Drac
la Chapelle-en-Valgaudemar
l'Argent
St-Alban-sur-Limagnole
Montpezat-sous-Bauzon
Thueyts
Vals
Chomérac
Vivarais
F
Drôme
Lalley
Col de la Croix Haute
Orcières
Châteauneuf-de-Randon
Aubenas
Luc-en-Diois
Col Bayard
Mende
Bagnols-les-Bains
le Bleymard
Largentière
Villeneuve-de-Berg
le Teil
Montélimar
Bourdeaux
Dieulefit
Bouvières
Gap
Chorges
L. du Serre-Ponçon
Savine
Veynes
Aspres-sur-Buëch
Tallard
Remollen
Lauze
Joyeuse
Viviers
Villefort
Donzère
Grignan
Rémuzat
Serres
Seyne-les-Al
Florac
le Pont-de-Montvert
les Vans
Vallon
Pierrelatte
Valréas
Nyons
Eyguians
Laragne
la Motte-du-Caire
Bessèges
Barjac
Bourg-St-Andéol
Aygues
St-Auban-sur-l'Ouvèze
Buëch
les Vanels
St. Ambroix
la Grand' Combe
Pont-St-Esprit
Bollène
Mondragon
Vaison
Sisteron
Cévennes
Lussan
Bagnols-sur-Cèze
Malaucène
Mt. Ventoux
Séderon
Montbrun-les-Bains
la Javie
Alès
St-Jean-du-Gard
Valleraugue
Orange
Ouvèze
Sault
Château-Arnoux
Digne
le Vigan
St-Hippolyte-du-Fort
Ganges
Lédignan
Boucoiran
Uzès
Roquemaure
Châteauneuf
Carpentras
St-Étienne-les-Orgues
la Bastie
Bléone
Chabrières
Quissac
Pont du Gard
Remoulins
Villeneuve-lès-Avignon
Sorgues
le Pontet
Avignon
Forcalquier
Barrême
St-Maurice
Fons
Nîmes
l'Isle-sur-la-Sorgue
Gordes
Apt
la Brillanne
la Bégude-Blanche
Valflaunès
St-Martin-de-Londres
Sommières
Beaucaire
Tarascon
Châteaurenard
Plan-d'Orgon
Cavaillon
Manosque
Moustiers-Ste-Marie
Riez
St-Guilhem-le-Desert
St-Gély-du-Fesc
Bellegarde
St-Rémy
les Baux
Sénas
Cadenet
Gréoux-les-Bains
Verdon
la Palud
Canyon du Verdon
Gignac
Lunel
Vauvert
St-Gilles
Arles
Eyguières
Vinon
Pertuis
Durance
Montpellier
Fréjorgues
Aigues-Mortes
Camargue
Salon
St. Cannat
Peyrolles
Rians
Aups
Villecroze
Palavas
le Grau-du-Roi
Étang de Vaccarès
Istres
Étang de Berre
Aix en Provence
Varages
Barjols
Salernes
Frontignan
Sète
Étang de Thau
Balaruc
Mèze
les Saintes-Maries
Martigues
Marignane
Gardanne
Trets
St. Zacharie
St-Maximin
Tourves
Brignoles
Carcès
Argens
Vidau
Port-St-Louis
Port-de-Bouc
Septèmes
Roquevaire
le Luc
Cap d'Agde
C. Couronne
Marseille
Gémenos
Aubagne
Ste Baume
St. daume
Méounes
Puget-Ville
Collobrières
Pierrefeu
Cuers
C. Croisette
Cassis
St-Cyr-sur-Mer
la Ciotat
Solliès-Pont
Bormes
le Canadel
Golfe du Lion
Bandol
Sanary
Toulon
Hyères
le Lavandou
Cavalière
Six-Fours-la-Plage
la Seyne-sur-Mer
Carqueiranne
Giens
Tanger
Tunis
Ajaccio & Bastia
C. Sicié
Iles d'Hyères
Porto Torres & Calvi

1:1 250 000

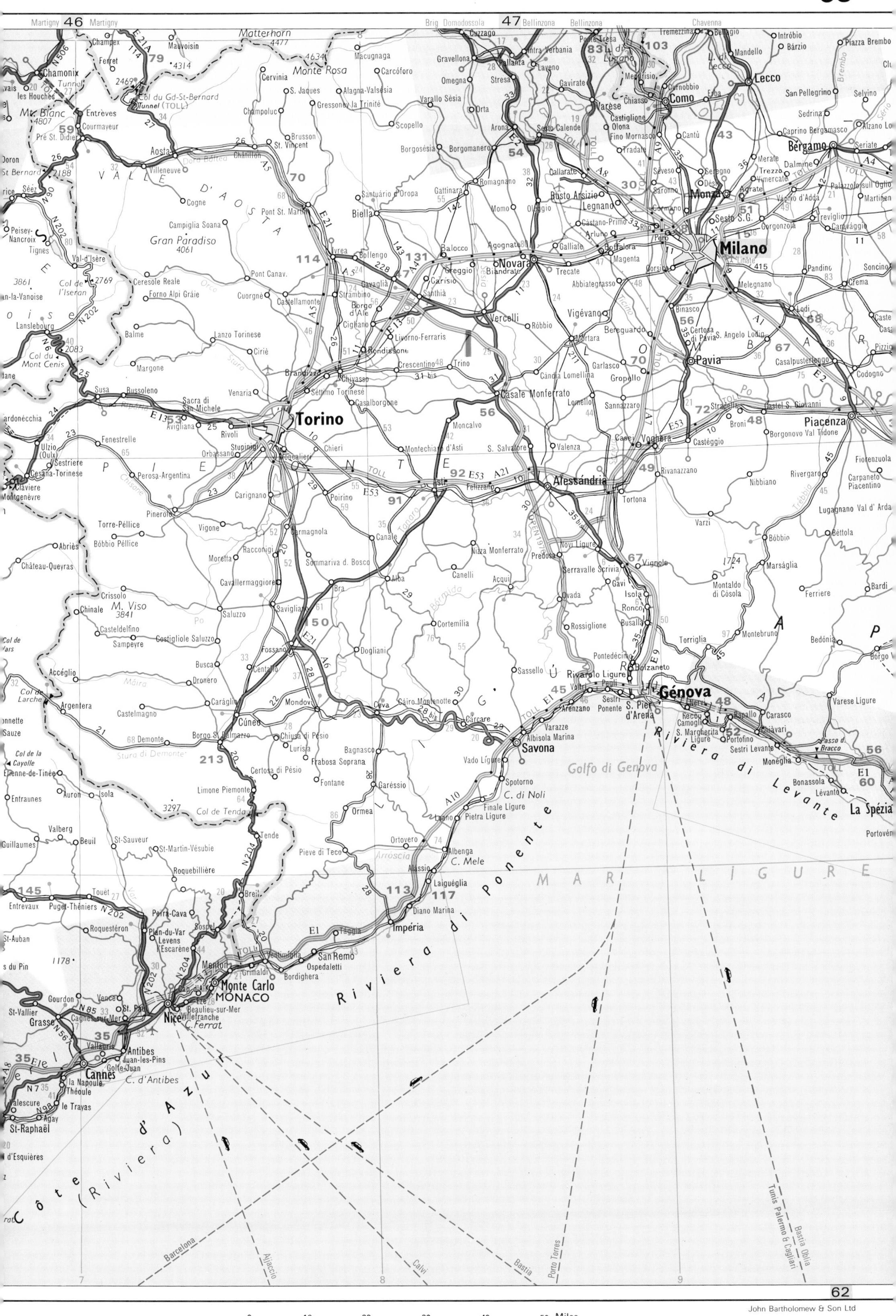

John Bartholomew & Son Ltd

0	10	20	30	40	50	Miles
0	10	20	30	40	50	60	70	80	Kilometres

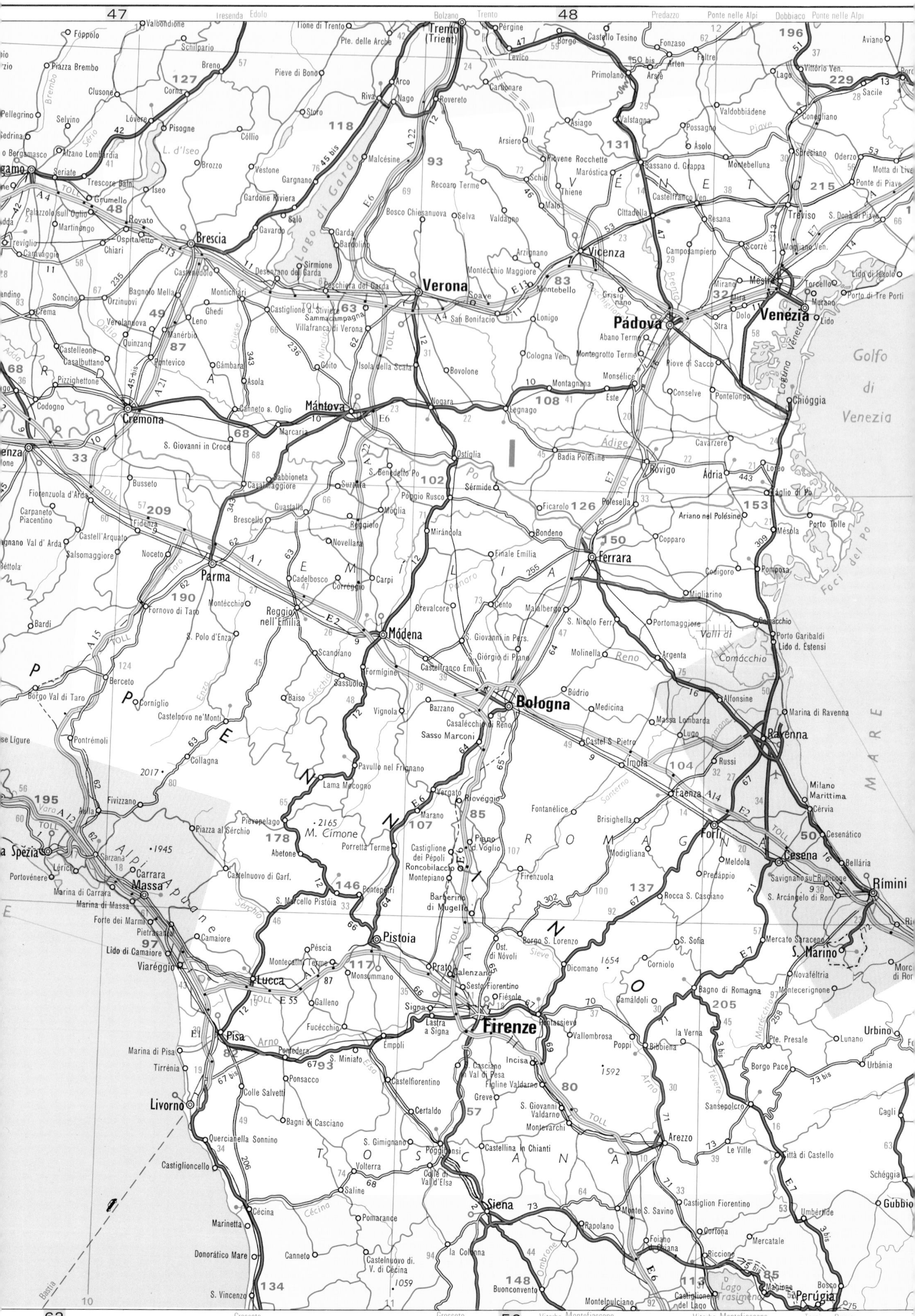

47
48
Trento
Bolzano
Verona
Brescia
Cremona
Mántova
Vicenza
Pádova
Venezia
Mestre
Treviso
Chióggia
Golfo di Venezia
Rovigo
Ferrara
Parma
Reggio nell'Emilia
Módena
Bologna
Ravenna
Forlì
Cesena
Rímini
S. Marino
Imola
Faenza
La Spezia
Carrara
Massa
Lucca
Pisa
Livorno
Pistoia
Prato
Firenze
Siena
Arezzo
Perúgia
Urbino
Gubbio
Lago di Garda
L. d'Iseo
Valli di Comácchio
Foci del Po
Laguna Veneta
M. Cimone
Alpi Apuane
VENETO
EMILIA
ROMAGNA
TOSCANA
APPENNINO
MARE
62
56

1:1 250 000

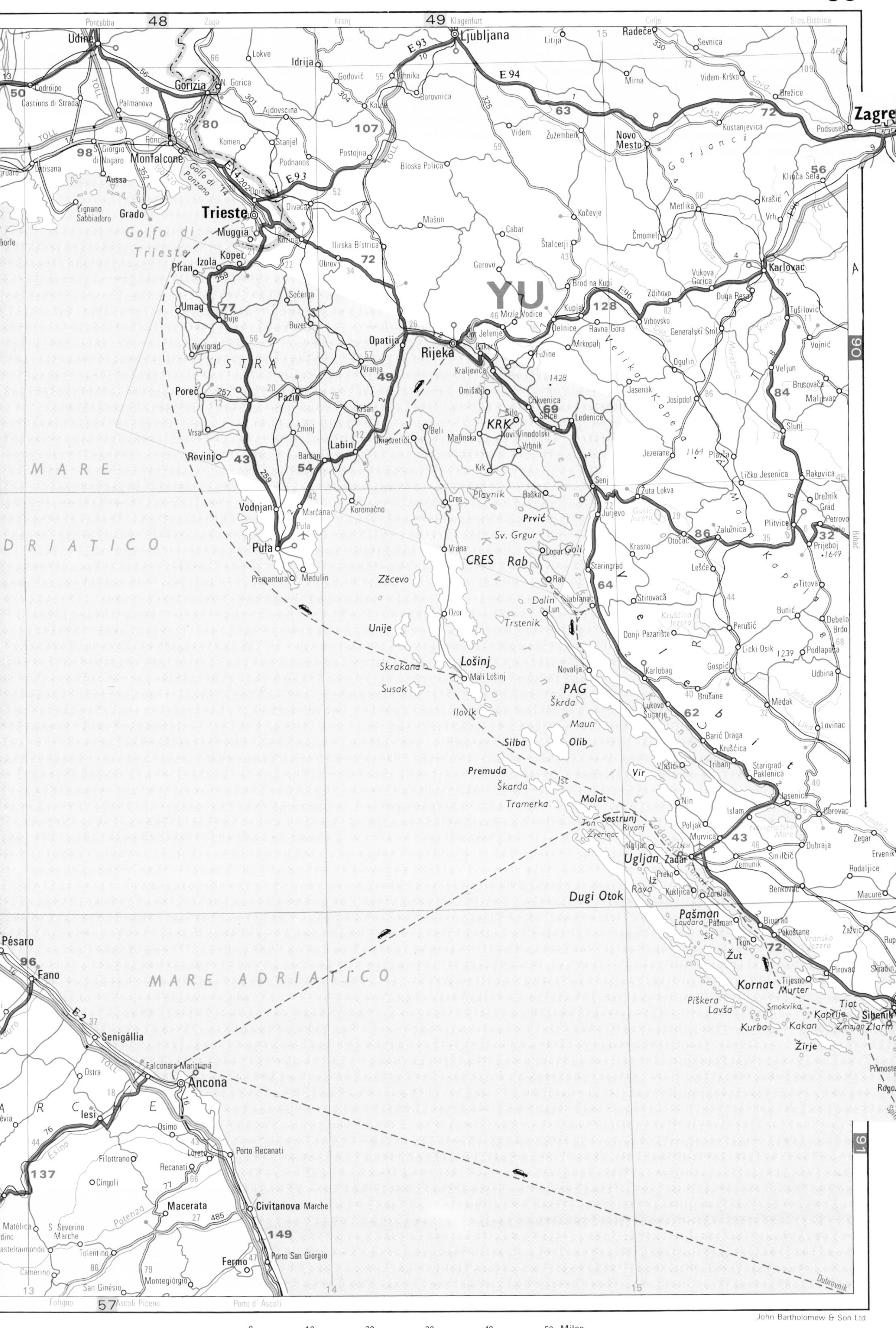

0   10   20   30   40   50 Miles
0 10 20 30 40 50 60 70 80 Kilometres

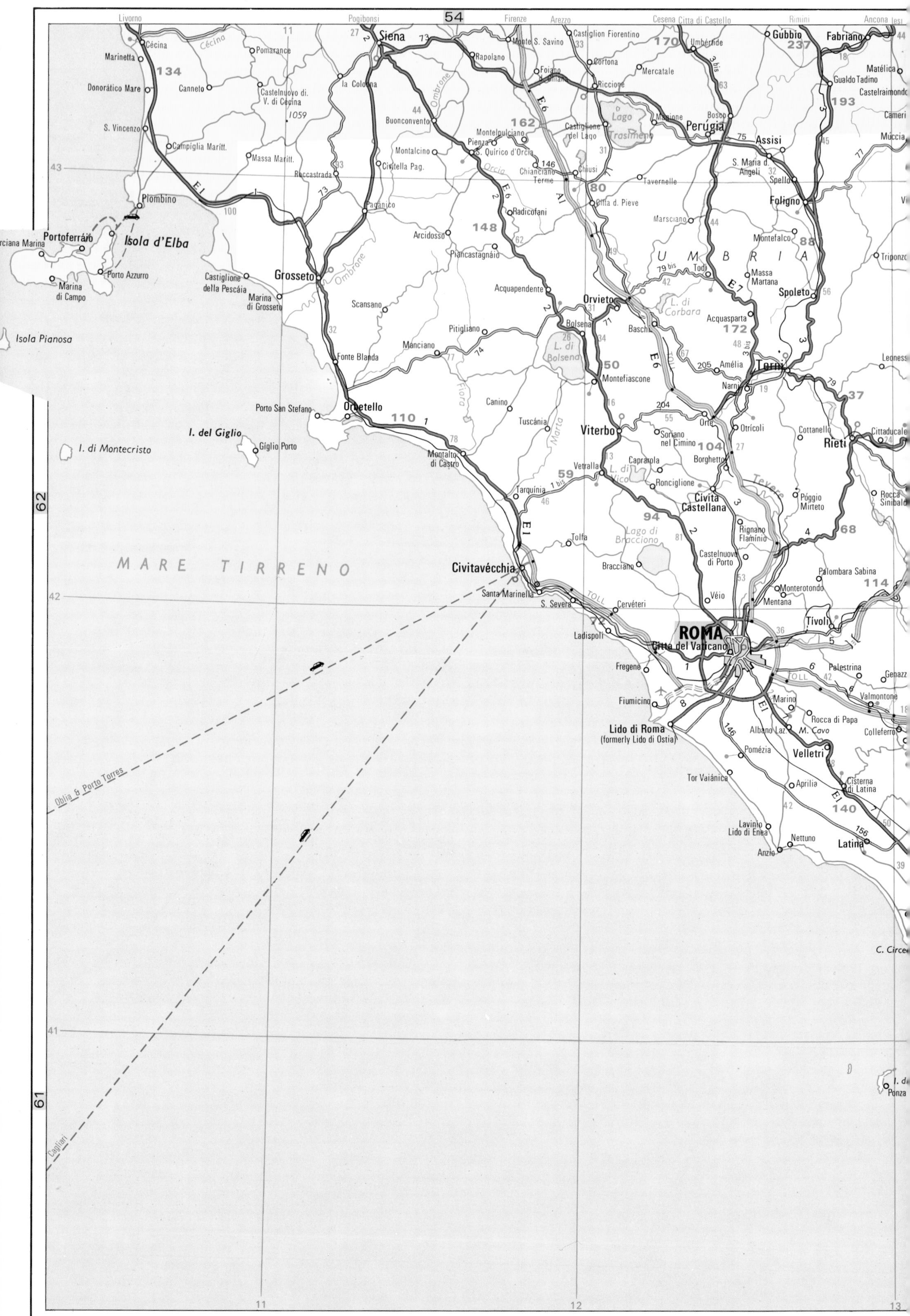

MARE TIRRENO

Isola d'Elba

Isola Pianosa

I. del Giglio

I. di Montecristo

Livorno
Cécina
Marinetta
Donorático Mare
S. Vincenzo
Campíglia Marítt.
Massa Maritt.
Pomarance
Canneto
Castelnuovo di. V. di Cécina
1059
la Colonna
Pogibonsi
SIENA
Rapolano
Monte S. Savino
Castiglion Fiorentino
Cortona
Arezzo
Firenze
Mercatale
Riccione
Foiano
Cesena Citta di Castello
UMBERTIDE
170
Bosco
PERÚGIA
Mastione
S. Maria d. Angeli
Spello
Rimini
GUBBIO
Fabriano
Matélica
Gualdo Tadino
Castelraimondo
193
Cameri
Muccia
Ancona Iesi
44

ARCIAna Marina
Portoferráio
Porto Azzurro
Marina di Campo
134
Plombino
Castiglione della Pescáia
Marina di Grosseto
GROSSETO
Scansano
Roccastrada
Buonconvento
Montalcino
Paganico
Arcidosso
Pitigliano
Mánciano
Fonte Blanda
Montepulciano
Pienza
S. Quírico d'Orcia
Civitella Pag.
162
148
Chianciano Terme
146
Chiusi
80
Città d. Pieve
Marsciano
Montefalco
85
FOLIGNO
UMBRIA
Todi
Massa Martana
SPOLETO
Triponzo
Leonessa
172
Acquasparta
Amélia
Narni
TERNI
37
205
204
Orte
Otrícoli
Cottanello
Rieti
Cittaducale
79

Porto San Stefano
Orbetello
110
Montalto di Castro
Tarquínia
46
Tuscánia
Acquapendente
Bolsena
L. di Bolsena
Montefiascone
Baschi
ORVIETO
L. di Corbara
VITERBO
Soriano nel Cimino
Vetralla
Capránica
Ronciglione
Borghetto
Póggio Mirteto
Rocca Sinibalda
68
114
59
94
Civita Castellana
Rignano Flaminio
Castelnuovo di Porto
Palombara Sabina
Monterotondo
TOLL
Mentana
TIVOLI

CIVITAVÉCCHIA
Santa Marinella
S. Severa
Ladispoli
Fregene
Fiumicino
Tolfa
Bracciano
Lago di Bracciano
Cervéteri
Véio
ROMA
Città del Vaticano
Palestrina
Genazz
Marino
Valmontone
Rocca di Papa
M. Cavo
Colleferro
36

LIDO DI ROMA
(formerly Lido di Ostia)
Pomézia
Tor Vaiánica
Lavinio
Lido di Enea
Aprilia
Nettuno
Anzio
Albano Laz.
Velletri
Cisterna di Latina
140
LATINA
156

Obía o Porto Torres

Cagliari

C. Circeo

I. di Ponza

54

1:1 250 000

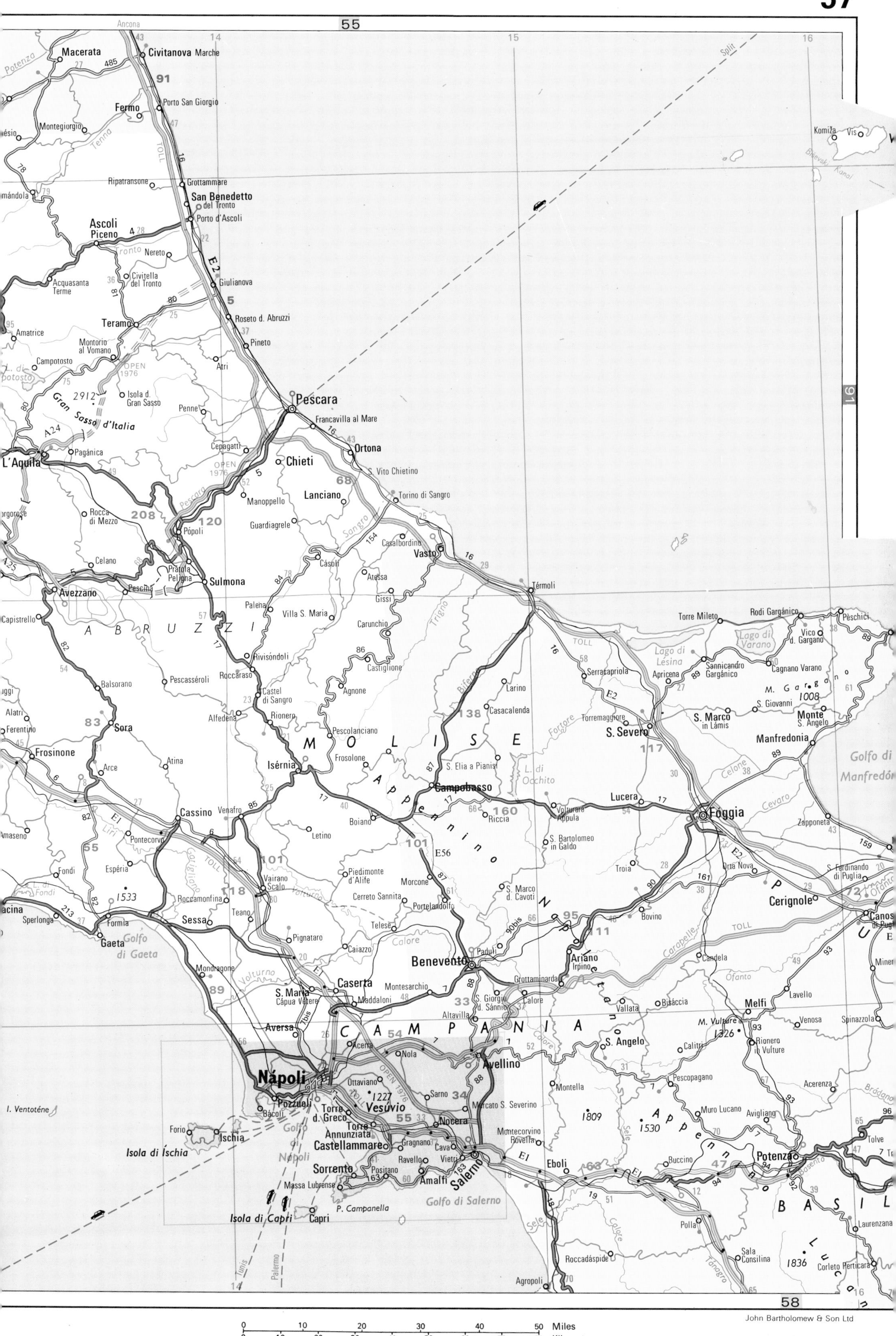

55

15

91

16

Ancona
Macerata
Civitanova Marche
14
485
91
Potenza
78
Fermo
Montegiorgio
Porto San Giorgio
147
mándola
79
Ripatransone
Grottammare
16
San Benedetto
del Tronto
Porto d'Ascoli
Ascoli
Piceno
4 28
E2
Nereto
Acquasanta
Terme
36
Civitella
del Tronto
81
Giulianova
5
95
Amatrice
Teramo
Roseto d. Abruzzi
37
Montorio
al Vomano
Pineto
OPEN
1976
Campotosto
2912
Isola d.
Gran Sasso
Atri
L. di
otosto
Gran Sasso d'Italia
75
Penne
Pescara
A24
Francavilla al Mare
L'Aquila
Pagánica
16
43
49
OPEN
1976
Ortona
Cepagatti
5
S. Vito Chietino
Chieti
68
Rocca
di Mezzo
208
52
Pópoli
120
Manoppello
Lanciano
Torino di Sangro
Celano
Guardiagrele
154
Casalbordino
Avezzano
Pescina
Pratola
Peligna
Sulmona
84 78
Cásoli
Vasto
16
29
Capistrello
Palena
Atessa
Térmoli
57
Villa S. Maria
Gissi
82
Balsorano
Carunchio
TOLL
Torre Mileto
Rodi Gargánico
Péschici
Pescasséroli
Rivisóndoli
86
Roccaráso
Castiglione
16
58
Lago di
Varano
38
Vico
d. Gargano
89
54
A B R U Z Z I
17
Castel
di Sangro
Agnone
Biferno
Lago di
Lésina
Sannicandro
Gargánico
89
M. Garga no
1008
Monte
61
uggi
83
Sora
Alfedena
Rionero
Pescolanciano
Larino
E2
27
Apricena
S. Marco
in Lámis
S. Giovanni
S. Angelo
Alatri
Ferentino
Frosinone
Arce
Atina
M O L I S E
Isérnia
Frosolone
87
L. di
Occhito
Casacalenda
138
Torremaggiore
S. Severo
Manfredonia
45
Golfo di
Manfredón
6
Cassino
Venafro
85
17
40
Boiano
S. Elia a Pianisi
Campobasso
717
Riccia
160
Celone
Lucera
30
89
117
Fóggia
Caserta
Vairano
Scalo
101
E56
Morcone
87
S. Bartolomeo
in Galdo
Troia
28
Zapponeta
43
159
55
Espéria
Pontecorvo
64
101
Piedimonte
d'Alife
Cerreto Sannita
61
S. Marco
d. Cavoti
90bis
66
161
161
Orta Nova
72
Fondi
1533
Roccamonfina
Telese
Portelandolfo
90
Cerignole
Canos
L. di
Fondi
213
82
118
Teano
Caiazzo
Calore
Paduli
95
Bovino
Candela
49
93
Miner
U
E
Sessa
Pignataro
20
Benevento
Ariano
Irpino
711
Formia
Gaeta
Golfo
di Gaeta
89
Mondragone
S. Maria
Cápua Vetere
Montesarchio
7
Grottaminarda
Lavello
Melfi
Venosa
Spinazzola
Aversa
26
C A M P A N I A
54
S. Giorgio
d. Sánnio
Calore
S. Angelo
Bisáccia
M. Vulture
1326
Rionero
in Vulture
Acerenza
Acerra
Nola
52
7
Altavilla
Nápoli
Ottaviano
1227
Sarno
34
Avellino
88
Montella
Calitri
Pescopagano
93
Muro Lucano
Avigliano
Pozzuoli
Vesuvio
Mercato S. Severino
Montecorvino
Rovella
1809
1530
65
96
Bácoli
Torre
d. Greco
55
33
Nocera
Golfo
di
Torre
Annunziata
Castellammare
Gragnano
Cava
S. Angelo
Buccino
47
Tolve
Potenza
Forio
Ischia
Ravello
Vietri
Eboli
94
94
Sorrento
Positano
Amalfi
63
60
Salerno
E1
92
39
Isola di Ischia
Massa Lubrense
Golfo di Salerno
19
51
Polla
B A S I L
I. Ventotène
P. Campanella
Sele
65
Laurenzana
Isola di Capri
Capri
Roccadáspide
Sala
Consilina
1836
Corleto Perticara
Agropoli
70
65
16

0   10   20   30   40   50 Miles
0   10   20   30   40   50   60   70   80 Kilometres

John Bartholomew & Son Ltd

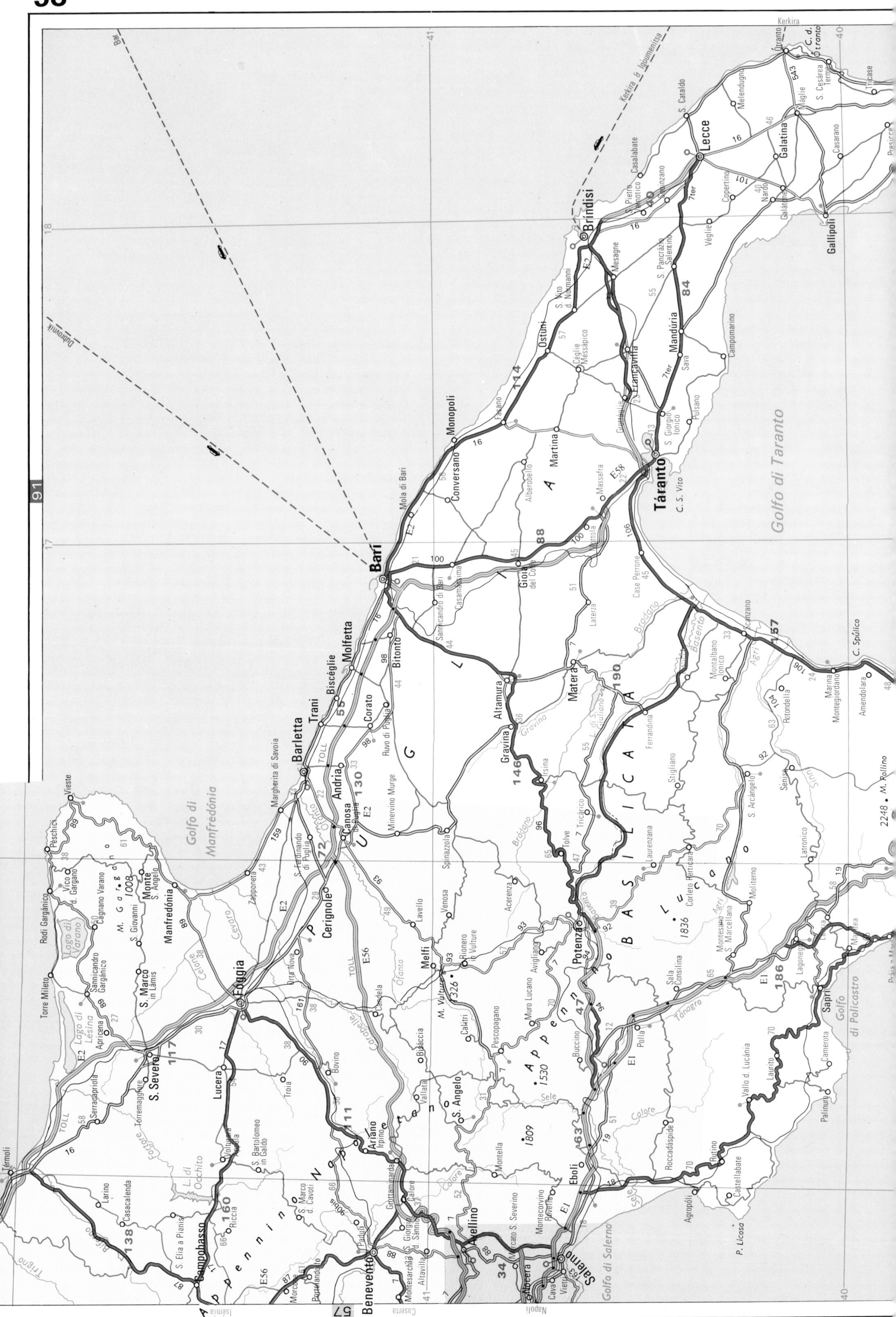

1:1 250 000

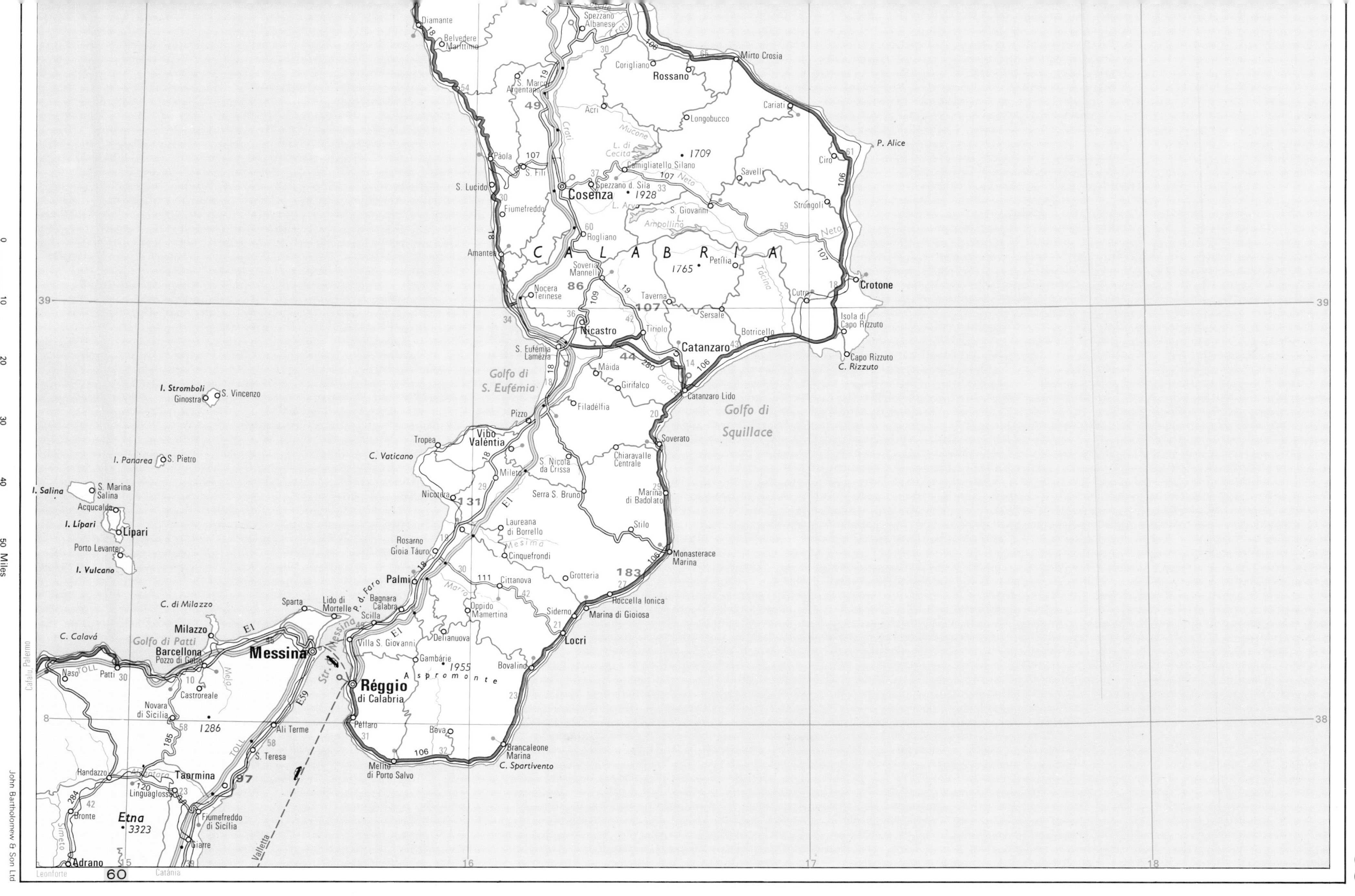

0	10	20	30	40	50 Miles
0	10	20	30 40	50 60	70 80 Kilometres

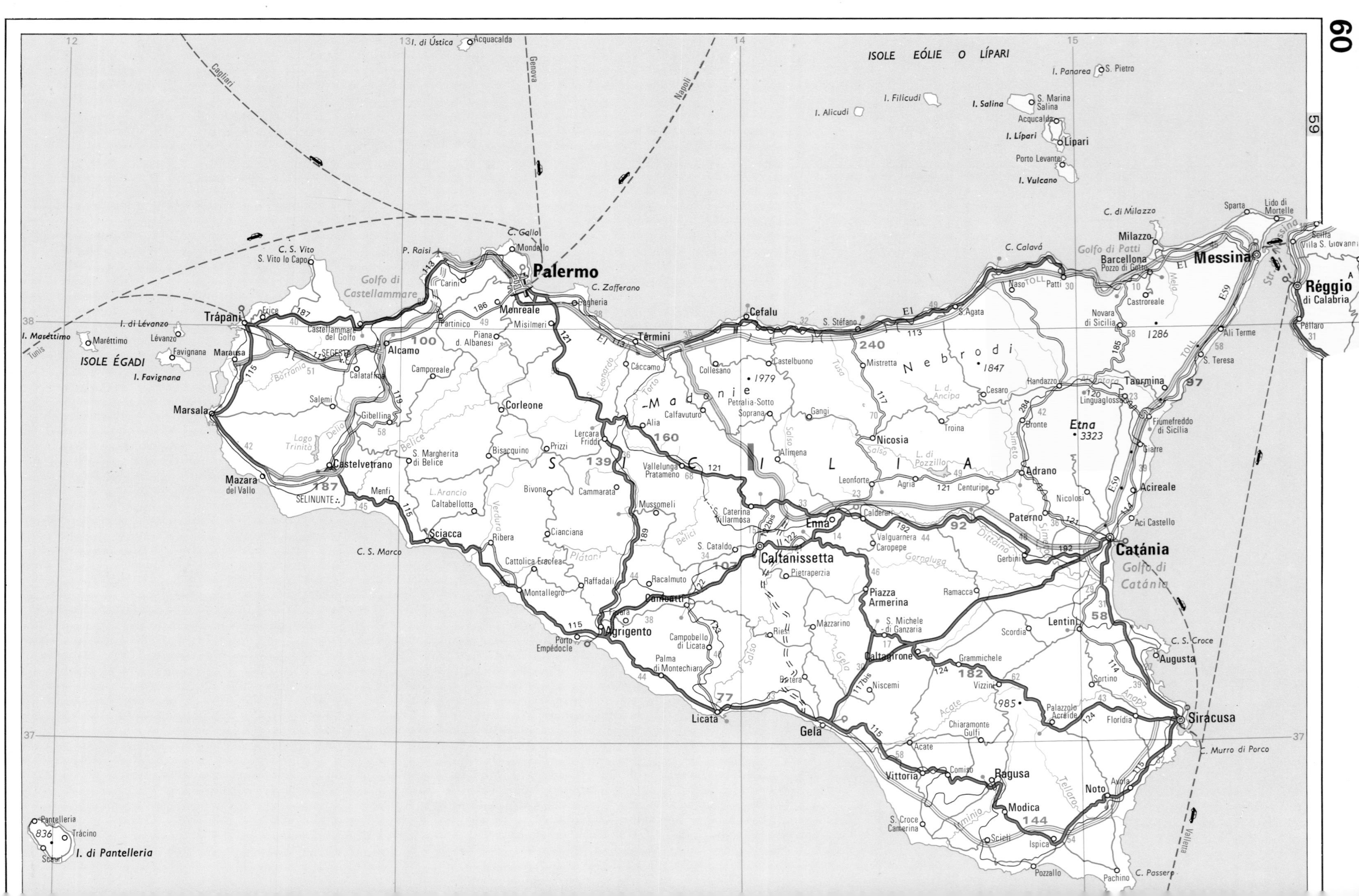
59
ISOLE EÓLIE O LÍPARI
I. Panarea
S. Pietro
I. Filicudi
I. Alicudi
I. Salina
S. Marina Salina
Acquacalda
I. Lípari
Lipari
Porto Levante
I. Vulcano
I. di Ústica
Acquacalda
Cagliari
Genova
Napoli
C. di Milazzo
Milazzo
Golfo di Patti
Barcellona Pozzo di Gotto
Messina
Sparta
Lido di Mortelle
Scilla
Villa S. Giovanni
Réggio di Calabria
Pellaro
C. Calavá
Patti
Naso
S. Agata
Castroreale
Novara di Sicilia
1286
Ali Terme
S. Teresa
Taormina
Linguaglossa
Fiumefreddo di Sicilia
Giarre
Etna
3323
Acireale
Aci Castello
Catánia
Golfo di Catánia
Nicolosi
Adrano
Bronte
Handazzo
Cesaro
Troina
Nebrodi
1847
Mistretta
S. Stéfano
Cefalu
Castelbuono
Collesano
Madonie
1979
Petralia-Sotto
Soprana
Gangi
Nicosia
Alimena
Leonforte
Agria
Centuripe
Paterno
Gerbini
Calderari
Enna
Valguarnera
Caropepe
Caltanissetta
S. Caterina Villarmosa
S. Cataldo
Pietraperzia
Piazza Armerina
Ramacca
Lentini
Scordia
C. S. Croce
Augusta
Sortino
Siracusa
C. Murro di Porco
Floridia
Palazzolo Acreide
985
Vizzini
Grammichele
S. Michele di Ganzaria
Caltagirone
Mazzarino
Riesi
Butera
Niscemi
Chiaramonte Gulfi
Acate
Vittoria
Comiso
Ragusa
Modica
Scicli
Ispica
Noto
Avola
Pozzallo
Pachino
C. Passero
S. Croce Camerina
Gela
Licata
Palma di Montechiaro
Campobello di Licata
Favara
Agrigento
Porto Empédocle
Racalmuto
Raffadali
Montallegro
Cattolica Eraclea
Ribera
Cianciana
Bivona
Cammarata
Mussomeli
Vallelunga Pratameno
Lercara Friddi
Alia
Calfavuturo
Cáccamo
Términi
Bagheria
C. Zafferano
Palermo
Mondello
C. Gallo
P. Raisi
Carini
Monreale
Misilmeri
Piana d. Albanesi
Partinico
Corleone
Prizzi
Bisacquino
S. Margherita di Belice
Caltabellotta
Sciacca
C. S. Marco
Menfi
Selinunte
Castelvetrano
Mazara del Vallo
Marsala
Salemi
Gibellina
Calatafimi
Segesta
Alcamo
Camporeale
Castellammare del Golfo
Golfo di Castellammare
C. S. Vito
S. Vito lo Capo
Erice
Trápani
Marausa
I. di Lévanzo
Lévanzo
Favignana
I. Favignana
I. Marettimo
Maréttimo
ISOLE ÉGADI
Tunis
Valletta
Pantelleria
Trácino
Scauri
836
I. di Pantelleria

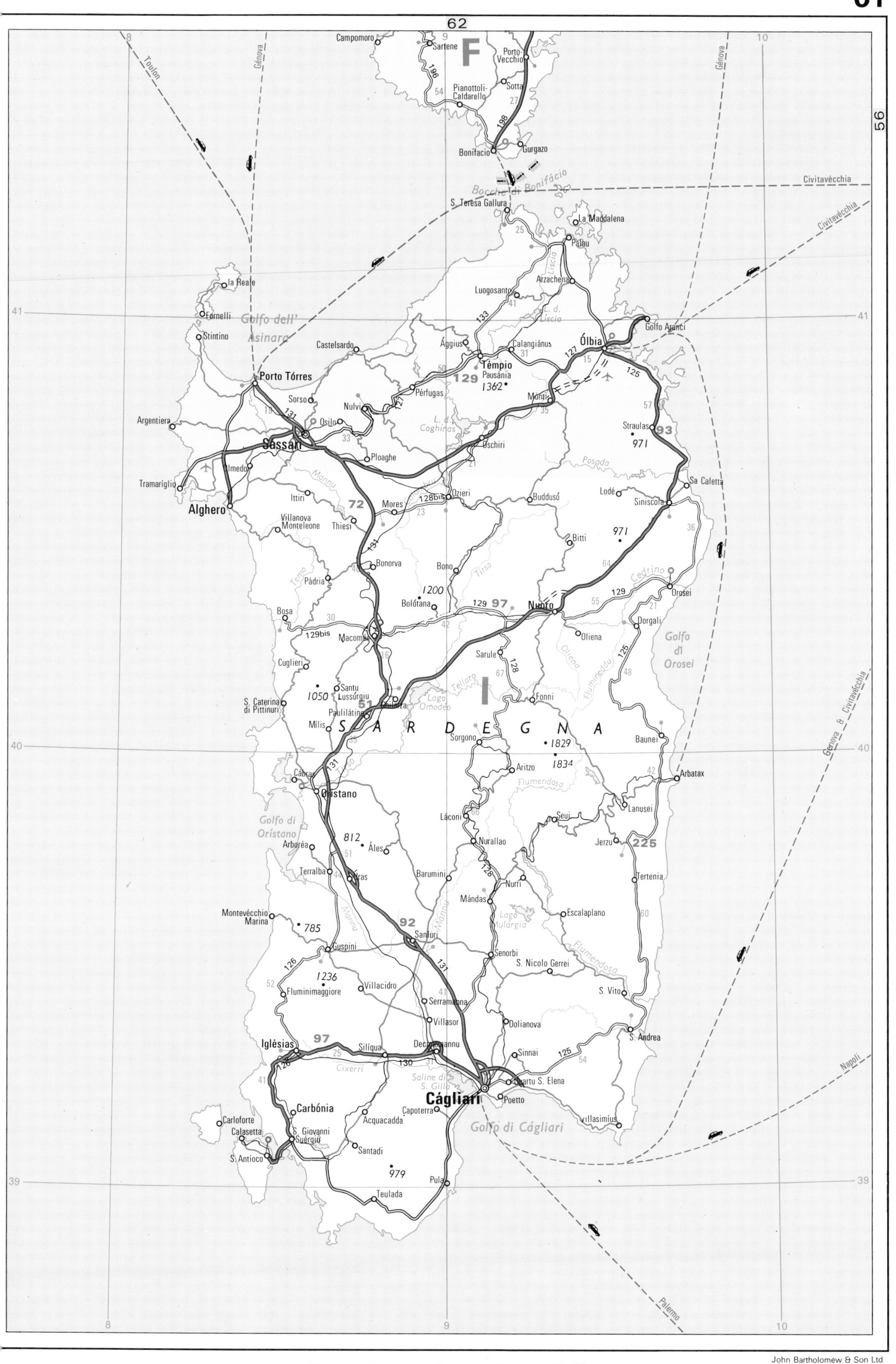

Map of Sardegna (Sardinia).

Places and labels shown on the map include:

Campomoro, Sartene, Porto-Vecchio, Pianottoli-Caldarello, Sotta, Bonifacio, Gurgazo, Bocche di Bonifácio, S. Teresa Gallura, La Maddalena, Palau, la Reale, Fornelli, Luogosanto, Arzachena, Golfo Aranci, Stintino, Golfo dell' Asinara, Castelsardo, Ággius, Calangiánus, Ólbia, Porto Tórres, Tempio Pausánia, 1362, Mónti, 125, Sorso, Pérfugas, 57, Nulvi, Argentiera, Osilo, Óschiri, Straulas, 971, Sássari, Ploaghe, Sa Caletta, Tramaríglio, Imedos, Ittiri, Mores, Ozieri, Buddusó, Lodé, Siniscola, Alghero, Villanova Monteleone, Thiesi, 128bis, Bitti, 971, Bonorva, Bono, Pádria, Tirso, Bosa, 1200, Bolótana, 129, 97, Núoro, Orosei, 129bis, Macomer, Oliena, Dorgali, Cuglieri, Sarule, Golfo di Orosei, S. Caterina di Píttinuri, 1050, Santu Lussúrgiu, 128, Fonni, Milis, SARDEGNA, Sorgono, 1829, Baunei, Áritzo, 1834, Cábras, Flumendosa, Arbatax, Oristano, Láconi, Seui, Lanusei, Golfo di Oristano, 812, Áles, Nurallao, Jerzu, 225, Arboréa, Terralba, Úras, Barumini, Nurri, Tertenia, Mándas, Montevécchio Marina, 785, Escalaplano, 92, Sánluri, Senorbi, S. Nicolo Gerri, Guspini, 1236, Villacidro, Serramanna, S. Vito, Fluminimaggiore, Villasor, Dolianova, S. Andrea, Iglésias, 97, Silíqua, Decimomannu, Sinnai, Cixerri, Carbónia, Acquacadda, Cágliari, Poetto, Carloforte, Calasetta, S. Giovanni Suérgiu, Santadi, Capoterra, Golfo di Cágliari, villasimíus, S. Antíoco, 979, Pula, Teulada

Toulon, Génova, Civitavécchia, Napoli, Palermo

61

0	10	20	30	40	50	Miles
0	10	20	30	40	50 60 70	80	Kilometres

John Bartholomew & Son Ltd

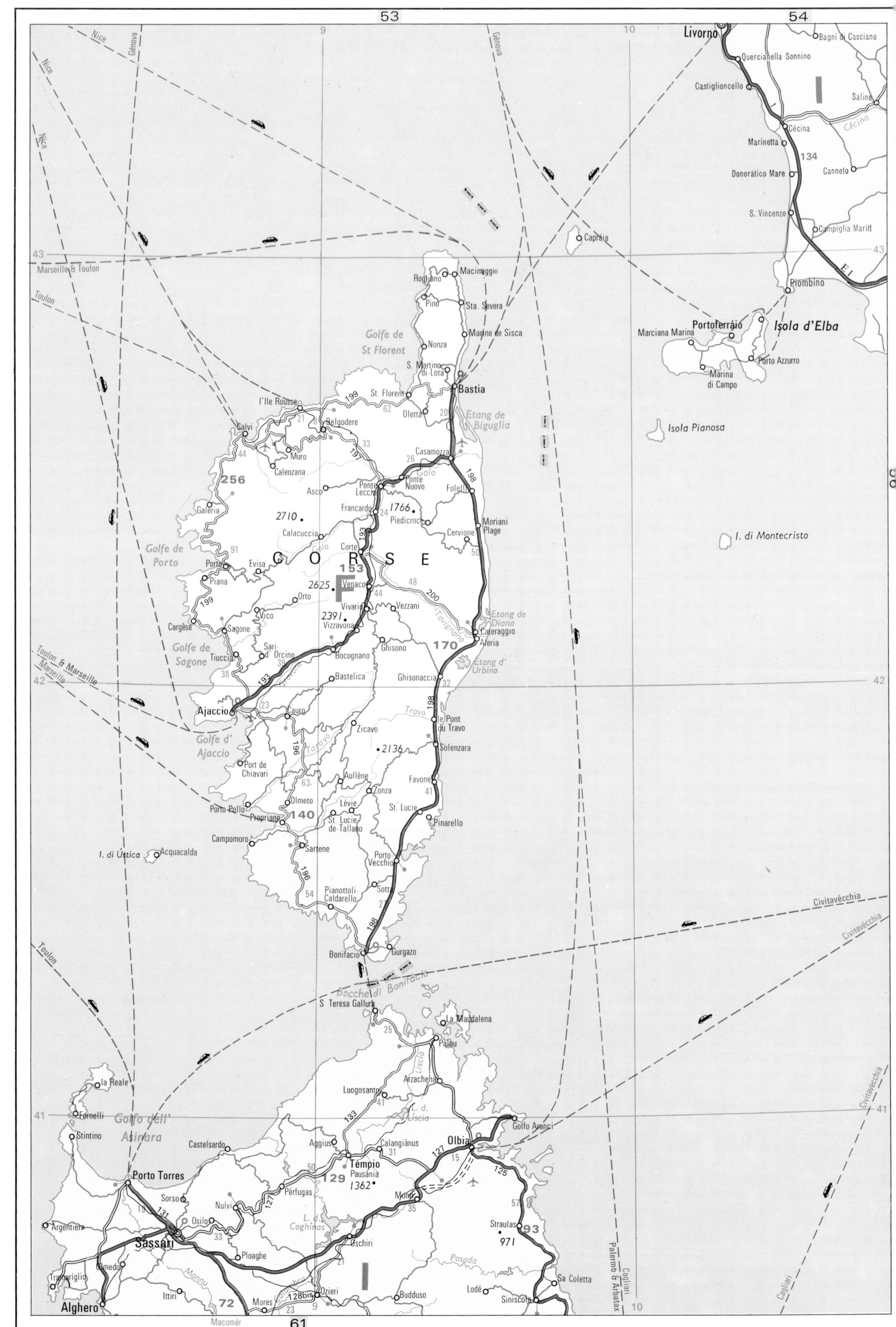

53   54

Livorno
Bagni di Casciano
Quercianella Sonnino
Castiglioncello
Saline
Cécina   Cécina
Marinetta
Canneto
134
Donorático Mare
S. Vincenzo
Campiglia Maritt

43   43
Marseille & Toulon
Toulon
Piombino
Portoferráio   Isola d'Elba
Marciana Marina
Capraia   Porto Azzurro
Marina
di Campo

Rogliano
Macinaggio
Pino
Sta. Severa
Golfe de
St Florent
Nonza
Marine de Sisca
S. Martino
di-Lota
Isola Pianosa

l'Ile Rousse   199   St. Florent
Oletta   Bastia
Calvi   21   Belgodere   62
33   Casamozza   Etang de
Muro   197   Biguglia
44   Caleozana   26   Ponte
Nuovo
256   Asco   Golo
Ponte   198
Leccia   Folelli
Galeria   Francardo   1766   I. di Montecristo
2710   24   Piedicroce   Cervione   Moriani
Calacuccia   193   Plage
Corte   50
G O R S E
91   Evisa   153
Porto   Venaco   48
Golfe de   Piana   2625   Vivario   Vezzani   200
Porto   Orto   44   Etang de
199   2391   Diana
Vico   Vizzavona   Cateraggio
Cargèse   Sagone   Ghisoni   170   Aleria
Sari-   Bocognano   Etang d'
Golfe de   d'Orcino   Urbino
Sagone   Tiuccia   38   193   Bastelica   Ghisonaccia
42   Marseille   42
Ajaccio   23   Cauro   Travo
Zicavo   198   le Pont
du Travo
Golfe d'   196   2136   Solenzara
Ajaccio   Aullène   Favone   41
Port de   Zonza
Chiavari   63   St. Lucie
Olmeto   Lévie   Pinarello
I. di Ústica   Acquacalda   Porto Pollo   Propriano   140   St. Lucie
de-Tallano
Campomoro   Sartene   Porto
196   Vecchio
Pianottoli-
Caldarello   Sotta
54   2
Bonifacio   Gurgazo
198
Bocche di Bonifacio
S Teresa Gallura
La Maddalena
25
Palau   Civitavecchia   Civitavecchia
Arzachena
Luiscia
Luogosanto
41   L. di
41   Liscia   Golfo Aranci   41
la Reale   133   Olbia
Fornelli   Aggius   Calangiánus   127
Golfo dell'   Castelsardo   31   15   125
Stintino   Asinara   Témpio   127
Porto Torres   129   Pausánia   57
Sorso   1362   Monti   Straulas   93
50   Pérfugas   35   971
Argentiera   Nulvi   127   Monti
Osilo   Oschiri   Sa Coletta
Sassari   33   Posada
Tramariglio   21   Lodé
Ozieri   Buddúso   Siniscola
Olmedo   128bis   10
Alghero   72   Ittiri   Mores
Macomér   61

1:1 250 000

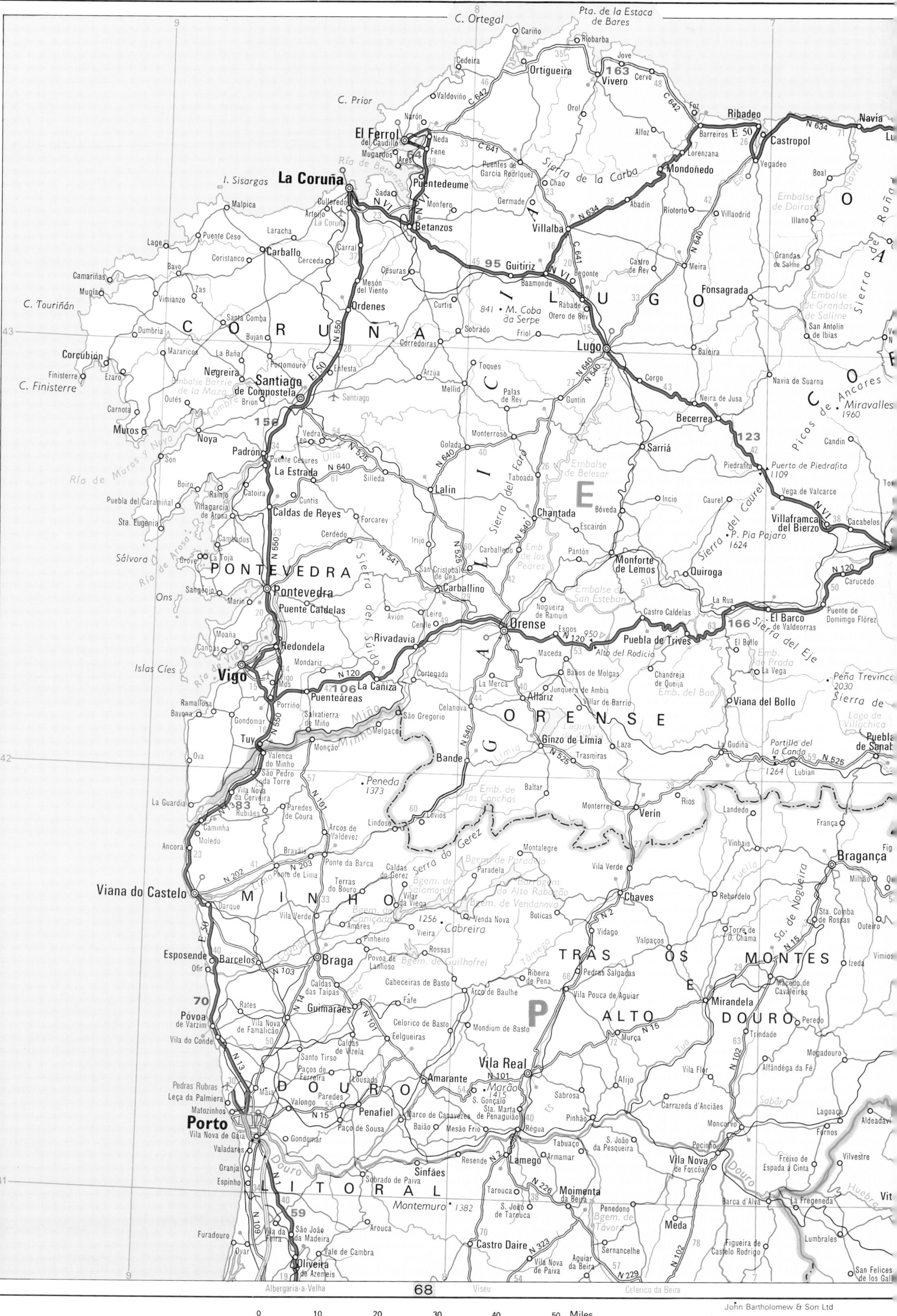

John Bartholomew & Son Ltd

| 0 | 10 | 20 | 30 | 40 | 50 | Miles |
| 0 | 10 | 20 | 30 | 40 | 50 | 60 | 70 | 80 | Kilometres |

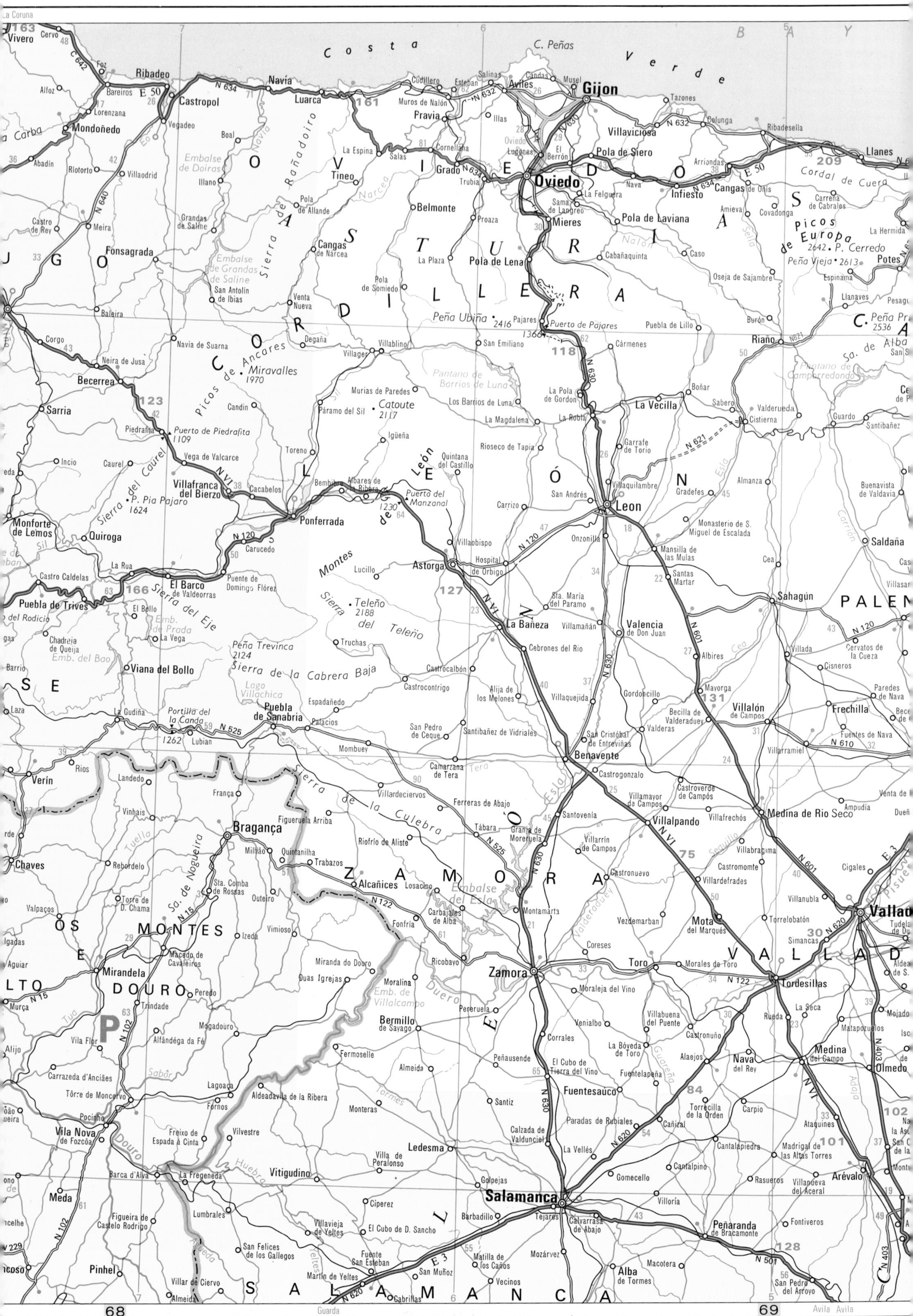
La Coruna
Costa Verde
BAY
C. Peñas
Vivero
Cervo
Ribadeo
Castropol
Navia
Luarca
Cudillero
Avilés
Gijon
Tazones
Colunga
Ribadesella
Llanes
Mondoñedo
Vegadeo
Pravia
Villaviciosa
Pola de Siero
Grado
Oviedo
Infiesto
Cangas de Onís
Tineo
Belmonte
Mieres
Pola de Laviana
Fonsagrada
Cangas de Narcea
Pola de Lena
Picos de Europa
2642 P. Cerredo
Peña Vieja 2613
Potes
CORDILLERA CANTABRICA
ASTURIAS
Peña Ubiña 2416
Puerto de Pajares
1366
Puebla de Lillo
Riaño
C. Peña Prieta 2536
Lugo
Becerrea
Sarria
Picos de Ancares
Miravalles 1970
Villablino
Catoute 2117
La Pola de Gordón
La Vecilla
Boñar
Cistierna
Guardo
Puerto de Piedrafita 1109
Villafranca del Bierzo
Sierra del Caurel
P. Pía Pájaro 1624
Ponferrada
Puerto del Manzanal 1230
LEÓN
Leon
Monforte de Lemos
Quiroga
Astorga
Hospital de Órbigo
Mansilla de las Mulas
Saldaña
El Barco de Valdeorras
Puebla de Trives
Montes de León
Teleño 2188
Sierra del Teleño
La Bañeza
Valencia de Don Juan
Sahagún
PALENCIA
Viana del Bollo
Peña Trevinca 2124
Sierra de la Cabrera Baja
Lago Villachica
Puebla de Sanabria
Portilla del la Canda 1262
Mayorga
Villalón de Campos
Frechilla
ORENSE
Verín
Benavente
Bragança
Sierra de la Culebra
Villalpando
Medina de Rio Seco
Chaves
ZAMORA
Alcañices
Embalse del Esla
MONTES
Mirandela
DOURO
P
Zamora
Toro
Tordesillas
Mota del Marqués
VALLADOLID
Valladolid
Simancas
Bermillo de Sayago
Fermoselle
Fuentesaúco
Medina del Campo
Olmedo
Vila Nova de Fozcôa
Ledesma
Vitigudino
Salamanca
Peñaranda de Bracamonte
Arévalo
Meda
Pinhel
Alba de Tormes
SALAMANCA
68
Guarda
69
Avila Avila

1:1 250 000

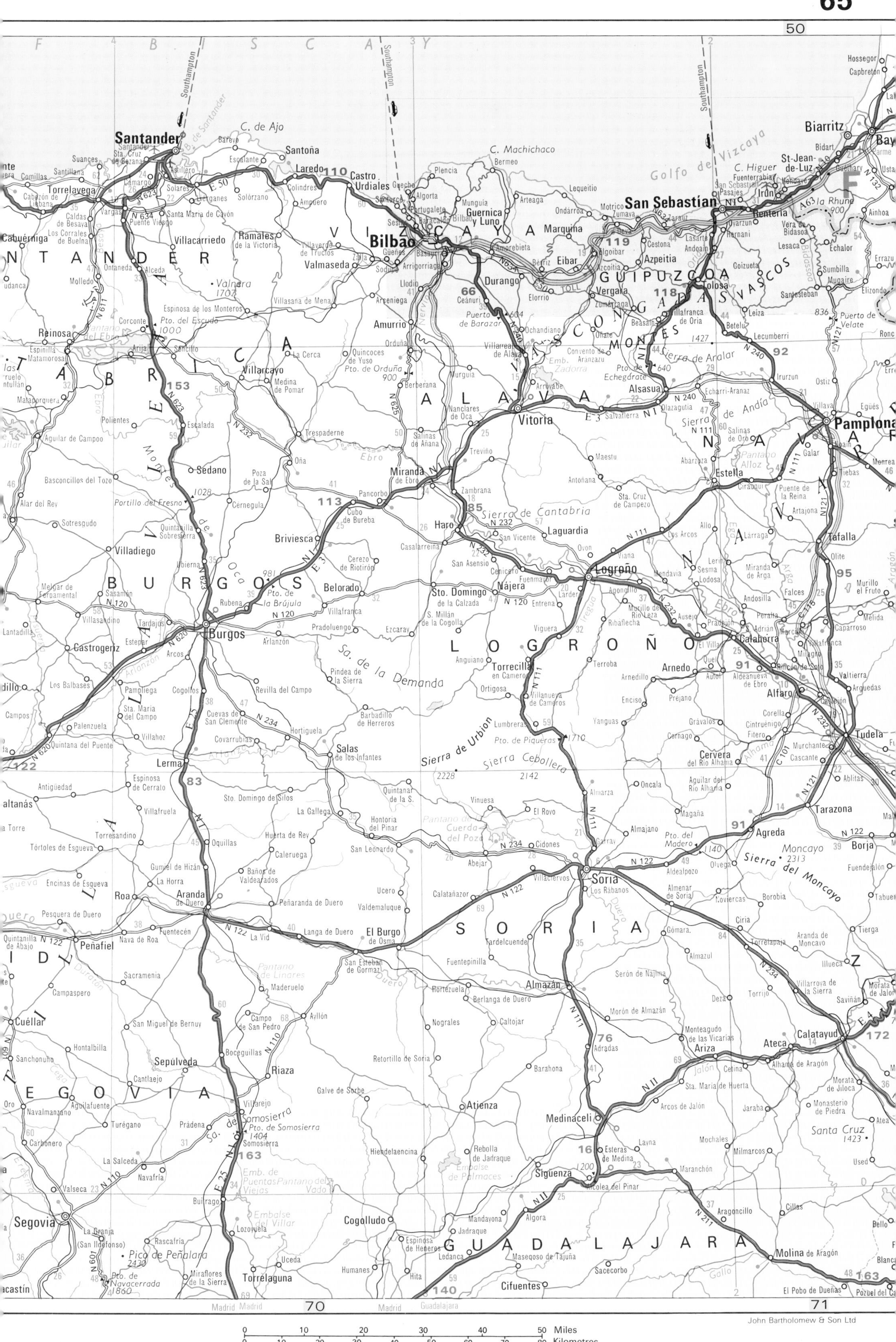

John Bartholomew & Son Ltd

0   10   20   30   40   50 Miles

0 10 20 30 40 50 60 70 80 Kilometres

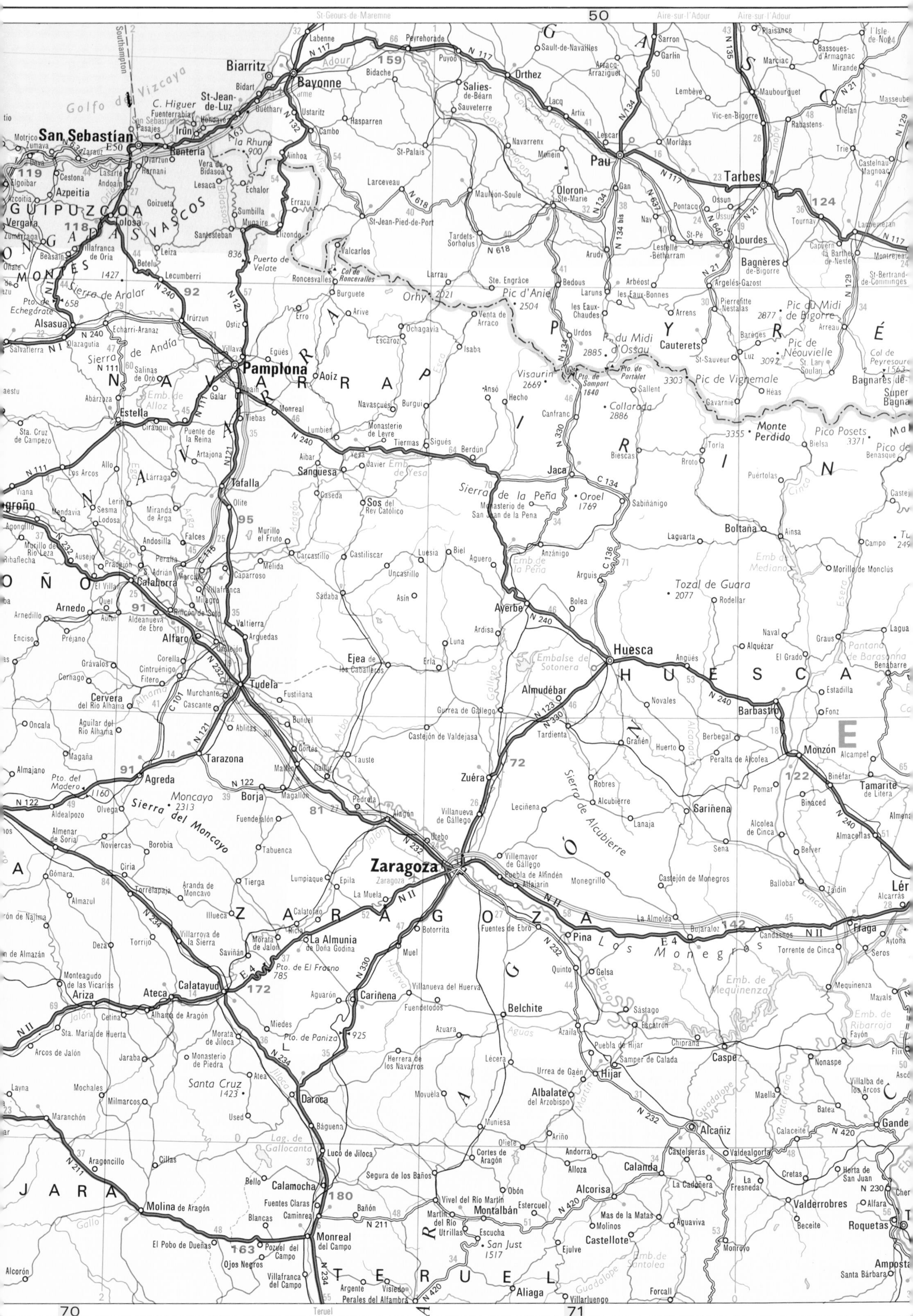
San Sebastián
Biarritz
Bayonne
Golfo de Vizcaya
St-Jean-de-Luz
Irún
Fuenterrabia
Hendaye
Renteria
Tolosa
Azpeitia
GUIPUZCOA
Vergara
Alsasua
Sierra de Aralar
Pamplona
NAVARRA
Estella
Tafalla
Olite
Sangüesa
Tudela
Tarazona
Agreda
Moncayo
2313
Sierra del Moncayo
Borja
Calahorra
Arnedo
Alfaro
Cervera del Río Alhama
Orthez
Pau
Tarbes
Lourdes
Oloron-Ste-Marie
Mauléon-Soule
St-Jean-Pied-de-Port
Roncesvalles
Pic d'Anie
2504
Pic du Midi d'Ossau
2885
Pic du Midi de Bigorre
2877
Pic de Néouvielle
3092
Pic de Vignemale
3303
Monte Perdido
3355
Pico Posets
3371
Cauterets
Bagnères-de-Bigorre
PYRÉNÉES
Jaca
Sierra de la Peña
Oroel
1769
Sabiñánigo
Biescas
Ayerbe
Huesca
HUESCA
Tozal de Guara
2077
Boltaña
Barbastro
Monzón
Binéfar
Tamarite de Litera
Fraga
Ejea de los Caballeros
Sos del Rey Católico
Zuéra
Almudébar
Sierra de Alcubierre
Sariñena
Zaragoza
ZARAGOZA
Los Monegros
Pina
Caspe
Alcañiz
Calatayud
Ateca
Ariza
La Almunia de Doña Godina
Cariñena
Belchite
Daroca
Hijar
Albalate del Arzobispo
Calamocha
Molina de Aragón
Monreal del Campo
Montalbán
Alcorisa
Calanda
Castellote
Valderrobres
Aliaga
TERUEL
ARAGÓN
Pto. de Somport
1640
Collarada
2886
Emb. de Mequinenza
Emb. de Yesa
Embalse de Sotonera
Lag. de Gallocanta
E 4
N II
N 232
N 240
N 330
N 234
N 121
N 111
N 117
N 134
50
70
71

1:1 250 000

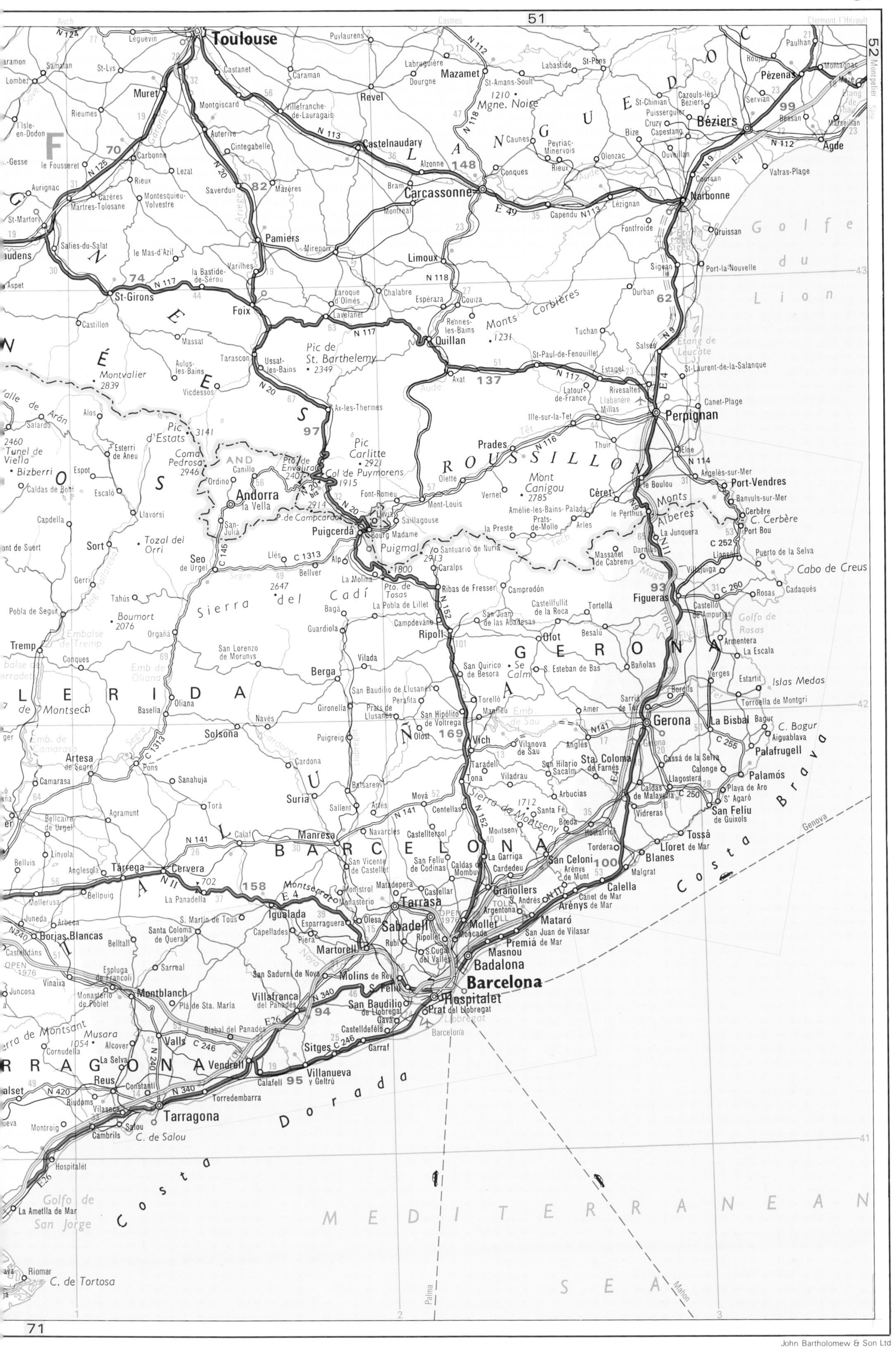

51

52

Toulouse

Léguevin
St-Lys
Castanet
Muret
Montgiscard
Villefranche-de-Lauragais
Puylaurens
Castres
Clermont-l'Hérault
Paulhan
Mazamet
Mgne. Noire
St-Pons
Labastide
Roujan
Montagnac
Pézenas
Béziers

Samatan
Lombez
St-Martin
le Fousseret
Carbonne
Rieux
Lezat
Auterive
Cintegabelle
Caraman
Revel
Castelnaudary
Alzonne
Bram
Caunes
Peyriac-Minervois
Rieux
Olonzac
Cazouls-lès-Béziers
Servian
Abéssan
Marseillan
Agde

l'Isle-en-Dodon
Gesse
Rieumes
Aurignac
Cazères
Martres-Tolosane
Saverdun
Mazères
Montesquieu-Volvestre
Montréal
Carcassonne
Capendu
Lézignan
Narbonne
Cruzy
Capestang
Coursan
Vatras-Plage

Salies-du-Salat
Aspet
le Mas-d'Azil
Pamiers
Mirepoix
Limoux
Espéraza
Couiza
Fontfroide
Sigean
Gruissan
Port-la-Nouvelle

St-Girons
Foix
Varilhes
la Bastide-de-Sérou
Chalabre
Laroque d'Olmès
Lavelanet
Rennes-les-Bains
Monts Corbières
Tuchan
Durban
Golfe du Lion

Castillon
Massat
Tarascon
Ussat-les-Bains
Pic de St. Barthelemy
Quillan
St-Paul-de-Fenouillet
Estagel
Salses
Étang de Leucate
St-Laurent-de-la-Salanque

Montvalier
Aulus-les-Bains
Vicdessos
Ax-les-Thermes
Axat
Latour-de-France
Millas
Rivesaltes
Canet-Plage

Valle de Arán
Alos
Esterri de Aneu
Espot
Pic d'Estats
Coma Pedrosa
Ordino
Canillo
Pic Carlitte
Font-Romeu
Mont-Louis
Olette
Prades
Thuir
Perpignan

Tunel de Viella
Bizberri
Caldas de Bohi
Llavorsí
Andorra la Vella
St-Julià
P. de Campcardos
Bourg Madame
Puigcerdá
Saillagouse
la Preste
Amélie-les-Bains-Palada
Prats-de-Mollo
Arles
le Boulou
Céret
le Perthus
Argelès-sur-Mer
Port-Vendres
Banyuls-sur-Mer
C. Cerbère
Port Bou

Capdella
Sort
Tozal del Orri
Seo de Urgel
Lés
Bellver
Alp
La Molina
Pto. de Tosas
Puigmal
Santuario de Nuria
Caralps
Mont Canigou
Vernet
La Junquera
Llançà
Puerto de la Selva

Pobla de Segur
Tahús
Boumort
Organá
Bagá
La Pobla de Lillet
Ribas de Fresser
Camprodón
Massanet de Cabrenys
Darnius
Vilajuiga
Cabo de Creus

Tremp
Conques
Emb. de Oliana
Basella
Guardiola
Castellfullit de la Roca
Tortellá
Besalú
Figueras
Castelló de Ampurias
Rosas
Cadaqués
Golfo de Rosas

Sierra del Cadí
Campdevánol
Ripoll
San Juan de las Abadesas
Olot
Bañolas
Armentera
La Escala

Pobla de Segur
San Lorenzo de Morunys
Vilada
Berga
San Quirico de Besora
S. Esteban de Bas
Verges
Estartit
Islas Medas

Artesa de Segre
Pons
Solsona
Oliana
Gironella
Prats de Llusanès
San Hipólito de Voltregá
Manlleu
Sarriá de Ter
Bordils
Torroella de Montgri
Gerona
La Bisbal
C. Bagur

Camarasa
Sanahuja
Cardona
Navès
Puigreig
Olost
Vich
Vilanova de Sau
San Hilario Sacalm
Sta. Coloma de Farnés
Cassá de la Selva
Aiguablava
Palafrugell

Agramunt
Torá
Calaf
Súria
Sallent
Artés
Moyá
Centellas
Taradell
Tona
Viladrau
Arbucias
Santa Fè
Caldas de Malavella
Llagostera
Calonge
Palamós
Playa de Aro
S' Agaró

Bellcaire de Urgel
Bellvis
Linyola
Tárrega
Cervera
Manresa
Navarcles
Castelltersol
Montseny
Sierra de Montseny
Breda
Hostalric
Vidreras
Tossa
San Feliu de Guixols

Bellpuig
La Panadella
Igualada
Castellbell
San Feliu de Codinas
La Garriga
Cardedeu
San Celoni
Arbucias
Lloret de Mar
Blanes
Costa Brava

Mollerusa
Juneda
Arbeca
Santa Coloma de Queralt
Capellades
Esparraguera
Olesa
Terrassa
Matadepera
Caldas de Montbuy
Granollers
Arénys de Munt
Calella
Canet de Mar
Arénys de Mar

Borjas Blancas
Belltall
Sarreal
Sta. María de Tous
Piera
Rubí
Sabadell
Ripollet
San Cugat del Vallés
Mollet
Montcada
S. Andrés
Argentona
Mataró
San Juan de Vilasar
Premiá de Mar

Vinaixa
Espluga de Francolí
Montblanch
Monasterio de Poblet
Plá de Sta. María
Martorell
Molins de Rey
Villafranca del Panadés
S. Feliu
San Baudilio de Llobregat
Gavá
Masnou
Badalona
Barcelona
Hospitalet
Prat del Llobregat

Juncosa
Musara
Alcover
Valls
Bisbal del Panadés
Castelldeféls
Sitges
Garraf

Reus
Vilaseca
Riudoms
Constantí
Vendrell
Calafell
Villanueva y Geltrú

Tarragona
Cambrils
Salou
C. de Salou
Torredembarra

Sierra de Montsant
Montroig
Hospitalet
La Ametlla de Mar
Golfo de San Jorge
Costa Dorada

Riomar
C. de Tortosa

MEDITERRANEAN SEA

N 124
N 125
N 20
N 113
N 112
N 9
E 4
E 49
N 117
N 118
N 116
N 114
N 152
N 141
N 340
N 240
N 420
N II
E 26

70
74
82
97
99
62
137
93
100
169
158
148
94
95

John Bartholomew & Son Ltd

0   10   20   30   40   50 Miles
0   10   20   30   40   50   60   70   80 Kilometres

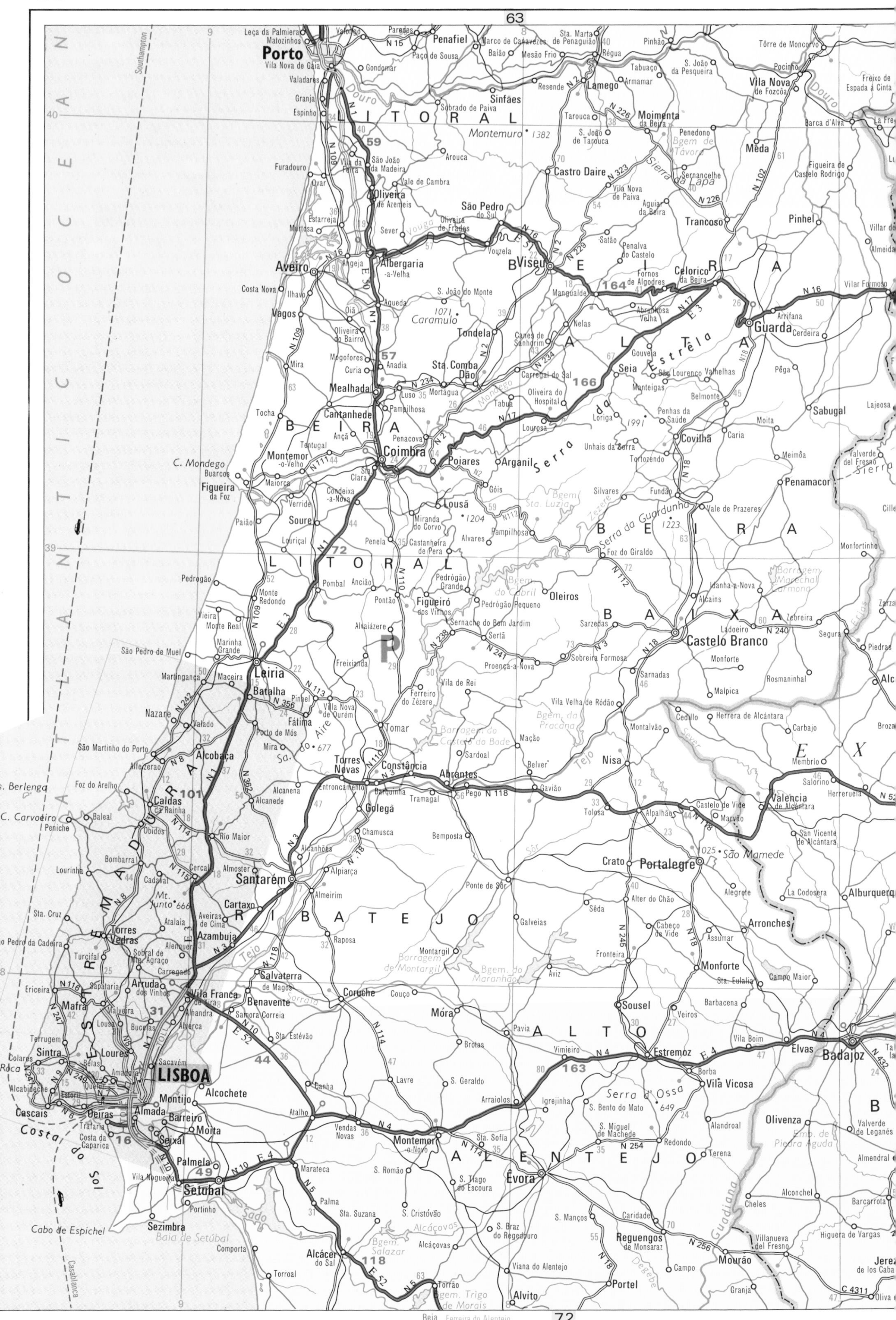

63

N O N

Southampton

Leça da Palmiera
Matozinhos
Porto
Vila Nova de Gaia
Valadares
Granja
Espinho

Paredes
Valongo
Gondomar
Sobrado de Paiva

Paredes
Paço de Sousa
Penafiel
Marco de Canavezes
Baião
Mesão Frio
Resende

Sta. Marta de Penaguião
Régua
Tabuaço
Lamego

Pinhão
S. João da Pesqueira
Armamar
Penedono
Bgem. de Tavor

Tôrre de Moncorvo
Pocinho
Vila Nova de Fozcôa
Barca d'Alva
La Fre

N 15
L I T O R A L
Sinfães
Montemuro 1382
Moimenta da Beira
Meda
Figueira de Castelo Rodrigo
Lu

N 1
Furadouro
Ovar
São João da Madeira
Vale de Cambra
Arouca
Castro Daire
N 323
Sernancelhe
Vila Nova de Paiva
Serra da Lapa
N 102
Almeida
Villar de

59
N 109
Estarreja
Murtosa
Oliveira de Azemeis
Oliveira de Frades
São Pedro do Sul
Aguiar da Beira
Satão
Penalva do Castelo
Abrolhosa Velha
Celorico da Beira
Trancoso
Pinhel

Aveiro
Angeja
Albergaria-a-Velha
Sever
Voutela
B **Viseu** E I R A
164
Fornos de Algodres
N 17
N 16
Vilar Formoso
Guarda
50

Costa Nova
Ilhavo
S. João do Monte
Caramulo 1071
Mangualde
Nelas
N 229
Arrifana
Cerdeira

Vagos
Oliveira do Bairro
Oiã
38
N 1
57
Tondela
N 2
Canas de Sanhrim
N 234
Carregal do Sal
Gouveia
S. Lourenço Valhelhas
Pêga
Sabugal

Mira
Megofores
Curia
Anadia
N 234
Sta. Comba Dão
166
Seia
Manteigas
Belmonte
Lajeosa

63
Mealhada
Luso
Mortágua
Tabua
Oliveira do Hospital
Loriga 1991
Penhas da Saúde
Moita
Caria

Tocha
Cantanhede
Pampilhosa
Penacova
N 17
Lourosa
Unhais da Serra
Covilhã
Meimoa

Tentugal
Ançã
Coimbra
Poiares
Arganil
Torozendo
Valverde del Fresno
Sierra

Montemor-o-Velho
N 111
Sta. Clara
Silvares
Fundão
Penamacor
Cille

C. Mondego
Buarcos
Condeixa-a-Nova
Góis
Bgem. Sta. Luzia
N 18
Vale de Prazeres

Figueira da Foz
Maiorca
Verride
Miranda do Corvo
1204
Lousã
Zezere
1223
Monfortinho

Paião
Soure
Louriçal
Castanheira de Pera
Pampilhosa
Alvares
Foz do Giraldo
B E I R A

Pedrógão
Pombal
Anção
Penela
Pedrógão Grande
Oleiros
N 112
Idanha-a-Nova
Bgem Marechal Carmona

Monte Redondo
Figueiró dos Vinhos
Pedrógão Pequeno
N 112
Alcains
Zebreira

Vieira
Monte Real
Pontão
Alvaiázere
Sernache do Bom Jardim
Sertã
Sarzedas
N 240
Segura

Marinha Grande
Freixianda
Ferreira do Zêzere
N 241
Proença-a-Nova
Sobreira Formosa
N 18
Castelo Branco
Alc

S. Pedro de Muel
Leiria
N 113
22
Vila de Rei
Sarnadas
Monforte
Rosmaninhal

Martingança
Maceira
Batalha
Pinhel
Vila Velha de Ródão
Malpica

Nazaré
Valado
N 356
Porto de Mós
Mira
Sa. 677
Mação
Belver
Montalvão
Cedillo
Herrera de Alcántara
Carbajo
Broza

São Martinho do Porto
Alfezerao
N 8
Alcobaça
Fátima
Torres Novas
Constância
Sardoal
Tejo
Nisa
Membrio
E X

Foz do Arelho
101
Sa.
Tomar
Barragem do Castelo do Bode
Ahrantes
Gavião
N 118
Salorino
Herreruela
N 52

Is. Berlenga
Baleal
Caldas da Rainha
Rio Maior
Alcanena
Entroncamento
Barquinha
Tramagal
Pego
N 118
Tolosa
Alpalhão
Castelo de Vide
San Vicente de Alcántara
Valencia de Alcántara

C. Carvoeiro
Peniche
Óbidos
N 114
Golegã
Chamusca
Sôr
1025 São Mamede
Marvão

Lourinha
Bombarral
Cadaval
N 115
Almoster
Alpiarça
Bemposta
Crato
Portalegre
La Codosera
Alburquerq

Sta. Cruz
Atalaia
Aveiras de Cima
Raposa
Ponte de Sôr
Alter do Chão
Alegrete
Arronches

S. Pedro da Cadeira
Torres Vedras
Mt. Junto 666
Santarém
Cartaxo
Almeirim
R I B A T E J O
Galveias
N 245
Cabeço de Vide
Assumar
Campo Maior

Ericeira
N 116
Sapataria
Arruda dos Vinhos
Azambuja
Salvaterra de Magos
Coruche
Aviz
Fronteira
Monforte
Sta. Eulalia
Barbacena

Mafra
N 241
Malveira
31
Alenquer
Carregado
Vila Franca de Xira
Benavente
Samora Correia
Couço
Móra
Pavia
Sousel
Veiros
Vila Boim
N 4

Terrugem
Colares
Sintra
Lousa
Bucelas
Alhandra
Alverca
Sta. Estevão
N 114
Brotas
A L T O
Estremoz
E 4
Elvas
Badajoz

de Roca
Belas
Queluz
Amadora
Loures
Sacavém
44
Vimieiro
N 4
163
Borba
Vila Viçosa
N 4
Ta

Alcabideche
Estoril
LISBOA
Alcochete
Canha
Lavre
S. Geraldo
Arraiolos
Igrejinha
S. Bento do Mato
Alandroal
Olivenza
B

Cascais
N 6
Trafaria
Almada
Montijo
Atalho
Vendas Novas
S. Tiago do Escoura
649
Serra d'Ossa
Redondo
Terena
Emb. de Piedra Aguda

Costa do Sol
Costa da Caparica
16
Barreiro
Moita
Seixal
N 4
Montemor-o-Novo
N 254
Alconchel

Cabo de Espichel
Palmela
N 10 E 4
Marateca
S. Romão
Sta. Sofia
Évora
Cheles
Barcarrota

Sezimbra
49
Setúbal
Portinho
Palma
Sta. Suzana
S. Cristóvão
S. Manços
Caridade
Villanueva del Fresno
Higuera de Vargas
Jere de los Caba

Baia de Setúbal
Comporta
Sado
Bgem. Salazar
118
Alcácovas
S. Bráz do Regeduro
Reguengos de Monsaraz
N 256
Campo
Mourão
Oliv

Casablanca
Torroal
Alcácer do Sal
Torrão
Bgem. Trigo de Morais
Alvito
Viana do Alentejo
N 18
Portel
Granja
C 4311
47

Beja
Ferreira do Alentejo
72

1:1 250 000

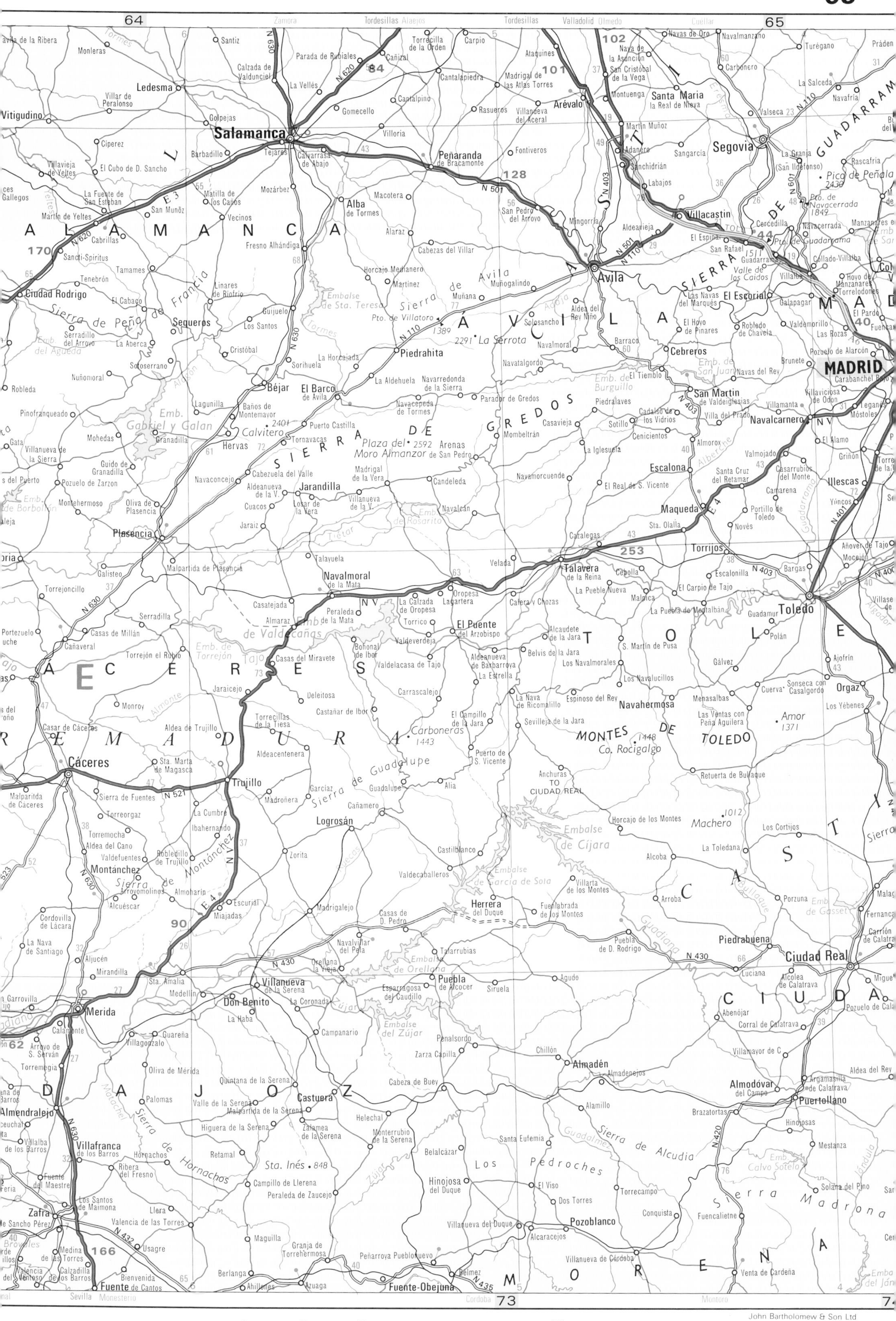
64
65
Zamora
Tordesillas Alaejos
Tordesillas
Valladolid Olmedo
Cuellar
Navas de Oro
Navalmanzano
Turégano
Práden
Monleras
Santiz
Parada de Rubiales
Cañizal
Torrecilla de la Orden
Carpio
Ataquines
Nava de la Asunción
San Cristóbal de la Vega
Carbonero
Calzada de Valdunciel
La Vellés
Cantalapiedra
Madrigal de las Altas Torres
Arévalo
Montuenga
Santa María la Real de Nieva
La Salceda
Navafría
Ledesma
Villar de Peralonso
Gomecello
Cantalpino
Rasueros
Villanueva del Aceral
Valseca
Vitigudino
Golpejas
Villoria
Martín Muñoz
Salamanca
Ciperez
Barbadillo
Tejares
Calvarrasa de Abajo
Peñaranda de Bracamonte
Fontiveros
Adanero
Sangarcía
Segovia
La Granja (San Ildefonso)
Rascafría
Villavieja de Yeltes
El Cubo de D. Sancho
Sanchidrián
Pico de Peñalara 2430
Gallegos
La Fuente de San Esteban
Matilla de los Caños
Mozárbez
Macotera
Labajos
Pto. de Navacerrada 1849
San Muñoz
Alba de Tormes
San Pedro del Arroyo
Martín de Yeltes
Vecinos
Mingorría
Villacastín
Navacerrada
Manzanares el Real
Cercedilla
Pto. de Guadarrama
Aldeavieja
El Espinar
Cabrillas
Fresno Alhándiga
Alaraz
San Rafael
Collado-Villalba
Sancti-Spiritus
Cabezas del Villar
Guadarrama
Tamames
Horcajo Medianero
Ávila
Valle de los Caídos
Villalba
Hoyo de Manzanares
Torrelodones
Tenebrón
Martínez
Muñogalindo
Ciudad Rodrigo
Linares de Riofrío
Las Navas del Marqués
El Escorial
Galapagar
El Cabaco
Embalse de Sta. Teresa
Muñana
Aldea del Rey Niño
El Pardo
Sierra de Peña de Francia
Sequeros
Guijuelo
Los Santos
Pto. de Villatoro
Solosancho
El Hoyo de Pinares
Robledo de Chavela
Valdemorillo
Las Rozas
Serradillo del Arroyo
La Alberca
1389
2291 La Serrota
Navalmoral
Barraco
Emb. del Agueda
Cristóbal
Piedrahita
Cebreros
Pozuelo de Alarcón
Sotoserrano
La Horcajada
Navatalgordo
Emb. de San Juan
Navas del Rey
Brunete
MADRID
Carabanchel Bajo
Robleda
Nuñomoral
Sorihuela
El Tiemblo
Béjar
El Barco de Avila
La Aldehuela
Navarredonda de la Sierra
Emb. de Burguillo
San Martín de Valdeiglesias
Villaviciosa de Odón
Lagunilla
Baños de Montemayor
Navacepeda de Tormes
Parador de Gredos
Piedralaves
Villamanta
Leganés
Móstoles
Pinofranqueado
Emb. Gabriel y Galan
Calvitero 2401
Puerto Castilla
Casavieja
Cadalso de los Vidrios
Villa del Prado
Navalcarnero
Mohedas
Hervas
Tornavacas
Plaza del Moro Almanzor 2592
Arenas de San Pedro
Mombeltrán
Sotillo
Cenicientos
Almorox
El Alamo
Gata
Villanueva de la Sierra
Granadilla
Sierra de Gredos
La Iglesuela
Valmojado
Griñón
Guido de Granadilla
Cabezuela del Valle
Madrigal de la Vera
Candeleda
Escalona
Casarrubios del Monte
Pozuelo de Zarzon
Navaconcejo
Aldeanueva de la V.
Jarandilla
Navamorcuende
El Real de S. Vicente
Santa Cruz del Retamar
Illescas
Emb. de Borbollón
Montehermoso
Oliva de Plasencia
Cuacos
Losar de la Vera
Villanueva de la V.
Navalcán
Camarena
Yuncos
Jaraiz
Emb. de Rosarito
Maqueda
Portillo de Toledo
Plasencia
Tiétar
Cazalegas
Sta. Olalla
Novés
Torrijos
Talayuela
Añover de Tajo
Mocejón
Galisteo
Malpartida de Plasencia
Velada
Talavera de la Reina
Cebolla
Escalonilla
Bargas
Torrejoncillo
Navalmoral de la Mata
Oropesa
Lagartera
Calera y Chozas
La Pueble Nueva
Malpica
El Carpio de Tajo
Casatejada
La Calzada de Oropesa
La Puebla de Montalbán
Guadamur
Toledo
Serradilla
Almaraz
Peraleda de la Mata
Torrico
El Puente del Arzobispo
Portezuelo
Casas de Millán
Emb. de Valdecañas
Valdeverdeja
Alcaudete de la Jara
Polán
Cañaveral
Emb. de Torrejón
Bohonal de Ibor
Aldeanueva de Barbarroya
Belvis de la Jara
S. Martín de Pusa
Torrejón el Rubio
Casas del Miravete
Valdelacasa de Tajo
Los Navalmorales
Gálvez
Ajofrín
CÁCERES
La Estrella
Los Navalucillos
Jaraicejo
Carrascalejo
La Nava de Ricomalillo
Espinoso del Rey
Menasalbas
Cuerva
Sonseca con Casalgordo
Orgaz
Monroy
Deleitosa
Navahermosa
Castañar de Ibor
Torrecillas de la Tiesa
El Campillo de la Jara
Las Ventas con Peña Aguilera
Amor 1371
Los Yébenes
Casar de Cáceres
Aldea de Trujillo
Carboneras 1443
Sevilleja de la Jara
MONTES DE TOLEDO
1448 Co. Rocigalgo
EXTREMADURA
Cáceres
Sta. Marta de Magasca
Aldeacentenera
Puerto de S. Vicente
Sierra de Guadalupe
Anchuras TO CIUDAD REAL
Retuerta de Bullaque
Malpartida de Cáceres
Sierra de Fuentes
Trujillo
Garciaz
Guadalupe
Alia
Madroñera
Torreorgaz
La Cumbre
Cañamero
Horcajo de los Montes
Machero 1012
Los Cortijos
Torremocha
Ibahernando
Logrosán
Embalse de Cijara
Aldea del Cano
Valdefuentes
Robledillo de Trujillo
Castilblanco
La Toledana
Montánchez
Sierra de Montánchez
Zorita
Alcoba
Sierra
Arroyomolinos
Almoharín
Valdecaballeros
Embalse de García de Sola
Villarta de los Montes
Arroba
Porzuna
Alcuéscar
Escurial
Madrigalejo
Herrera del Duque
Fuenlabrada de los Montes
Emb. de Gasset
Cordovilla de Lácara
Miajadas
Casas de D. Pedro
La Nava de Santiago
Navalvillar de Pela
Talarrubias
Puebla de D. Rodrigo
Piedrabuena
Ciudad Real
Aljucén
Orellana la Vieja
Emb. de Orellana
Luciana
Mirandilla
Sta. Amalia
Villanueva de la Serena
Puebla de Alcocer
Esparragosa del Caudillo
Siruela
Agudo
Alcolea de Calatrava
Medellín
Don Benito
La Coronada
La Garrovilla
Merida
Calamonte
La Haba
Embalse del Zújar
Campanario
Abenójar
Corral de Calatrava
Pozuelo de Calatrava
Villagonzalo
Guareña
Peñalsordo
Arroyo de S. Serván
Zarza Capilla
Chillón
Villamayor de C.
Torremegía
Oliva de Mérida
Almadén
Almadenejos
Aldea del Rey
Quintana de la Serena
Cabeza de Buey
Almodóvar del Campo
Argamasilla de Calatrava
Almendralejo
Palomas
Valle de la Serena
Castuera
Puertollano
Malpartida de la Serena
Alamillo
Brazatortas
Helechal
Higuera de la Serena
Zalamea de la Serena
Hinojosas
Villafranca de los Barros
Monterrubio de la Serena
Santa Eufemia
Sierra de Alcudia
Mestanza
Hornachos
Retamal
Belalcázar
Ribera del Fresno
Sierra de Hornachos
Sta. Inés 848
Los Pedroches
Emb. Calvo Sotelo
Fuente del Maestre
Campillo de Llerena
Hinojosa del Duque
El Viso
Torrecampo
Solana del Pino
Peraleda de Zaucejo
Dos Torres
Sierra Madrona
Los Santos de Maimona
Zafra
Llera
Conquista
Fuencaliente
Valencia de las Torres
Pozoblanco
Villanueva del Duque
Maguilla
Alcaracejos
Usagre
Medina de las Torres
Granja de Torrehermosa
Peñarroya Pueblonuevo
Villanueva de Córdoba
Calzadilla de los Barros
Bienvenida
Berlanga
Belmez
Venta de Cardeña
Valencia del Ventoso
Fuente de Cantos
Ahillones
Azuaga
Fuente-Obejuna
SIERRA MORENA
Sevilla Monesterio
Córdoba
73
Montoro
CIUDAD REAL
CASTILLA
CASTILLA LA NUEVA
SALAMANCA
ÁVILA
SIERRA DE AVILA
SIERRA DE GUADARRAMA
TOLEDO
BADAJOZ
N 630
N 620
N 501
N 110
N 403
N 601
N VI
N V
N 401
N 521
N 430
N 432
N 435
N 420
E 3
E 4
84
101
102
128
170
253
90
166
62
40
44

John Bartholomew & Son Ltd

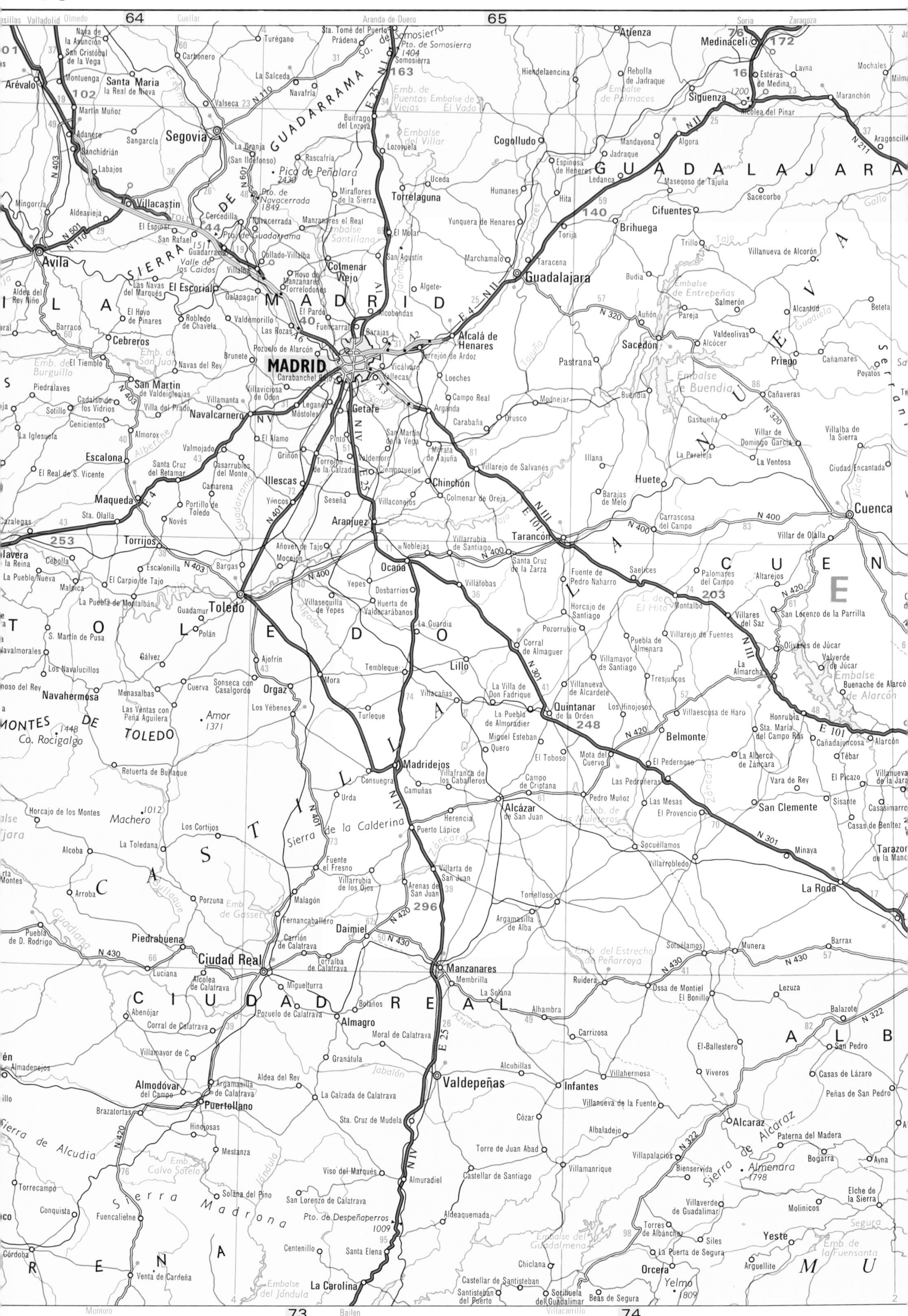
64
65
73
74
Arévalo
Santa Maria la Real de Nieva
Segovia
Villacastín
Avila
S I E R R A D E G U A D A R R A M A
Pico de Peñalara 2430
Pto. de Navacerrada 1849
Pto. de Somosierra 1404
Somosierra
Buitrago del Lozoya
Embalse de Puentes Viejas
Embalse El Vado
Embalse del Villar
Torrelaguna
Cogolludo
Atienza
Medinaceli
Sigüenza
Alcolea del Pinar
Maranchón
G U A D A L A J A R A
Cifuentes
Brihuega
Guadalajara
Alcalá de Henares
Torrejón de Ardoz
Colmenar Viejo
El Escorial
Valle de los Caidos
Cercedilla
Guadarrama
Navacerrada
Manzanares el Real
M A D R I D
MADRID
El Pardo
Fuencarral
Barajas
Las Rozas
Pozuelo de Alarcón
Carabanchel Bajo
Vallecas
Vicálvaro
Leganés
Móstoles
Getafe
Navalcarnero
Cebreros
San Martin de Valdeiglesias
Escalona
Maqueda
Illescas
Aranjuez
Chinchón
Arganda
Pastrana
Sacedón
Embalse de Entrepeñas
Embalse de Buendia
Priego
Huete
Cuenca
Tarancón
Ocaña
Toledo
Torrijos
Talavera la Reina
T O L E D O
Orgaz
Mora
Tembleque
Lillo
Quintanar de la Orden
Belmonte
Madridejos
Consuegra
Alcázar de San Juan
Campo de Criptana
El Toboso
Mota del Cuervo
San Clemente
La Roda
Villarrobledo
Tomelloso
Embalse de Alarcón
Motilla del Palancar
C U E N C A
L A N U E V A
C A S T I L L A
MONTES DE TOLEDO
Amor 1371
Machero 1012
Sierra de la Calderina
Puerto Lápice
Herencia
Villarta de San Juan
Arenas de San Juan
Daimiel
Manzanares
Membrilla
La Solana
Ciudad Real
Piedrabuena
Almagro
C I U D A D R E A L
Valdepeñas
Infantes
Almodóvar del Campo
Puertollano
Sierra de Alcudia
Sierra Madrona
Pto. de Despeñaperros 1009
Santa Elena
La Carolina
Almuradiel
Viso del Marqués
Alcaraz
Sierra de Alcaraz
Almenara 1798
Munera
Ossa de Montiel
Ruidera
Villarrobledo
Yeste
Orcera
Yelmo 1809
Embalse de Guadalmena
Embalse del Jándula
Montoro
Bailen
Villacarrillo
Cuellar
Aranda-de-Duero
Soria
Zaragoza
Valladolid
Olmedo
A L B
M U
E

1:1 250 000

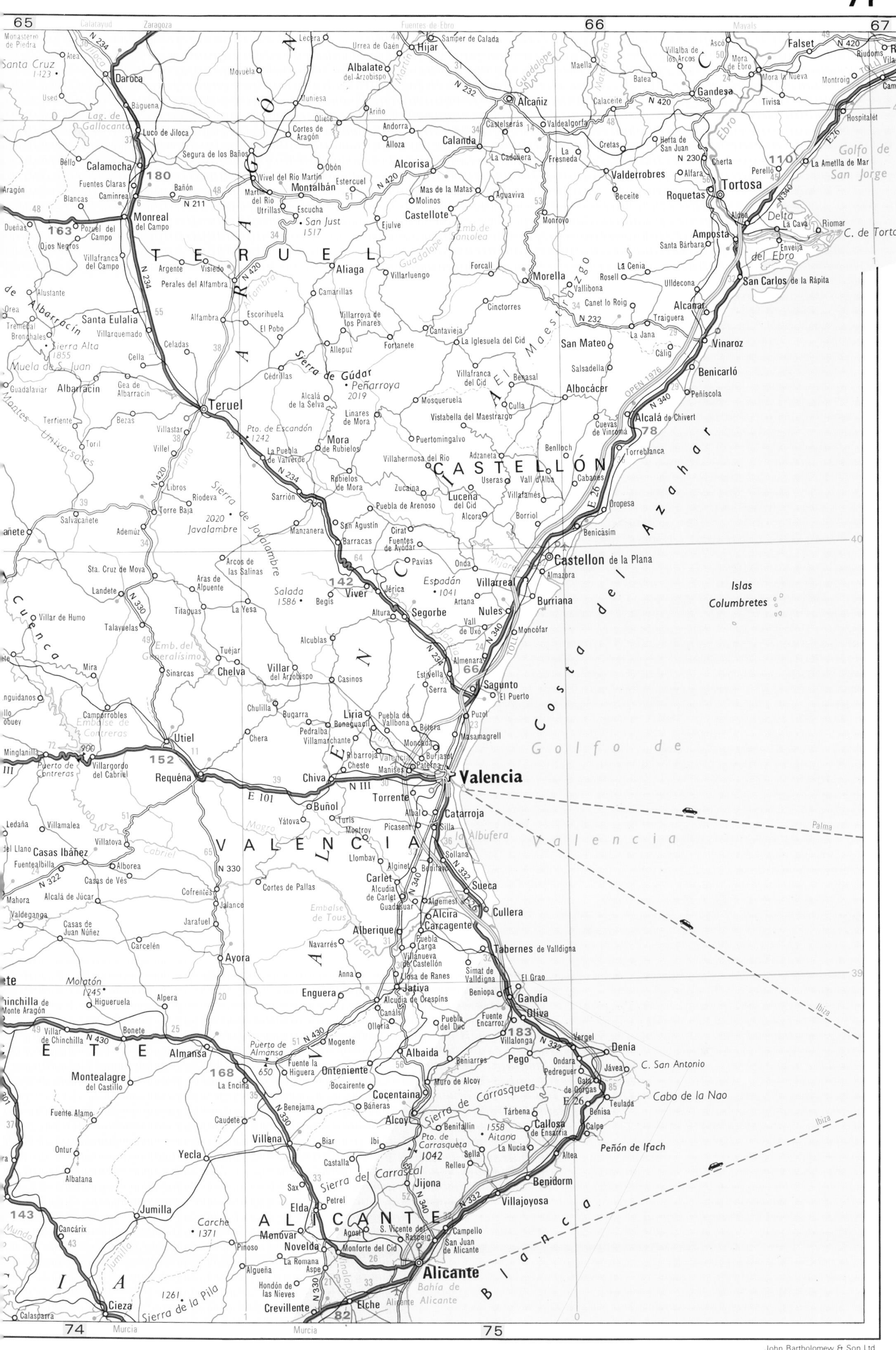

65 · Calatayud · Zaragoza · N 234 · 66 · Mayals · 67

Santa Cruz 1123
Monasterio de Piedra · N 234 · Atea · Daroca · Movuela · Lécera · Urrea de Gaén · Híjar · Samper de Calada · Villalba de los Arcos · Asco · Mora de Ebro · Falset · N 420 · Reus · Vilaseca · Riudoms
Used · Báguena · Montroig · Cambrils
Luco de Jiloca · Albalate del Arzobispo · Maella · Batea · Gandesa · Tivisa · Hospitalet
Calamocha · 180 · Segura de los Baños · Alcañiz · Calaceite · N 420 · E 26
Fuentes Claras · Bañón · Obón · Andorra · Alloza · Calanda · La Cadonera · La Fresneda · Cretas · Horta de San Juan · N 230 · Cherta · 110 · La Ametlla de Mar
Blancas · Caminreal · N 211 · 48 · Martín del Río · Estercuel · Mas de las Matas · Valdealgorfa · Beceite · Perelló · 49 · San Jorge
Dueñas · 163 · Pozuel del Campo · Monreal del Campo · Utrillas · Escucha · Ejulve · Monroyo · Valderrobres · Santa Bárbara · Amposta · La Cava · Riomar
Ojos Negros · Villafranca del Campo · N 234 · San Just 1517 · Castellote · Emb. de Santolea · Forcall · Rossell · Lt Cenia · Roquetas · Delta del Ebro · Enveija · C. de Tortosa
Argente · Visiedo · Aliaga · Villarluengo · Morella · Ulldecona · San Carlos de la Rápita
TERUEL · Perales del Alfambra · Camarillas · Cantavieja · Cinctorres · Canet lo Roig · N 232 · Alcanar
Alfambra · Escorihuela · Allepuz · Villarroya de los Pinares · La Iglesuela del Cid · Benasal · Salsadella · Traiguera · La Jana · Vinaroz
Santa Eulalia · El Pobo · Fortanete · Villafranca del Cid · Culla · San Mateo · Cálig · Benicarló
Villarquemado · Celadas · Cédrillas · Mosqueruela · Vistabella del Maestrazgo · Albocácer · Peñíscola
Albarracín · Gea de Albarracín · Teruel · Pto. de Escandón 1242 · Puertomingalvo · Adzaneta · Cuevas de Vinromá · N 340 · Alcalá de Chivert
Cella · Villastar · Bezas · La Puebla de Valverde · Mora de Rubielos · Villahermosa del Río · Useras · Vall d'Alba · Cabanes · Torreblanca · 78
Teriente · Villel · Linares de Mora · Rubielos de Mora · Zucaina · Villafamés · Benlloch · Oropesa
Albentosa · Manzanera · Puebla de Arenoso · Lucena del Cid · Alcora · Borriol · Benicásim
142 · Viver · Jérica · Espadán 1041 · Villarreal · Castellón de la Plana · Almazora
Salada 1586 · Begis · San Agustín · Barracas · Cirat · Fuentes de Ayodar · Pavias · Onda · Artana · Nules · Burriana
Javalambre 2020 · Arcos de las Salinas · Aras de Alpuente · Altura · Segorbe · Vall de Uxó · Moncófar
Islas Columbretes
CASTELLÓN
Costa del Azahar
Golfo de Valencia
ARAGÓN
Guadalope
Ebro
Golfo de San Jorge
Tortosa

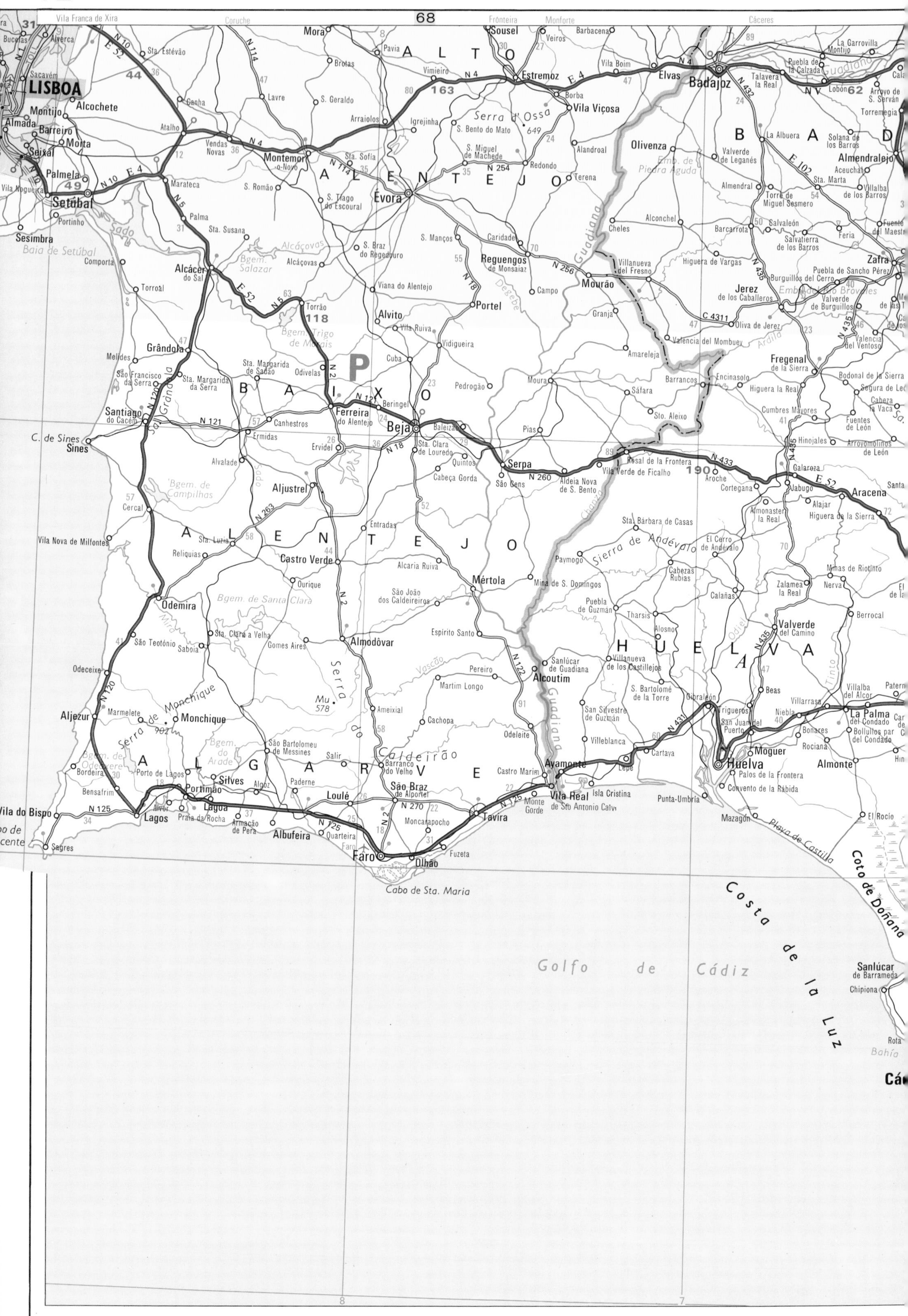
Vila Franca de Xira
Coruche
68
Fronteira
Monforte
Cáceres
LISBOA
Almada
Barreiro
Montijo
Alcochete
Seixal
Moita
Palmela
Setúbal
Sesimbra
Baía de Setúbal
Sado
Mora
Brotas
Pavia
Lavre
S. Geraldo
Canha
Sta. Estêvão
Vendas Novas
Montemor-o-Novo
Arraiolos
Igrejinha
Vimieiro
Sousel
Estremoz
Veiros
Barbacena
Vila Boim
Elvas
Badajoz
Borba
Vila Viçosa
Serra d'Ossa
S. Bento do Mato
S. Miguel de Machede
Redondo
Alandroal
Terena
Olivenza
Emb. de Piedra Aguda
Valverde de Leganés
Almendral
Torre de Miguel Sesmero
A L T O
A L E N T E J O
B A D
Évora
Sta. Sofia
S. Tiago do Escoural
S. Romão
Marateca
Palma
Sta. Susana
Alcácer do Sal
Bgem. Salazar
Alcáçovas
S. Braz do Regedouro
S. Manços
Caridade
Reguengos de Monsaraz
Mourão
Campo
Portel
Viana do Alentejo
Alvito
Vila Ruiva
Vidigueira
Torrão
Bgem. Trigo de Morais
Comporta
Torroal
Grândola
Melides
São Francisco da Serra
Sta. Margarida da Serra
Sta. Margarida de Sadão
Odivelas
Cuba
P
B A I X O
A L E N T E J O
Beringel
Ferreira do Alentejo
Beja
Baleizão
Pedrogão
Moura
Sáfara
Sto. Aleixo
Barrancos
Encinasolo
Santiago do Cacém
C. de Sines
Sines
Canhestros
Ermidas
Ervidel
Sta. Clara de Louredo
Quintos
Cabeça Gorda
Serpa
São Gens
Pias
Aldeia Nova de S. Bento
Vila Verde de Ficalho
Rosal de la Frontera
Aroche
Cortegana
Alvalade
Bgem. de Campilhas
Cercal
Aljustrel
Entradas
Vila Nova de Milfontes
Sta. Luzia
Reliquias
Castro Verde
Alcaria Ruiva
Ourique
São João dos Caldeireiros
Mértola
Mina de S. Domingos
Odemira
Bgem. de Santa Clara
Mira
São Teotónio
Saboia
Sta. Clara a Velha
Gomes Aires
Almodôvar
Serra do Caldeirão
Espírito Santo
Pereiro
Martim Longo
Alcoutim
Sanlúcar de Guadiana
Odeceixe
Aljezur
Marmelete
Monchique
Serra de Monchique
Mu 578
Ameixial
Cachopa
Odeleite
Guadiana
Bordeira
Porto de Lagos
Silves
São Bartolomeu de Messines
Salir
Barranco do Velho
A L G A R V E
Castro Marim
Ayamonte
Vila Real de Sto António
Monte Gorde
Bensafrim
Lagos
Portimão
Lagoa
Alvor
Praia da Rocha
Armação de Pera
Algoz
Paderne
Albufeira
Loulé
São Braz de Alportel
Moncarapacho
Tavira
Quarteira
Faro
Olhão
Fuzeta
Vila do Bispo
Sagres
Cabo de Sta. Maria
Cheles
Alconchel
Villanueva del Fresno
Higuera de Vargas
Barcarrota
Salvaleón
Jerez de los Caballeros
Oliva de Jerez
Valencia del Mombuey
Granja
Amareleja
Fregenal de la Sierra
Higuera la Real
Cumbres Mayores
Hinojales
Galaroza
Jabugo
Aracena
Almonaster la Real
Alajar
Higuera de la Sierra
Sta. Bárbara de Casas
Sierra de Andévalo
El Cerro de Andévalo
Paymogo
Cabezas Rubias
Puebla de Guzmán
Tharsis
Alosno
Calañas
Zalamea la Real
Nerva
Minas de Riotinto
Berrocal
Valverde del Camino
H U E L V A
Villanueva de los Castillejos
S. Bartolomé de la Torre
San Silvestre de Guzmán
Villeblanca
Cartaya
Lepe
Isla Cristina
Gibraleón
Trigueros
San Juan del Puerto
Beas
Niebla
Villarrasa
La Palma del Condado
Bollullos par del Condado
Bonares
Rociana
Almonte
Moguer
Huelva
Palos de la Frontera
Convento de la Rábida
Punta-Umbría
Mazagón
El Rocío
Playa de Castilla
Coto de Doñana
Costa de la Luz
Golfo de Cádiz
Sanlúcar de Barrameda
Chipiona
Rota
Bahía
Talavera la Real
Puebla de la Calzada
Montijo
La Garrovilla
Lobón
Torremegía
La Albuera
Solana de los Barros
Almendralejo
Aceuchal
Sta. Marta
Villalba de los Barros
Feria
Salvatierra de los Barros
Zafra
Puebla de Sancho Pérez
Burguillos del Cerro
Valverde de Burguillos
Valencia del Ventoso
Bodonal de la Sierra
Segura de León
Cabeza la Vaca
Fuentes de León
Arroyomolinos de León
Villalba del Alcor
Guadiana
Ardila
Chança
Degebe
Odiel
Tinto
E 52
E 4
E 102
N 4
N 10
N 114
N 2
N 5
N 18
N 120
N 121
N 122
N 125
N 254
N 256
N 260
N 263
N 270
N 433
N 435
N 431
N V
C 4311
44
49
118
163
190
62
8
7

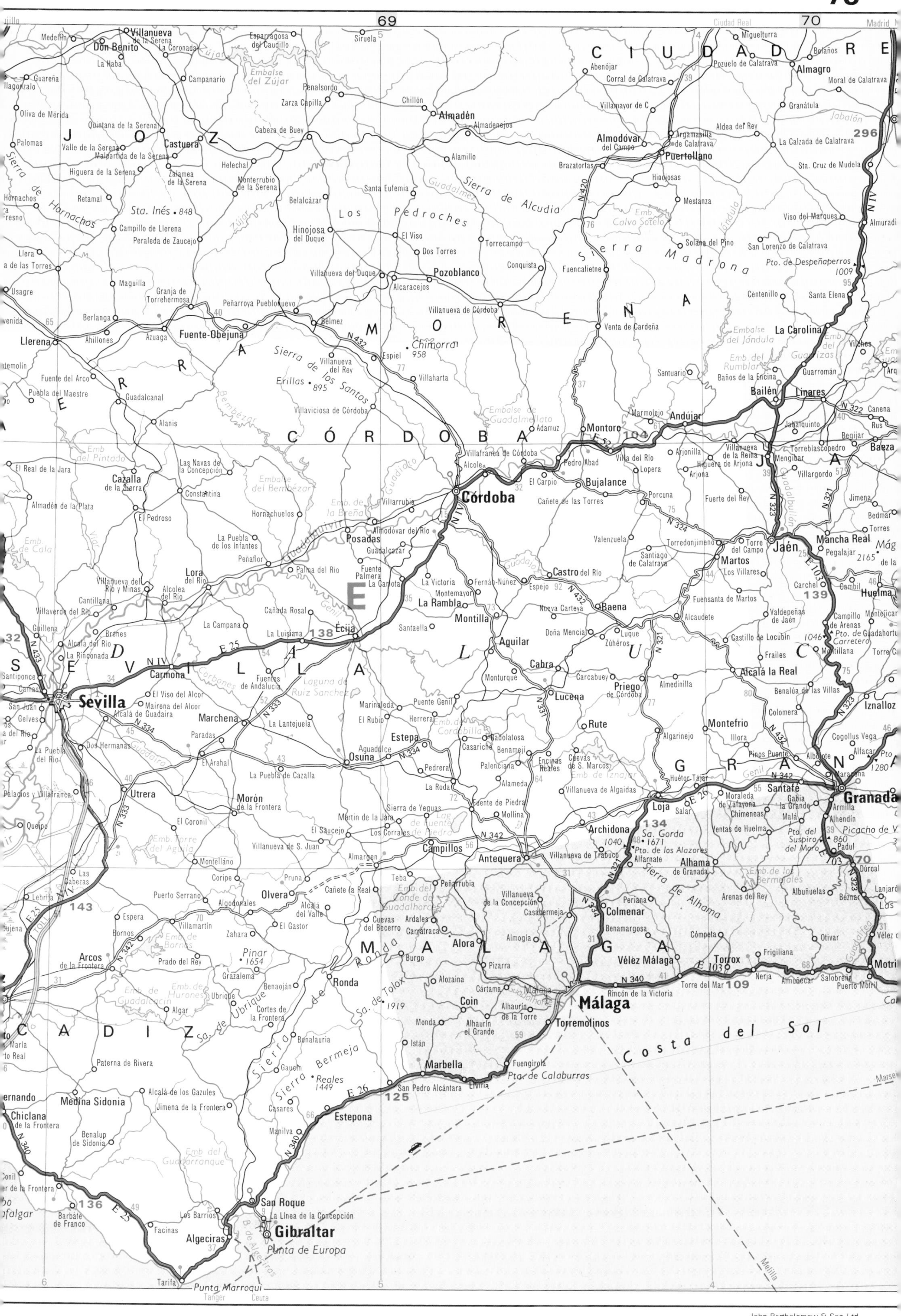

69
70

Madrid

C I U D A D   R E

Miguelturra
Botijos
Almagro
Moral de Calatrava

Medellín
Villanueva
de la Serena
Don Benito
La Coronada
Siruela
Esparragosa
del Caudillo
Abenojar
Corral de Calatrava
Pozuelo de Calatrava

La Haba
Campanario
Granátula
Aldea del Rey
La Calzada de Calatrava

Guareña
Quintana de la Serena
Zarza Capilla
Penalsordo
Chillón
Almadén
Almadenejos
Villamayor de C
Almodóvar
del Campo
Argamasilla
de Calatrava
Puertollano
Sta. Cruz de Mudela

296

Ilagonzalo
Oliva de Mérida
Valle de la Serena
Castuera
Cabeza de Buey
Helechal
Santa Eufemia
Brazatortas
Hinojosas
Mestanza
Viso del Marques
Almuradi

Palomas
J O Z
Malpartida de la Serena
Zalamea
de la Serena
Monterrubio
de la Serena
Belalcázar
Guadalme
Sierra   de   Alcudia
Calvo Sotelo
Emb.

Hornachos
resno
Higuera de la Serena
Retamal
Sta. Inés 848
Campillo de Llerena
Peraleda de Zaucejo
Hinojosa
del Duque
El Viso
Dos Torres
Torrecampo
Fuencaliente
San Lorenzo de Calatrava
Pto. de Despeñaperros
1009
Centenillo
Santa Elena
95

Usagre
Maguilla
Granja de
Torrehermosa
Villanueva del Duque
Pozoblanco
Conquista
Fuencalietne
Santuario
Baños de la Encina
Embalse
del Jándula
La Carolina
Vilches

Llera
de las Torres
Berlanga
Ahillones
Azuaga
Fuente-Obejuna
Peñarroya Pueblonuevo
Bélmez
Alcaracejos
Villanueva de Córdoba
Venta de Cardeña
Emb. del
Rumblar
Bailén
Guarromán
Linares
N 322
Canena
Arq

avenida
65
Llerena
40
M O R
S E N A
Marmolejo
Andújar
Baeza
Rus
Begijar
Arquillos

emolín
Fuente del Arco
Guadalcanal
Espiel
77
Villaharta
E R
Chimorra
958
Villanueva
del Rey
Erillas 895
Villaviciosa de Córdoba
Adamuz
Montoro
Arjonilla
Villanueva
de la Reina
Higuera de Arjona
Arjona
Torreblascopedro
J A
N 323
Villargordo
Mengibar
Jimena
Bedmar
Torres

Puebla del Maestre
Alanis
Emb.
del Pintado
C Ó R D O B A
Guadalm
Embalse de
Guadalmellato
Villafranca de Córdoba
Alcolea
Pedro Abad
Villa del Río
Lopera
Santiago
de Calatrava
Torre
del Campo
Jaén
Pegalajar
Mág
2165
Mancha Real
Carchel
Huelma
Cambil
Mág
de la

El Real de la Jara
Cazalla
de la Sierra
Las Navas de
la Concepción
Constantina
Emb. del
Bembézar
Villarrubia
Almodóvar del Río
Córdoba
32
El Carpio
Bujalance
Cañete de las Torres
Porcuna
Fuerte del Rey
75
Martos
Los Villares
Fuensanta de Martos
44
Castillo de Locubín
139
Valdepeñas
de Jaén
Campillo
de Arenas
Montejicar

Almadén de la Plata
El Pedroso
Hornachuelos
Emb. de
la Breña
Posadas
Guadalcázar
Fuente
Palmera
La Carlota
La Victoria
Fernán-Núñez
Espejo
92
N 432
Castro del Río
Baena
Nueva Carteya
Doña Mencía
Luque
Zúhéros
Alcaudete
1046
Pto. de Guadahortu
Carretero
Mancha Real
Montillana
Torre Ca

Guillena
Villanueva del
Río y Minas
Alcolea
del Río
Lora
del Río
Cañada Rosal
La Campana
La Luisiana
138
Écija
E
Santaella
La Rambla
Montilla
Aguilar
Cabra
Monturque
Carcabuey
Priego
de Córdoba
Almedinilla
Fraile
Alcalá la Real
Benalúa de las Villas
80
Colomera
Iznalloz

Villaverde del
Río
Brenes
Cantillana
Carmona
Fuentes
de Andalucía
Marinaleda
Herrera
Puente Genil
Lucena
Rute
Algarinejo
Illora
Montefrio
Pinos Puente
Cogollus Vega
Alfacar
N 323
46
Pto.
1280

SEVILLA
Santiponce
Cama
San Juan
Sevilla
Alcalá del Río
La Rinconada
N IV
34
Mairena del Alcor
El Viso del Alcor
Alcalá de Guadaira
Marchena
La Lantejuela
El Rubio
Estepa
Osuna
Badolatosa
Casariche
Benamejí
Cuevas
de S. Marcos
Emb. de Iznájar
Villanueva de Algaidas
Huétor-Tájar
Moraleda
de Zafayona
Santafé
Loja
Salar
Chimeneas
Gabia
la Grande
Malá
Armilla
Granada
Alhendín

a del Río
Gelves
Palacios y Villafranca
Dos Hermanas
El Arahal
Paradas
La Puebla de Cazalla
Aguadulce
Pedrera
La Roda
Alameda
Mollina
Fuente de Piedra
Las
de Fuente
Sierra de Yeguas
Martín de la Jara
Los Corrales de Piedra
N 342
Archidona
Sa. Gorda
1671
134
Pto. de los Alazores
Ventas de Huelma
Pto. del
Suspiro
del Moro
860
Padul
Dúrcal
Picacho de V

La Puebla
del Río
Utrera
Morón
de la Frontera
El Coronil
El Saucejo
Villanueva de S. Juan
Almargen
Campillos
56
Antequera
1040
Villanueva de Trabuco
Alfarnate
Alfarnatejo
Sierra de Alhama
Arenas del Rey
Albuñuelas
Béznar
Lanjaró
Las

Queipo
Emb. Torre
del Águila
Montellano
Coripe
Pruna
Teba
Peñarrubia
Villanueva
de la Concepción
Casabermeja
M Á L A G A
Periana
Colmenar
Cómpeta
Arenas
de Granada
Emb. de los
Bermejales
Otivar
Vélez
Las

Lebrija
143
Las
Cabezas
Espera
Bornos
Villamartín
Zahara
Olvera
Cañete la Real
Algodonales
Alcalá
del Valle
El Gastor
Emb. del
Conde de
Guadalhorce
Ardales
Cuevas
del Becerro
Carratraca
Almogía
N 331
Benamargosa
Colmenar
Vélez Málaga
Torrox
Frigiliana
68
Motril
E 103
N 340
109
Torre del Mar
Nerja
Almuñecar
Salobreña
Puerto Motril
Ca

Arcos
de la Frontera
Prado del Rey
N 342
Grazalema
Pinar
1654
Burgo
M Á L A G A
Alora
Pizarra
Cártama
Guadalhorce
Málaga
Rincón de la Victoria
G A

Espejo
Puerto Serrano
Algar
Ubrique
Benaoján
Ronda
Ronda
Alozaina
Coin
Alhaurín
de la Torre
Torremolinos
Costa   del   Sol

N 342
Bornos
Emb. de
Hurones
Emb. de
Guadalcacín
Sa. de Ubrique
Cortes de
la Frontera
Sa. de Tolox
1919
Monda
Alhaurín
el Grande
59

CÁDIZ
María
to Real
Chiclana
de la Frontera
N 340
Paterna de Rivera
Medina Sidonia
Benalup
de Sidonia
Alcalá de los Gazules
Jimena de la Frontera
M Á L A G A
Sierra   Bermeja
Reales
1449
Casares
Benalauria
Gaucin
Istán
Marbella
Fuengirola
Pto. de Calaburras
E 26
125
San Pedro Alcántara
Elviria

Emb. del
Guadarranque

Conil
r de la Frontera
bo
nfalgar
136
E 25
49
Barbate
de Franco
Facinas
Manilva
Estepona
N 340

Los Barrios
San Roque
La Línea de la Concepción
Tarifa
Algeciras
B. de Algeciras
Gibraltar
Punta de Europa
Punta Marroqui
Tanger
Ceuta

Costa   del   Sol

Marse
Melilla

6
5
4

0   10   20   30   40   50 Miles
0   10   20   30   40   50   60   70   80 Kilometres

John Bartholomew & Son Ltd

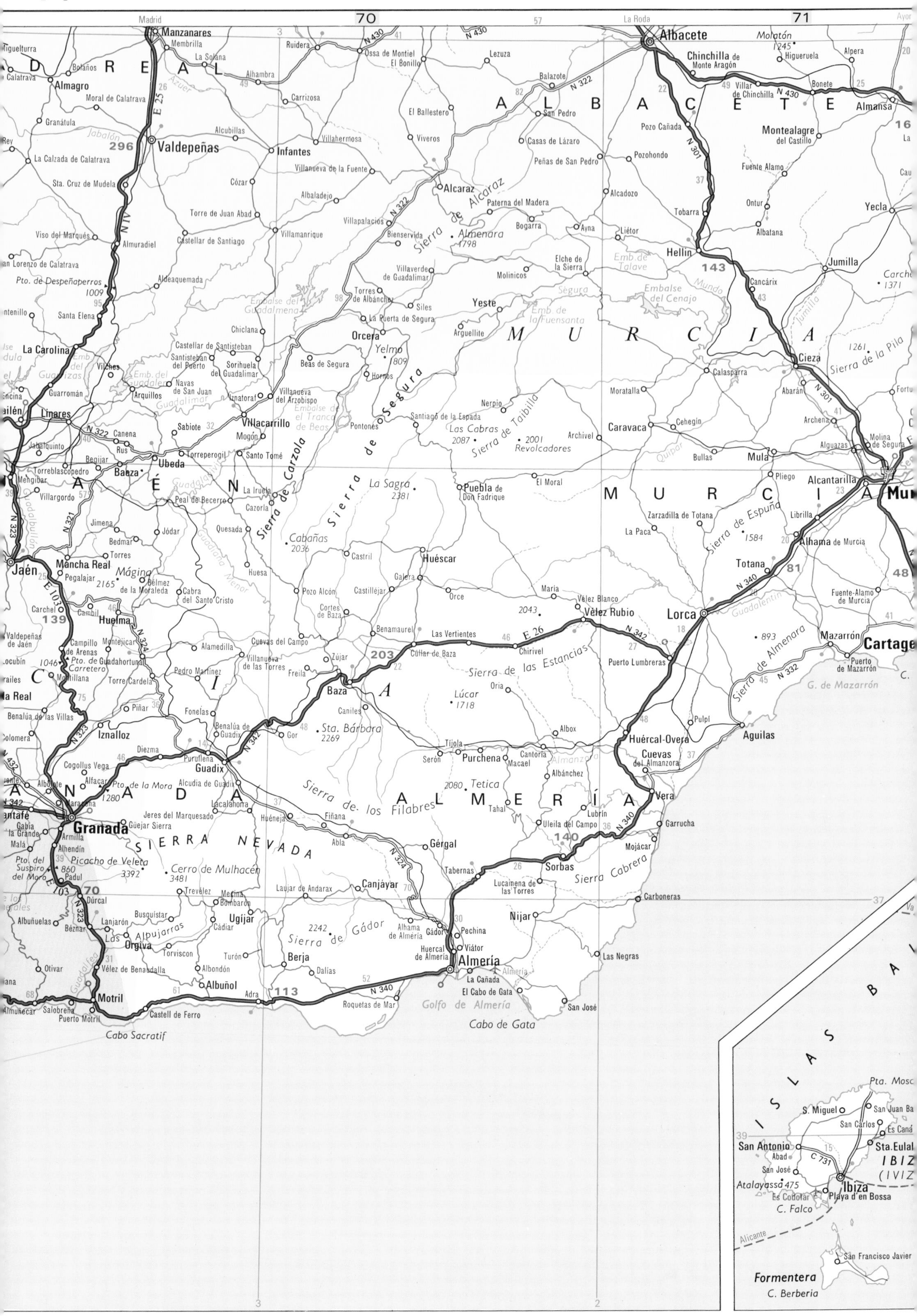

Madrid
70
La Roda
71
Manzanares
Membrillo
La Solana
Alhambra
Ruidera
Ossa de Montiel
El Bonillo
Lezuza
Albacete
Molatón
1245
Chinchilla de Monte Aragón
Higueruela
Alpera
Almagro
Bolaños
Moral de Calatrava
Granátula
Valdepeñas
La Calzada de Calatrava
Sta. Cruz de Mudela
Carrizosa
Alcubillas
Infantes
Villahermosa
Villanueva de la Fuente
El Ballestero
Viveros
Balazote
San Pedro
Casas de Lázaro
Peñas de San Pedro
Pozohondo
Pozo Cañada
Villar de Chinchilla
Bonete
Almansa
Montealegre del Castillo
Fuente Álamo
Ontur
Albatana
Yecla
Jumilla
Cózar
Albaladejo
Alcaraz
Sierra de Alcaraz
Paterna del Madera
Bogarra
Ayna
Liétor
Alcadozo
Tobarra
Hellín
Torre de Juan Abad
Villamanrique
Villapalacios
Bienservida
Almenara
1798
Castellar de Santiago
Viso del Marqués
Almuradiel
Villaverde de Guadalimar
Molinicos
Elche de la Sierra
Emb. de Talave
Embalse del Cenajo
Cancárix
Aldeaquemada
Pto. de Despeñaperros
1009
Santa Elena
Torres de Albanchez
Siles
La Puerta de Segura
Yeste
Arguellite
Emb. de la Fuensanta
Orcera
Chiclana
La Carolina
Castellar de Santisteban
Santisteban del Puerto
Sorihuela del Guadalimar
Vilches
Navas de San Juan
Arquillos
Guarromán
Beas de Segura
Yelmo
1809
Hornos
M U R C I A
Cieza
Calasparra
Sierra de la Pila
1261
Linares
Bailén
Villanueva del Arzobispo
Iznatoraf
Villacarrillo
Sabiote
Canena
Rus
Jabalquinto
Mogón
Santo Tomé
Torreperogil
Úbeda
Baeza
Begíjar
Torreblascopedro
Mengíbar
Villargordo
Pontones
Santiago de la Espada
Nerpio
Las Cabras
2087
Sierra de Taibilla
2001
Revolcadores
Archivel
Caravaca
Cehegín
Moratalla
Abarán
Archena
Molina de Segura
Alguazas
Bullas
Mula
Pliego
Alcantarilla
Murcia
J A É N
Sierra de Cazorla
Sierra de Segura
La Iruela
Peal de Becerro
Cazorla
La Sagra
2381
Puebla de Don Fadrique
El Moral
M U R C I A
Sierra de Espuña
1584
Zarzadilla de Totana
La Paca
Librilla
Alhama de Murcia
Totana
Jimena
Bedmar
Jódar
Torres
Quesada
Cabañas
2036
Mancha Real
Jaén
Pegalajar
Mágina
2165
Bélmez de la Moraleda
Cabra del Santo Cristo
Huesa
Castril
Huéscar
Galera
Castilléjar
Pozo Alcón
Orce
María
Vélez Blanco
Vélez Rubio
2043
Lorca
Fuente-Álamo de Murcia
Cartagena
Carchel
Cambil
Huelma
Cortes de Baza
Benamaurel
Las Vertientes
Cúllar de Baza
Chirivel
Puerto Lumbreras
Mazarrón
Puerto de Mazarrón
G. de Mazarrón
Sierra de Almenara
893
Valdepeñas de Jaén
Campillo de Arenas
Montejícar
Pto. de Carretero
Noalejo
Guadahortuna
Alamedilla
Cuevas del Campo
Villanueva de las Torres
Zújar
Freila
Baza
Pedro Martínez
Torre Cardela
Montillana
Sierra de las Estancias
Oria
Lúcar
1718
Alcalá la Real
Benalúa de las Villas
Colomera
Iznalloz
Piñar
Fonelas
Benalúa de Guadix
Gor
Caniles
Sta. Bárbara
2269
A L M E R Í A
Albox
Pulpí
Águilas
Huércal-Overa
Cuevas del Almanzora
Cogollos Vega
Diezma
Purullena
Guadix
Alfacar
Pto. de la Mora
1280
Alcudia de Guadix
Albolote
Granada
G R A N A D A
Santafé
Gabia la Grande
Malá
Armilla
Alhendín
Jeres del Marquesado
Lacalahorra
Huéneja
Fiñana
Güéjar Sierra
Sierra de los Filabres
Tíjola
Serón
Purchena
Macael
Cantoria
Albánchez
2080
Tetica
Tahal
Vera
Garrucha
Lubrín
Ulelia del Campo
S I E R R A N E V A D A
Abla
Gérgal
Tabernas
Sorbas
Mojácar
Sierra Cabrera
Pto. del Suspiro del Moro
860
Picacho de Veleta
3392
Cerro de Mulhacén
3481
Padul
Dúrcal
Trevélez
Mecina Bombarón
Laujar de Andarax
Canjáyar
Lucainena de las Torres
Carboneras
Níjar
Albuñuelas
Béznar
Lanjarón
Busquístar
Cádiar
Ugíjar
Las Alpujarras
Órgiva
Torviscón
Sierra de Gádor
2242
Alhama de Almería
Gádor
Pechina
Huércal de Almería
Viátor
Almería
Turón
Albondón
Berja
Dalías
Las Negras
Otívar
Vélez de Benaudalla
Albuñol
Adra
Roquetas de Mar
La Cañada
El Cabo de Gata
Golfo de Almería
Cabo de Gata
San José
Motril
Almuñécar
Salobreña
Puerto Motril
Castell de Ferro
Cabo Sacratif
ISLAS BALEARES
Pta. Mosca
S. Miguel
San Juan Bautista
San Carlos
Es Caná
San Antonio Abad
Sta. Eulalia
IBIZA (IVIZA)
San José
Atalayassa 475
Ibiza
Es Codolar
Playa d'en Bossa
C. Falco
Alicante
San Francisco Javier
Formentera
C. Berberia

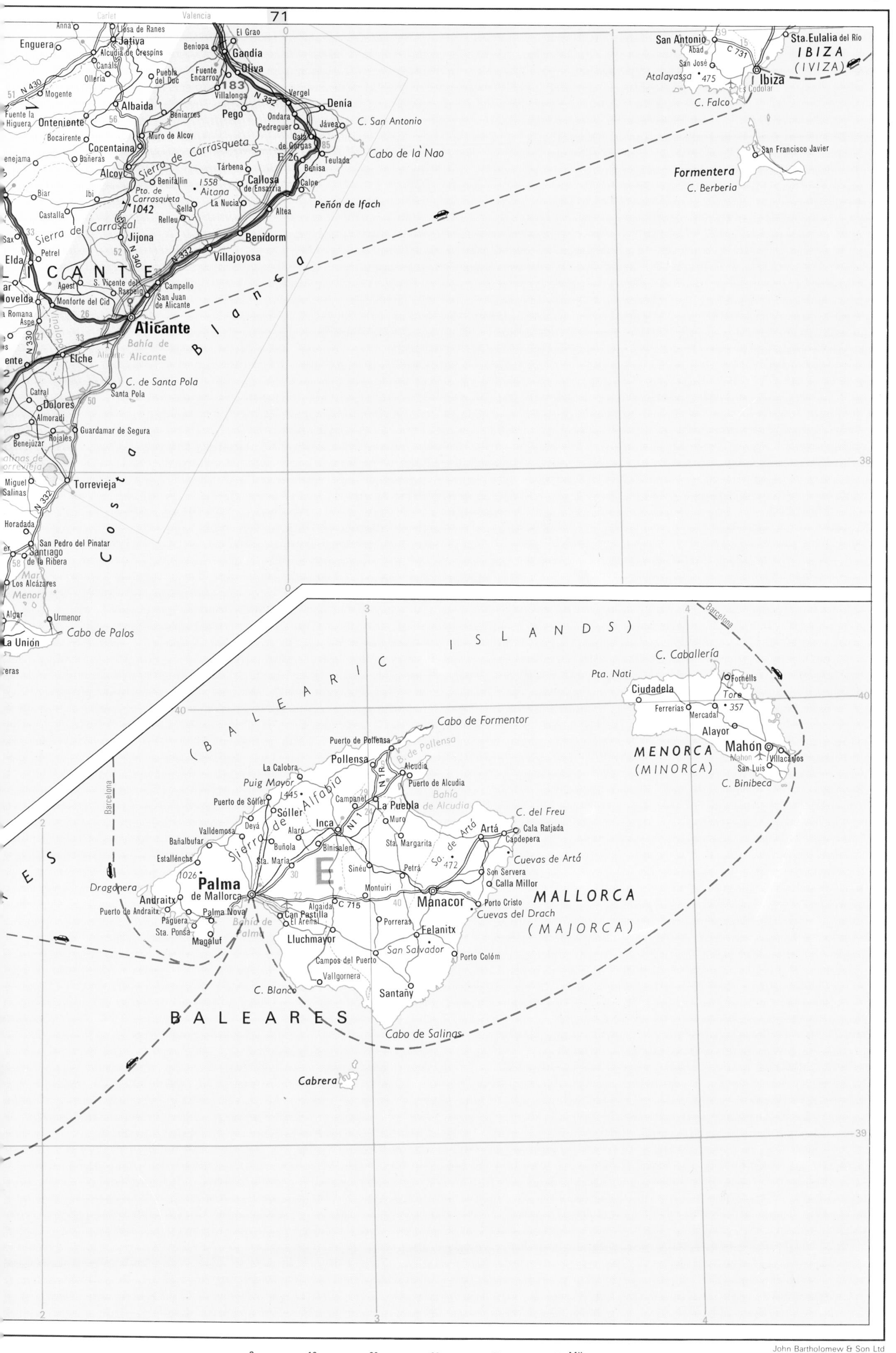
71
Valencia
Carlet
Anna
Llosa de Ranes
Játiva
Enguera
Alcudia de Crespins
Canals
Ollería
Puebla del Duc
Fuente Encarroz
Beniopa
El Grao
Gandia
Oliva
Mogente
N 430
Albaida
Villalonga
N 332
Vergel
Denía
Fuente la Higuera
Onteniente
Beniarrés
Pego
Ondara
Pedreguer
Jávea
C. San Antonio
Bocairente
Muro de Alcoy
Gata de Gorgas
Cocentaina
Sierra de Carrasqueta
Cabo de la Nao
Bañeras
E 26
Teulada
Alcoy
Tárbena
Benisa
Benifallim
1558
Aitana
Callosa de Ensarriá
Calpe
Biar
Ibi
Pto. de Carrasqueta
1042
Sella
La Nucia
Peñón de Ifach
Castalla
Relleu
Altea
Sierra del Carrascal
Jijona
Benidorm
Sax
N 340
Elda
Petrel
Villajoyosa
LICANTE
Costa Blanca
Agost
S. Vicente del Raspeig
Campello
San Juan de Alicante
Novelda
Monforte del Cid
Aspe
Alicante
Bahía de Alicante
N 330
Elche
C. de Santa Pola
Santa Pola
Catral
Dolores
Almoradí
Guardamar de Segura
Benejúzar
Rojales
Salinas de Torrevieja
Torrevieja
Miguel Salinas
N 332
Horadada
San Pedro del Pinatar
Santiago de la Ribera
Mar Menor
Los Alcázares
Algar
Urmenor
Cabo de Palos
La Unión
San Antonio Abad
C 731
Sta. Eulalia del Río
IBIZA
(IVIZA)
San José
Atalayassa
475
Ibiza
Es Codolar
C. Falco
San Francisco Javier
Formentera
C. Berberia
38
(BALEARIC ISLANDS)
Barcelona
C. Caballería
Pta. Nati
Fornélls
Ciudadela
Toro
357
Ferrerías
Mercadal
Alayor
MENORCA
(MINORCA)
Mahón
Villacarlos
San Luis
C. Binibeca
40
Cabo de Formentor
Puerto de Pollensa
B. de Pollensa
Pollensa
La Calobra
Alcudia
Puig Mayor
1445
Puerto de Alcudia
Puerto de Sóller
Sóller
Sierra de Alfabia
Campanet
La Puebla
Bahía de Alcudia
C. del Freu
Deyá
Valldemosa
Alaró
Inca
Muro
Artá
Cala Ratjada
Capdepera
Bañalbufar
Buñola
Sta. Margarita
So. de Artá
472
Cuevas de Artá
Binisalem
Estallénchs
Sta. María
Sineu
Petra
Son Servera
Calla Millor
1026
Palma de Mallorca
Montuiri
Dragonera
Andraitx
Manacor
Porto Cristo
MALLORCA
Puerto de Andraitx
Palma Nova
Algaida
C 715
Cuevas del Drach
Paguera
Can Pastilla
El Arenal
Porreras
(MAJORCA)
Sta. Ponsa
Bahía de Palma
Felanitx
Magaluf
Lluchmayor
San Salvador
Porto Colóm
Campos del Puerto
Vallgornera
C. Blanco
Santañy
BALEARES
Cabo de Salinas
Cabrera
39
E
0 10 20 30 40 50 Miles
0 10 20 30 40 50 60 70 80 Kilometres

John Bartholomew & Son Ltd

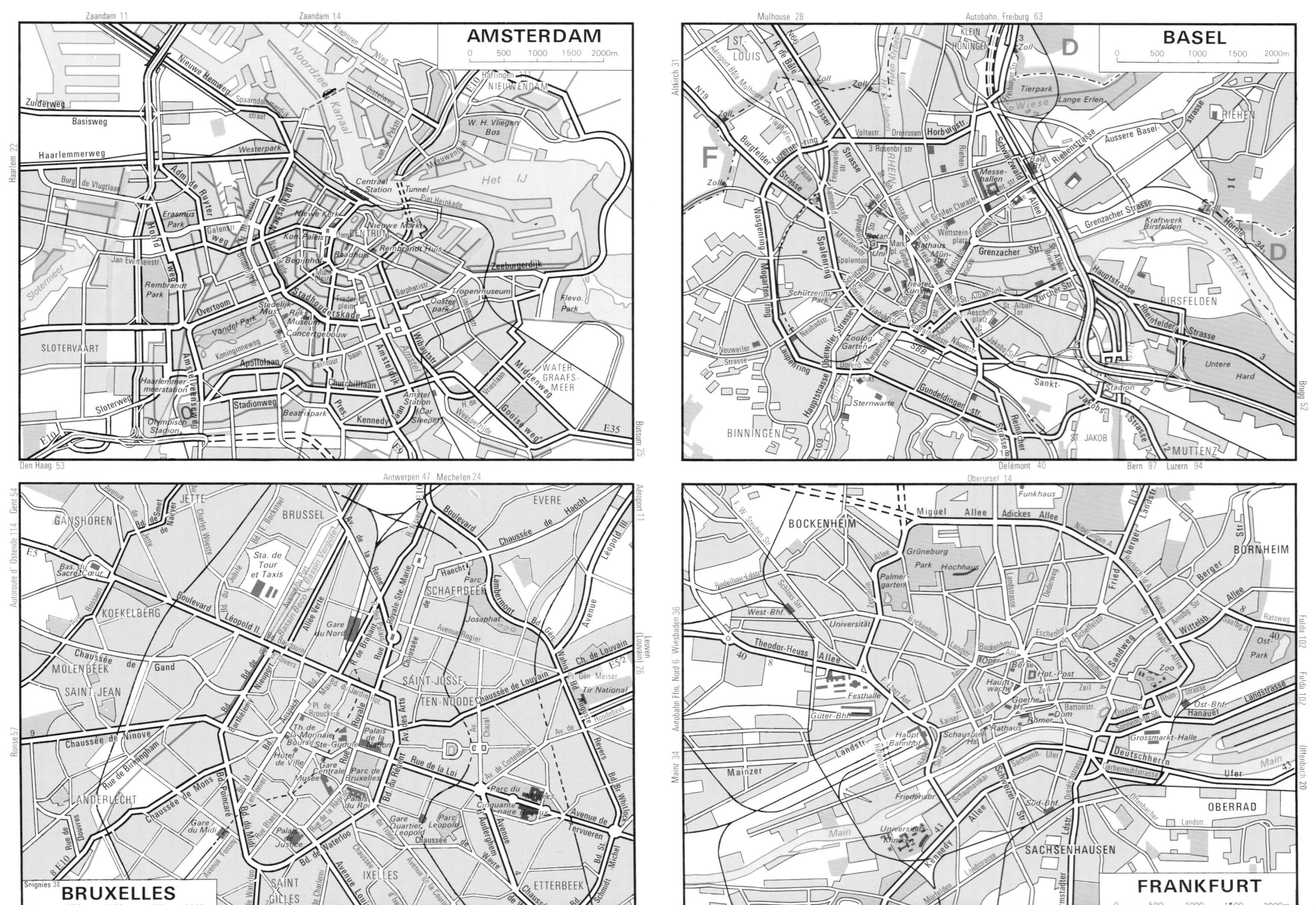
AMSTERDAM
Zaandam 11
Zaandam 14
Haarlem 22
Den Haag 53
Bussum 25
NIEUWENDAM
Het IJ
Centraal Station
Noordzee Kanaal
CENTRUM
Westerpark
Vondel Park
Oosterpark
Flevo. Park
WATERGRAAFSMEER
SLOTERVAART
Amstel
Slotermeer
BASEL
Mulhouse 28
Autobahn, Freiburg 63
Altkirch 31
Brugg 52
Delémont 40
Bern 97
Luzern 94
KLEIN HÜNINGEN
ST. LOUIS
RIEHEN
BIRSFELDEN
BINNINGEN
ST. JAKOB
MUTTENZ
RHEIN
Hauptstrasse
Spalenring
Wasgenring
Mogarten-ring
Burgfelder Strasse
Grenzacher Strasse
Sternwarte
BRUXELLES
Antwerpen 47
Mechelen 24
Aéroport 11
Leuven (Louvain) 26
Gent 54
Autoroute d'Ostende 114
Ronse 57
Soignies 38
GANSHOREN
JETTE
BRUSSEL
EVERE
SCHAERBEEK
KOEKELBERG
MOLENBEEK SAINT-JEAN
SAINT-JOSSE-TEN-NOODE
ANDERLECHT
IXELLES
SAINT GILLES
ETTERBEEK
Gare du Nord
Gare du Midi
Sta. de Tour et Taxis
Chaussée de Ninove
Chaussée de Mons
Chaussée de Louvain
Rue de la Loi
FRANKFURT
Oberursel 14
Mainz 34
Autobahn Flfh. Nord 6
Wiesbaden 36
Fulda 102
Offenbach 20
BOCKENHEIM
BORNHEIM
OBERRAD
SACHSENHAUSEN
Grüneburg Park
Palmengarten
Funkhaus
Festhalle
Haupt Bahnhof
Zoo
Ost-Park
Main
Miguel Allee
Adickes Allee
Theodor-Heuss Allee

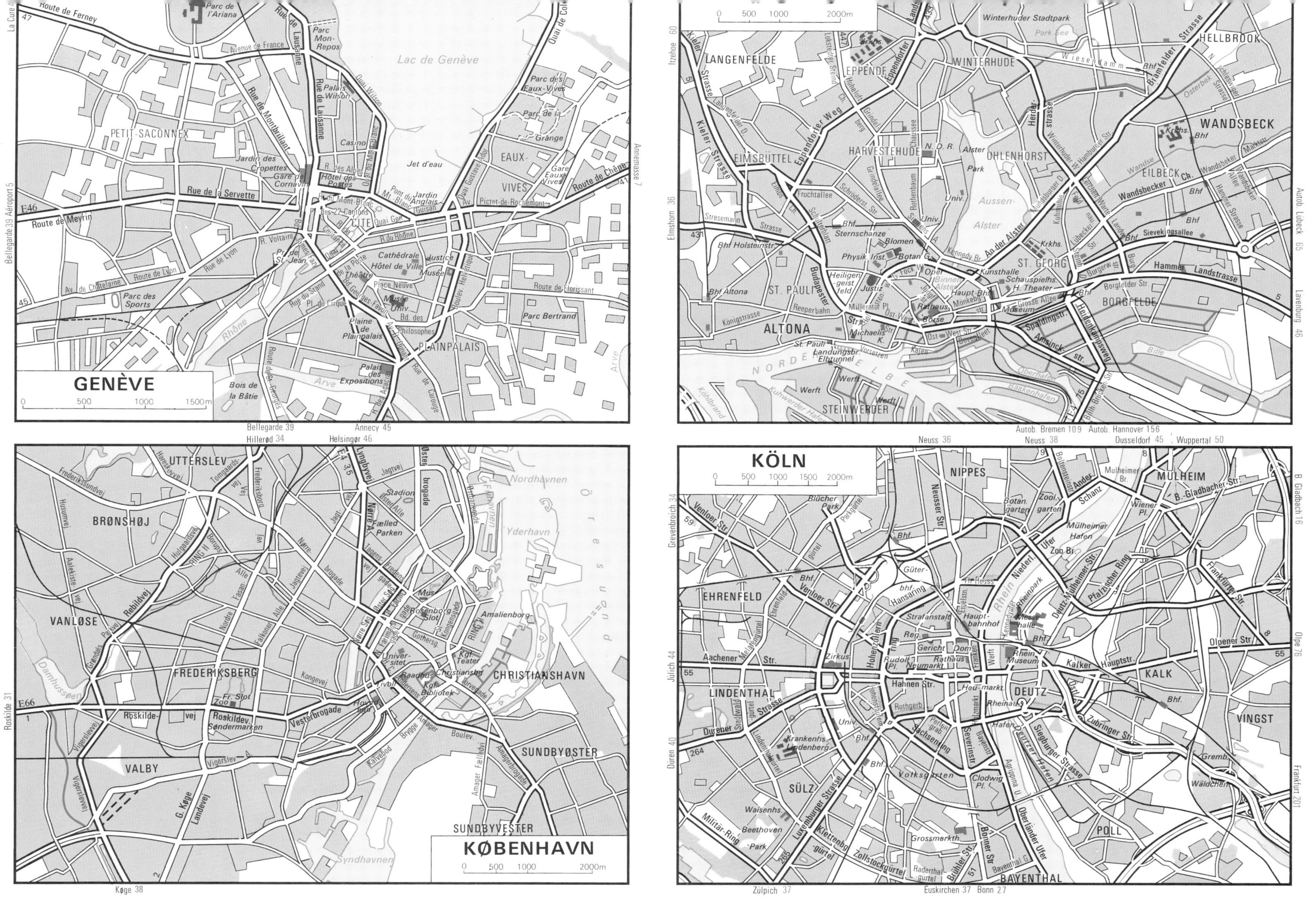
GENÈVE
Lac de Genève
PETIT-SACONNEX
EAUX-VIVES
CITÉ
PLAINPALAIS
Jet d'eau
Annemasse 7
Bellegarde 39 Aéroport 5
Bellegarde 39
Annecy 45
LANGENFELDE
EPPENDORF
WINTERHUDE
HELLBROOK
WANDSBECK
EILBECK
HARVESTEHUDE
UHLENHORST
EIMSBÜTTEL
Aussen-Alster
ST. GEORG
ST. PAULI
ALTONA
BORGFELDE
STEINWERDER
NORDERELBE
Itzehoe 60
Elmshorn 36
Autob. Lübeck 65
Lavenburg 46
UTTERSLEV
BRØNSHØJ
VANLØSE
FREDERIKSBERG
CHRISTIANSHAVN
SUNDBYØSTER
VALBY
SUNDBYVESTER
KØBENHAVN
Øresund
Hillerød 34
Helsingør 46
Roskilde 31
Køge 38
KÖLN
NIPPES
MÜLHEIM
EHRENFELD
LINDENTHAL
DEUTZ
KALK
VINGST
SÜLZ
POLL
BAYENTHAL
Neuss 36
Neuss 38
Autob. Bremen 109
Autob. Hannover 156
Dusseldorf 45
Wuppertal 50
B. Gladbach 16
Olpe 76
Frankfurt 201
Grevenbroich 34
Jülich 44
Düren 40
Zülpich 37
Euskirchen 37 Bonn 27

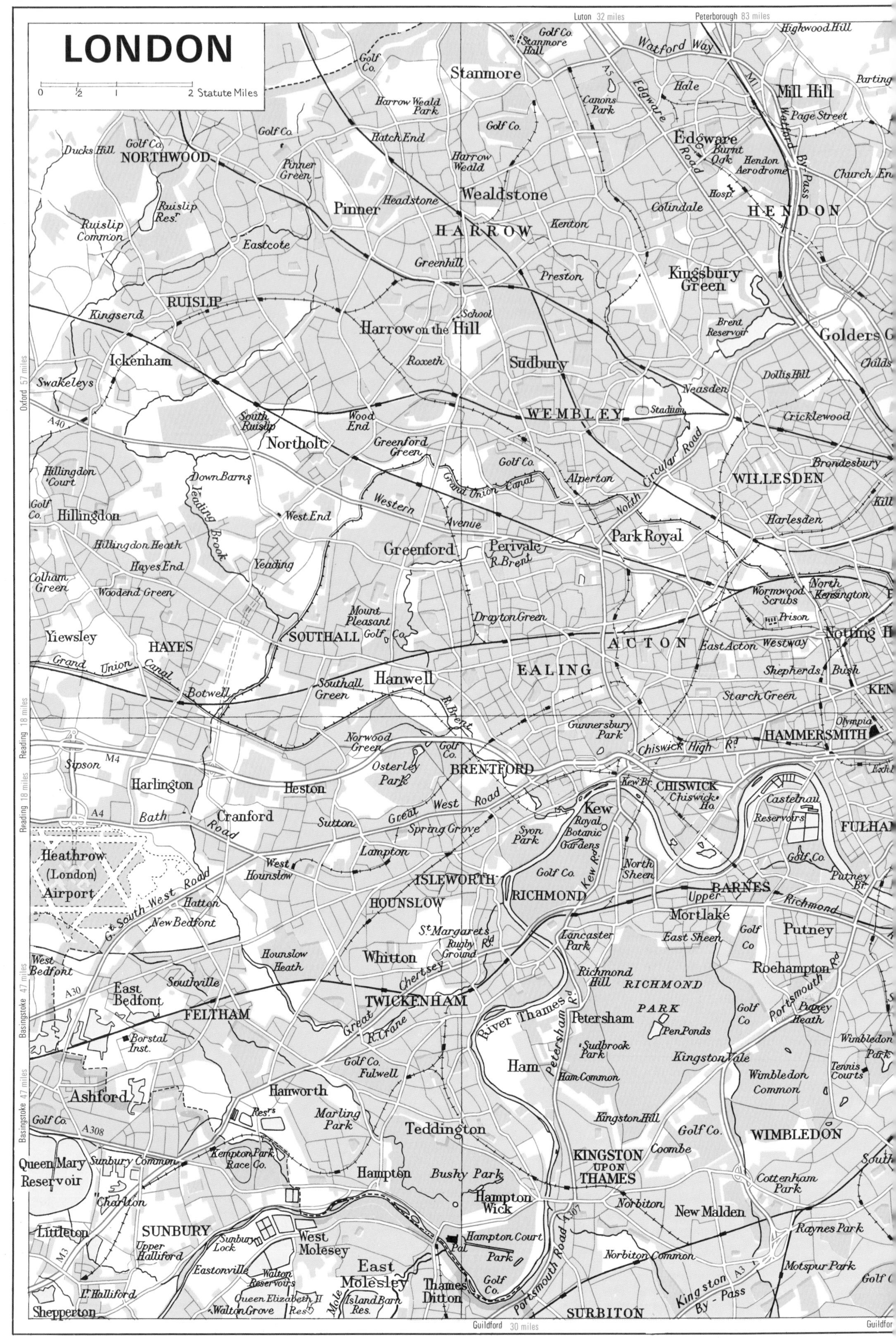

LONDON

0   ½   1   2 Statute Miles

Golf Co.
Stanmore Hill
Stanmore
Watford Way
Highwood Hill
Mill Hill
Parting

Harrow Weald Park
Canons Park
Hale
Edgware
Page Street
Church End

Ducks Hill
NORTHWOOD
Golf Co.
Golf Co.
Hatch End
Harrow Weald
Edgware Road
Burnt Oak
Hendon Aerodrome
HENDON

Pinner Green
Pinner
Headstone
Harrow Weald
Colindale
Hosp.

Ruislip Res.
Eastcote
HARROW
Kenton
Kingsbury Green
Brent Reservoir
Golders G

Ruislip Common
RUISLIP
Greenhill
Preston
Sudbury
Neasden
Dollis Hill
Childs

Kingsend
Harrow on the Hill
School
Stadium
WEMBLEY
Cricklewood

Ickenham
Roxeth
Sudbury
Golf Co.
North Circular Road
WILLESDEN
Brondesbury

Swakeleys
South Ruislip
Wood End
Greenford Green
Alperton
Harlesden
Kill

A40
Northolt
Great Union Canal
Golf Co.
Park Royal

Hillingdon Court
Down Barns
Western
Perivale
R. Brent
Wormwood Scrubs
North Kensington

Golf Co.
Hillingdon
West End
Avenue
Greenford
Prison
Notting H

Hillingdon Heath
Yeading
ACTON
East Acton
Westway
KEN

Colham Green
Hayes End
Woodend Green
Mount Pleasant
Drayton Green
EALING
Shepherds Bush
Starch Green

Yiewsley
HAYES
SOUTHALL
Golf Co.
Gunnersbury Park
Olympia
HAMMERSMITH

Grand Union Canal
Botwell
Southall Green
Hanwell
R. Brent
Chiswick High Rd.
Exh

Norwood Green
Golf Co.
BRENTFORD
Kew Br.
CHISWICK
Chiswick Ho.
Castelnau Reservoirs
FULHAM

Sipson
M4
Osterley Park
Spring Grove
West Road
Kew
Royal Botanic Gardens
Golf Co.
Putney Br.

Harlington
Heston
Sutton
Great
West Road
Syon Park
North Sheen
BARNES
Upper Richmond

A4
Cranford Road
Bath
Lampton
ISLEWORTH
RICHMOND
Mortlake
East Sheen
Putney

Heathrow (London) Airport
West Hounslow
HOUNSLOW
Golf Co.
Kew Road
Roehampton
Golf Co.
Putney Heath

Hatton
St Margarets
Rugby Ground
Lancaster Park
Richmond Hill
RICHMOND
Golf Co.
Portsmouth Rd.

New Bedfont
Whitton
Chertsey Rd
Petersham
PARK
Wimbledon Park

West Bedfont
Hounslow Heath
R. Crane
Petersham
PARK
Pen Ponds
Golf Co.
Tennis Courts

A30
East Bedfont
TWICKENHAM
Great
R. Crane
Petersham Rd
Sudbrook Park
Kingston Vale
Wimbledon Common

Southville
Borstal Inst.
Golf Co. Fulwell
Ham
Ham Common
Kingston Hill
Golf Co.
WIMBLEDON

Ashford
Hanworth
Marling Park
Teddington
Kingston Hill
Coombe
Cottenham Park

Golf Co.
A308
Res's
Hampton
Bushy Park
KINGSTON UPON THAMES
Norbiton
New Malden
Raynes Park

Queen Mary Reservoir
Sunbury Common
Kempton Park Race Co.
Hampton Wick
Norbiton Common
Motspur Park

Charlton
Sunbury Lock
West Molesey
Hampton Court Park
Norbiton
Kingston A.307
Golf Co.

Littleton
M3
SUNBURY
Upper Halliford
Eastonville
East Molesey
Pal
Hampton Court
Golf Co.
SURBITON
Kingston By-Pass

Shepperton
Lr. Halliford
Walton Grove
Queen Elizabeth II Res.
Mole
Island Barn Res.
Thames Ditton
Portsmouth Road A.307

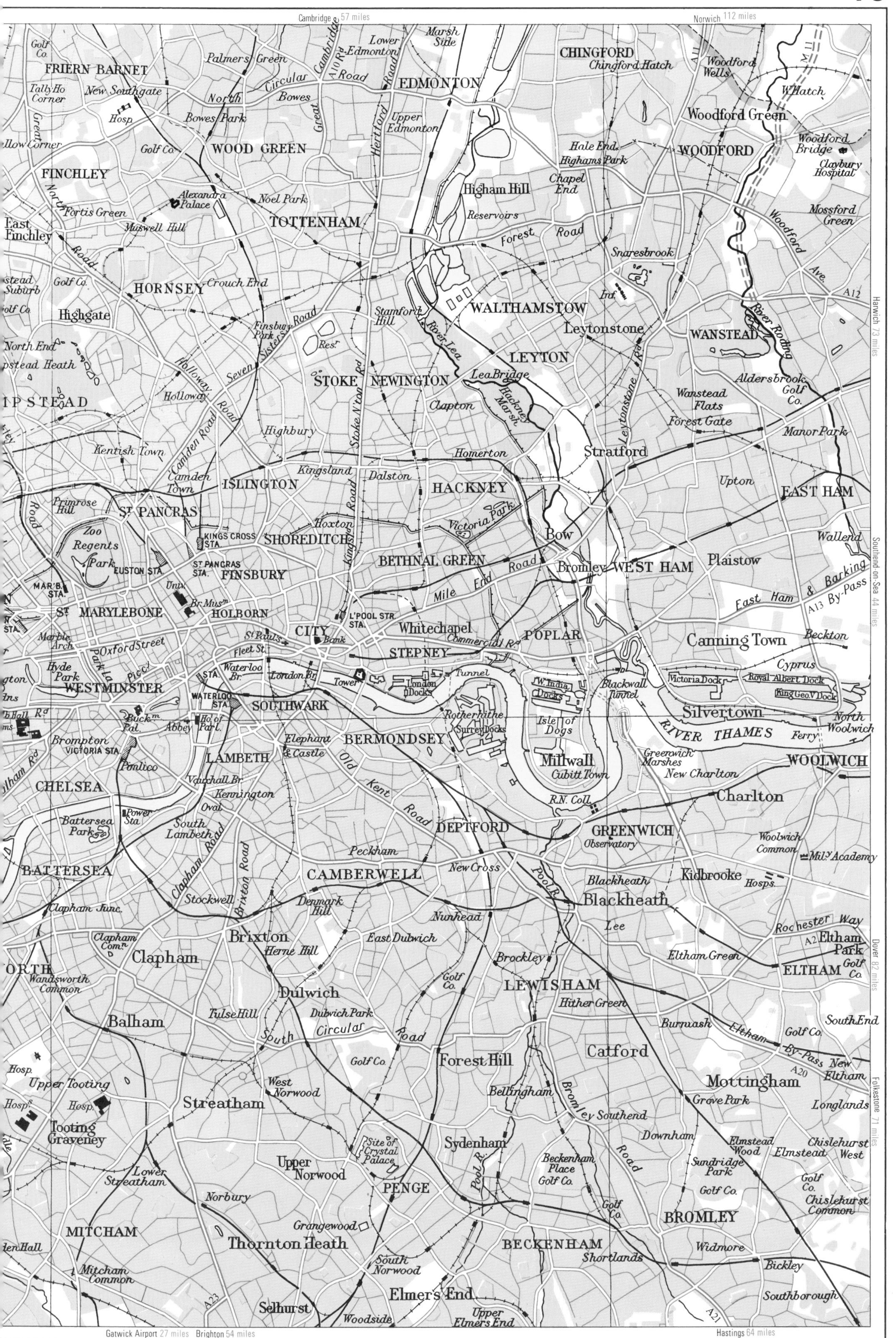

FRIERN BARNET

Palmers Green

Golf Co.

Tally Ho Corner

New Southgate

North Bowes

Circular Bowes

Bowes Park

Lower Edmonton

Marsh Side

CHINGFORD
Chingford Hatch

Woodford Wells

W. Hatch

WOOD GREEN

EDMONTON

Woodford Green

Golf Co.

Alexandra Palace

Noel Park

Upper Edmonton

Higham Hill

Hale End
Highams Park

WOODFORD

Woodford Bridge

Claybury Hospital

FINCHLEY

Fortis Green

Muswell Hill

TOTTENHAM

Chapel End

Mossford Green

East Finchley

Golf Co.

Reservoirs

Snaresbrook

Woodford Ave.

Ali2

stead Suburb

HORNSEY

Crouch End

Finsbury Park Seven Sisters Road

Stamford Hill

WALTHAMSTOW

Forest Road

River Roding

olf Co.

Highgate

Res.r

LEYTON

Leytonstone

WANSTEAD

Aldersbrook Golf Co.

North End
pstead Heath

Holloway Road

Highbury

STOKE NEWINGTON

Clapton

Lea Bridge

Hackney Marsh

Wanstead Flats

Forest Gate

IPSTEAD

Kentish Town

Holloway

Kingsland

Homerton

Stratford

Upton

Manor Park

Camden Road

ISLINGTON

Dalston

HACKNEY

EAST HAM

Camden Town

Hoxton

Victoria Park

Bow

St Pancras

SHOREDITCH

BETHNAL GREEN

Bromley WEST HAM

Plaistow

Wallend

Zoo Regents Park

Kings Cross Sta.

Euston Sta.

St Pancras Sta. FINSBURY

Mile End Road

East Ham & Barking By-Pass A13

Primrose Hill

MAR.B. STA.

Univ.

Br. Mus.m

HOLBORN

L'pool St.r Sta.

POPLAR

Canning Town

Beckton

St MARYLEBONE

Marble Arch

Oxford Street

CITY
Bank

St Paul's
Fleet St.

Whitechapel
Commercial Rd.

STEPNEY

Cyprus

Royal Albert Dock

Hyde Park

Park La.

STA.

Waterloo Br.

London Br.

Tower

Tunnel

W. India Docks

Blackwall Tunnel

Victoria Dock

King Geo. V Dock

WESTMINSTER

Picc.y

WATERLOO STA.

London Docks

Silvertown

North Woolwich

Buck.m Pal.

Ho. of Parl.

SOUTHWARK

Rotherhithe

Surrey Docks

Isle of Dogs

RIVER THAMES

Ferry

all Rd.

ms

Abbey

Elephant & Castle

BERMONDSEY

Millwall
Cubitt Town

Greenwich Marshes

New Charlton

WOOLWICH

Brompton

VICTORIA STA.

LAMBETH

R.N. Coll.

CHELSEA

Pimlico

Vauxhall Br.

Kennington Oval

Old Kent Road

Charlton

Power Sta.

South Lambeth

DEPTFORD

GREENWICH
Observatory

Woolwich Common

Mil.y Academy

Battersea Park

Clapham Road

Peckham

New Cross

Blackheath

Kidbrooke

Hosps.

BATTERSEA

Stockwell

Brixton Road

CAMBERWELL

Pool R.

Blackheath

Rochester Way

A2

Eltham Park

Clapham Junc.

Denmark Hill

Nunhead

Lee

Eltham Green

ELTHAM Golf Co.

ORTH

Clapham Com.

Brixton
Herne Hill

East Dulwich

Brockley

Wandsworth Common

Clapham

Dulwich

Golf Co.

LEWISHAM

Hither Green

Burnash

Eltham

Golf Co. SouthEnd

Balham

Tulse Hill

Dulwich Park

Circular Road

Catford

By-Pass A20 New Eltham

South

Golf Co.

Forest Hill

Bellingham

Mottingham

Hosp.

Upper Tooting

West Norwood

Grove Park

Longlands

Hosp.s

Streatham

Bromley Southend

Elmstead Wood

Chislehurst West

Tooting Graveney

Site of Crystal Palace

Sydenham

Beckenham Place Golf Co.

Downham

Sundridge Park

Elmstead

Lower Streatham

Upper Norwood

Pool R.

Golf Co.

Norbury

PENGE

Golf Co.

Chislehurst Common

MITCHAM

Grangewood

Thornton Heath

BECKENHAM
Shortlands

Widmore

BROMLEY

Bickley

en Hall

Mitcham Common

South Norwood

Woodside

Upper Elmers End

Elmers End

Southborough

A23

Selhurst

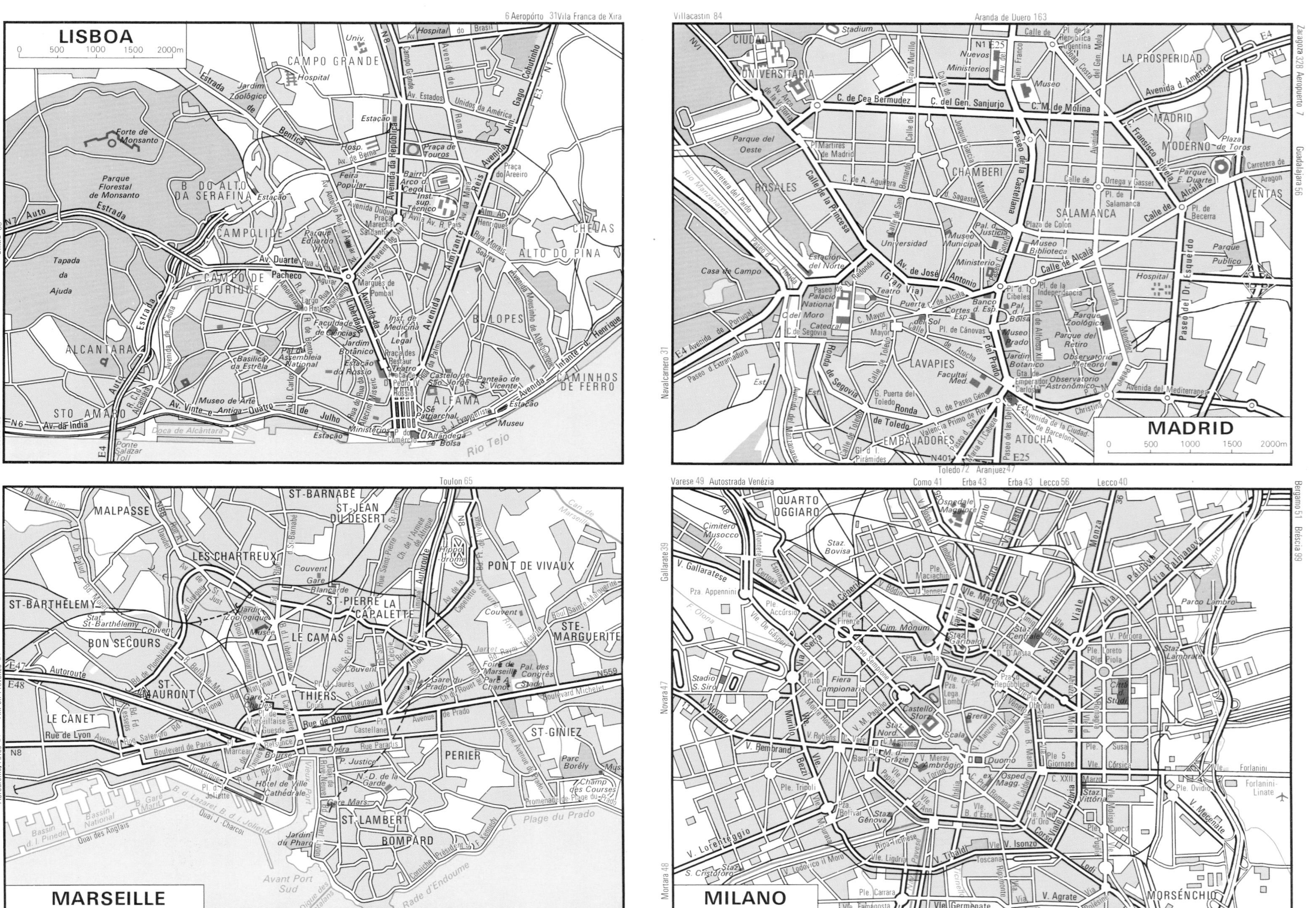
LISBOA
0 500 1000 1500 2000m
6 Aeropórto 31 Vila Franca de Xira
Sintra 30
Cascais 29
MADRID
0 500 1000 1500 2000m
Villacastín 84
Aranda de Duero 163
Zaragoza 328 Aeropuerto 7
Guadalajara 56
Navalcarnero 31
Toledo 72 Aranjuez 47
MARSEILLE
Toulon 65
Aix-en-Provence 31
Aérodrome/Arles 81
MILANO
Varese 49 Autostrada Venézia
Como 41 Erba 43 Erba 43 Lecco 56 Lecco 40
Bergamo 51 Bréscia 99
Gallarate 39
Novara 47
Mortara 48

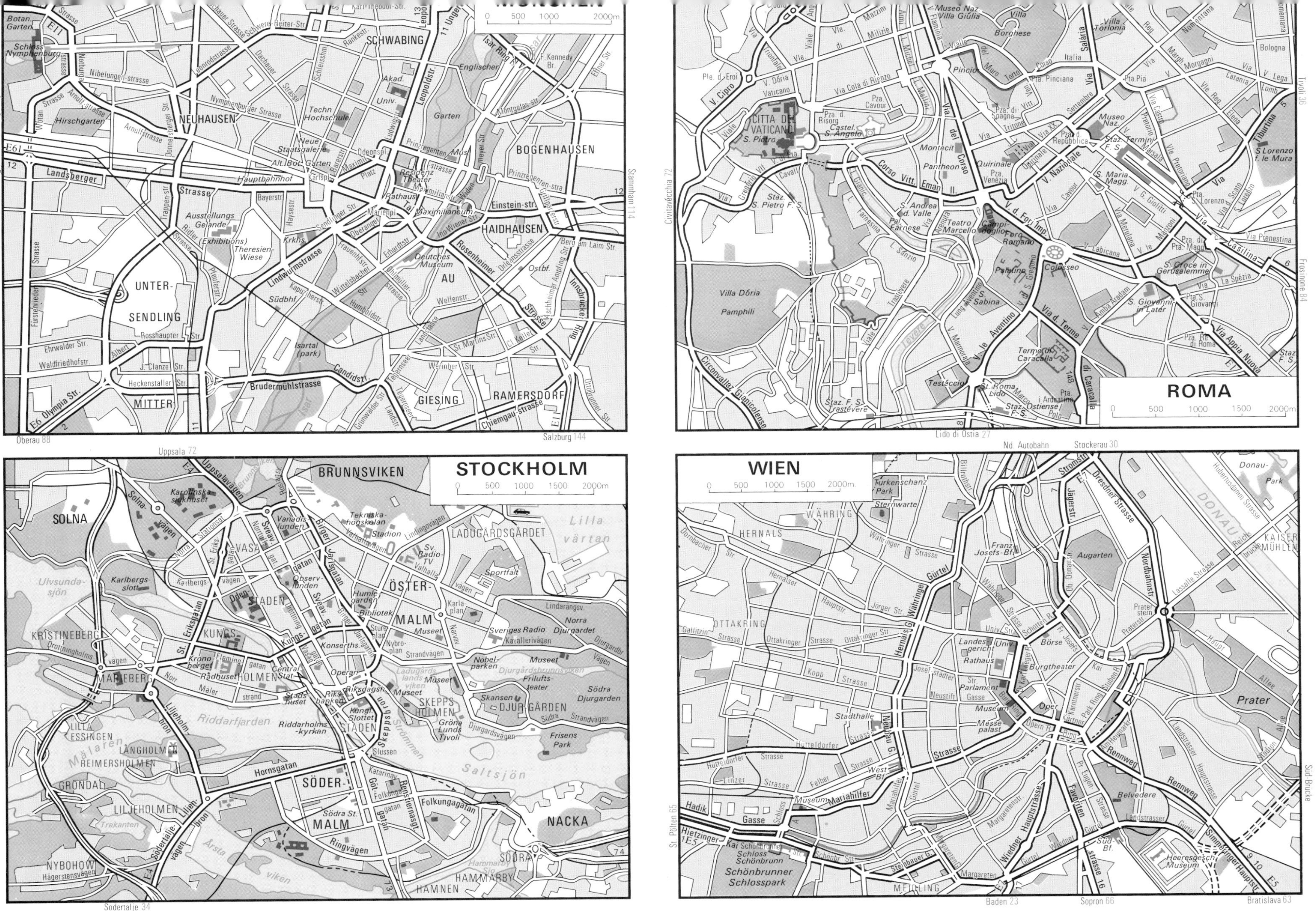
SCHWABING
NEUHAUSEN
BOGENHAUSEN
HAIDHAUSEN
AU
UNTER-
SENDLING
GIESING
RAMERSDORF
MITTER
0 500 1000 2000m
ROMA
0 500 1000 1500 2000m
CITTÀ DEL VATICANO
Villa Dória Pamphili
Colosseo
STOCKHOLM
0 500 1000 1500 2000m
BRUNNSVIKEN
SOLNA
LADUGÅRDSGÄRDET
ÖSTER-
MALM
KRISTINEBERG
SÖDER-
MALM
NACKA
Saltsjön
Riddarfjärden
WIEN
0 500 1000 1500 2000m
WÄHRING
HERNALS
OTTAKRING
Augarten
Prater
Belvedere
Schönbrunner Schlosspark
MEIDLING

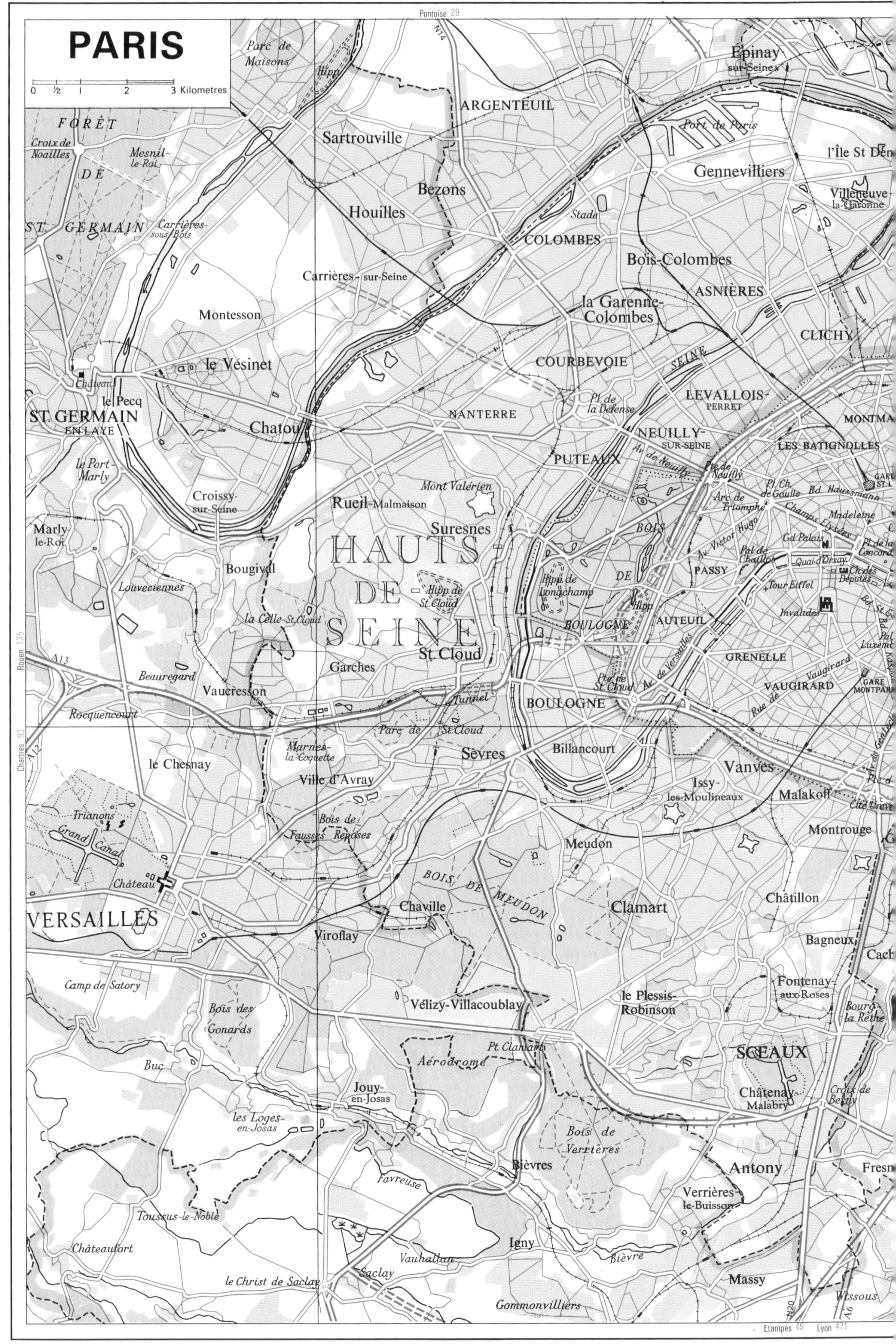

PARIS

0 ½ 1 2 3 Kilometres

FORÊT
DE
ST. GERMAIN

Pontoise 29
N14

Croix de
Noailles
Mesnil-
le-Roi
Parc de
Maisons
Hipp.

Sartrouville

ARGENTEUIL

Épinay
sur-Seine

l'Île St Den

Carrières-
sous-Bois

Bezons

Houilles

Stade

Port de Paris

Gennevilliers

Villeneuve-
la-Garonne

Montesson

Carrières - sur-Seine

COLOMBES

Bois-Colombes

ASNIÈRES

la Garenne-
Colombes

le Vésinet

COURBEVOIE

SEINE

CLICHY

le Pecq

Château

Chatou

NANTERRE

Pl. de
la Défense

LEVALLOIS-
PERRET

MONTMA

ST. GERMAIN
EN-LAYE

le Port
Marly

Croissy-
sur-Seine

Rueil-Malmaison

Mont Valérien

Suresnes

NEUILLY-
SUR-SEINE

Pte de
Neuilly

LES BATIGNOLLES

Av. de Neuilly

PUTEAUX

GARE
ST.

Pl. Ch.
de Gaulle

Bd. Haussmann

Marly-
le-Roi

HAUTS

Arc de
Triomphe

Champs Élysées

Madeleine

Av. Victor Hugo

Gd. Palais

Pl. de
la Concorde

Bougival

DE

BOIS

Pal. de
Chaillot

Quai-d'Orsay

Louveciennes

Hipp. de
St. Cloud

la Celle-St.Cloud

SEINE

Hipp. de
Longchamp

DE

PASSY

Tour Eiffel

Ch. des
Députés

Beauregard

Garches

St.Cloud

Hipp.

AUTEUIL

BOULOGNE

Invalides

GRENELLE

Bd. St. G

Pal. du
Luxemb

Vaucresson

Tunnel

Pte de
St. Cloud

Av. de Versailles

VAUGIRARD

GARE
MONTPARN

Rocquencourt

A13

Parc de

St. Cloud

BOULOGNE

Rue de

Bd.

Rouen 135

Marnes-
la-Coquette

Sèvres

Billancourt

Vanves

Pte de

Chartres 93   A12

le Chesnay

Ville d'Avray

Issy-
les-Moulineaux

Malakoff

Cité Unive

Trianons

Bois de
Fausses Reposes

Montrouge

Grand Canal

Meudon

Châtillon

Château

BOIS DE MEUDON

Clamart

Bagneux

Cach

VERSAILLES

Chaville

Viroflay

Fontenay-
aux-Roses

Camp de Satory

Vélizy-Villacoublay

le Plessis-
Robinson

Bourg
la Rethe

Bois des
Gonards

Pt. Clamart

SCEAUX

Buc

Aérodrome

Châtenay-
Malabry

Croix de
Berny

Jouy-
en-Josas

les Loges-
en-Josas

Bois de
Verrières

Bièvres

Antony

Fresn

Favreuse

Verrières
le-Buisson

Toussus-le-Noble

Igny

Bièvre

Massy

Châteaufort

Vauhallan

Saclay

le Christ de Saclay

Gommonvilliers

Wissous

N20   A6

Etampes 49   Lyon 471

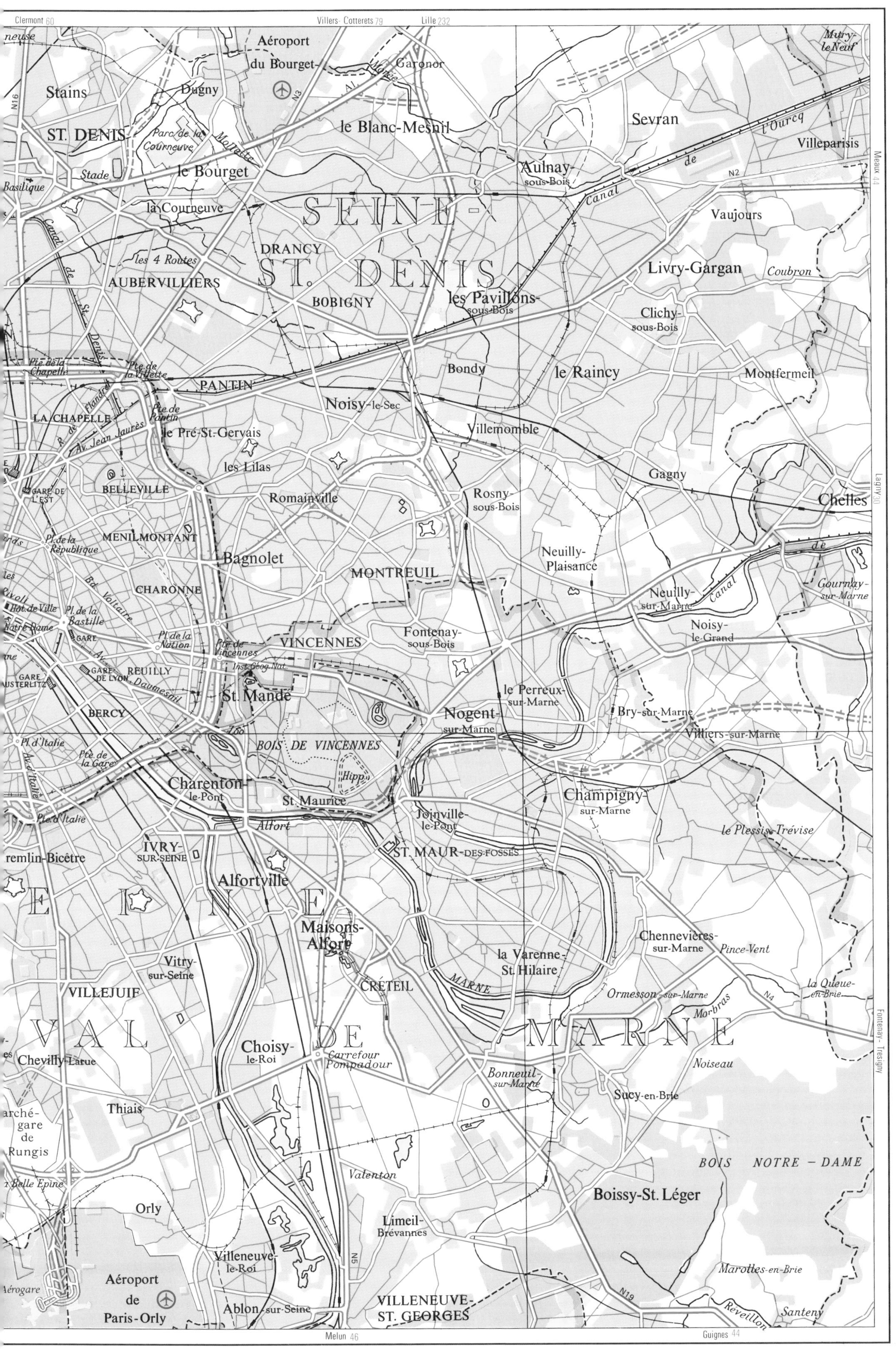

neuse   *Mury-le-Neuf*

Stains   Aéroport du Bourget-   Garonor

Dugny   N3

ST. DENIS   Parc de la Courneuve   le Blanc-Mesnil   Sevran   Villeparisis

le Bourget   Aulnay-sous-Bois

Stade   Basilique   la Courneuve   Vaujours

SEINE-

les 4 Routes   DRANCY   Livry-Gargan   Coubron

AUBERVILLIERS   **ST. DENIS**

BOBIGNY   les Pavillons-sous-Bois   Clichy-sous-Bois

Pte de la Chapelle   Pte de la Villette   Bondy   le Raincy   Montfermeil

PANTIN

LA CHAPELLE   Pte de Pantin   Noisy-le-Sec

le Pré-St-Gervais   Villemomble   Gagny   Chelles

GARE DE L'EST   BELLEVILLE   les Lilas   Romainville   Rosny-sous-Bois

Pl. de la République   MENILMONTANT   Neuilly-Plaisance   Gournay-sur-Marne

Bagnolet   Neuilly-sur-Marne

Bd. Voltaire   CHARONNE   MONTREUIL   Noisy-le-Grand

Hot. de Ville   Pl. de la Bastille   Notre-Dame

Pl. de la Nation   Pte de Vincennes   VINCENNES   Fontenay-sous-Bois

GARE   REUILLY   Inst. Géog. Nat.   le Perreux-sur-Marne

GARE DE LYON   Daumesnil   St. Mandé   Bry-sur-Marne

AUSTERLITZ   BERCY   Nogent-sur-Marne   Villiers-sur-Marne

Pl. d'Italie   Pte de la Gare   **BOIS DE VINCENNES**

Av d'Italie   Hipp.   Champigny-sur-Marne

Pte d'Italie   Charenton-le-Pont   St. Maurice   le Plessis-Trévise

Alfort   Joinville-le-Pont

remlin-Bicêtre   IVRY-SUR-SEINE   ST. MAUR-DES-FOSSES

Alfortville   Chennevières-sur-Marne   Pince-Vent

S E I N E   Maisons-Alfort

Vitry-sur-Seine   la Varenne-St. Hilaire

VILLEJUIF   CRÉTEIL   Ormesson-sur-Marne   Na   la Queue-en-Brie

V A L   **D E**   **M A R N E**   Mor-bras

Morbras

Chevilly-Larue   Choisy-le-Roi   MARNE   Noiseau

Carrefour Pompadour   Bonneuil-sur-Marne   Sucy-en-Brie

Thiais

arché-gare de Rungis   **BOIS NOTRE-DAME**

Belle Epine   Valenton   Boissy-St. Léger

Orly   Limeil-Brévannes

Villeneuve-le-Roi   Marolles-en-Brie

Aéroport de Paris-Orly   N5   N19   Reveillon   Santeny

Aérogare   Ablon-sur-Seine   **VILLENEUVE ST. GEORGES**

Meaux 44   Lagny 30   Fontenay- Tresigny

1:2 500 000

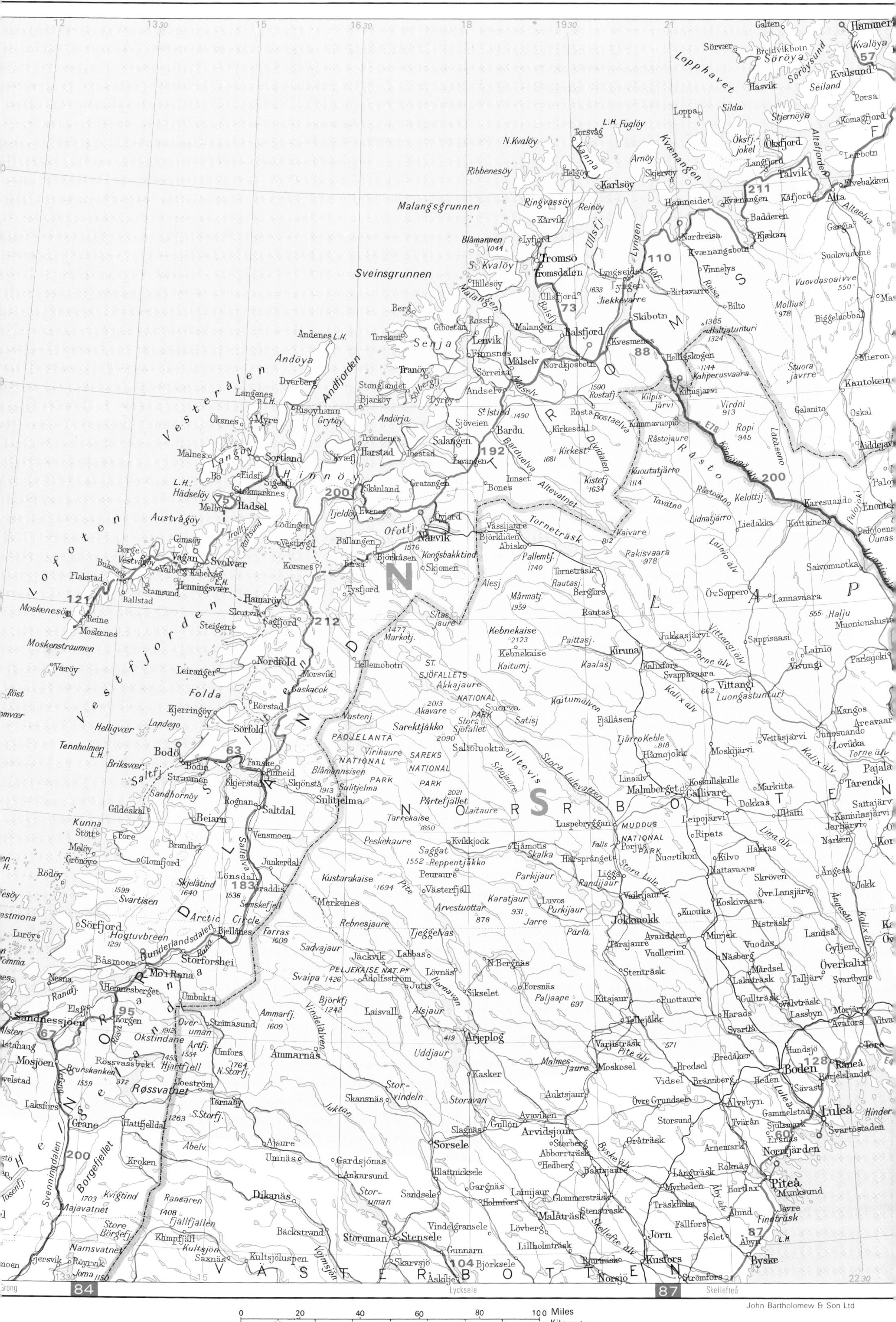
Lopphavet
Söröya
Hammerfest
Kvalöya
Kvalsund
Seiland
Loppa
Alta
Altafjorden
Kvænangen
Karlsöy
Ringvassöy
Tromsö
Tromsdalen
Malangsgrunnen
Sveinsgrunnen
Senja
Lyngen
Skibotn
Balsfjord
Nordreisa
Kautokeino
Andöya
Andenes
Vesterålen
Sortland
Hinnöy
Harstad
Lofoten
Austvågöy
Svolvær
Vestfjorden
Moskenesöy
Moskenstraumen
Værøy
Röst
Narvik
Ofotfj.
Bardu
Torneträsk
Abisko
Kebnekaise
Kiruna
Vittangi
Karesuando
Gällivare
Malmberget
Jokkmokk
Saltfj.
Bodö
Fauske
Sulitjelma
Saltdal
Arctic Circle
Mo i Rana
Sandnessjöen
Mosjöen
Rössvatnet
Arjeplog
Arvidsjaur
Sorsele
Storuman
Stensele
Boden
Luleå
Piteå
Älvsbyn
Överkalix
Kalix älv
Lule älv
Pite älv
Skellefte älv
Byske
Kusfors
Norsjö
Malå
Dikanäs
Grane
Majavatnet
Svartisen
Padjelanta National Park
Sareks National Park
St. Sjöfallets National Park
Muddus National Park
Peljekaise Nat. Pk.
Nordland
Troms
Finnmark
Norrbotten
Västerbotten
Lappland
E78
84
87
Lycksele
Skellefteå
Grong
0 20 40 60 80 100 Miles
0 20 40 60 80 100 120 140 160 Kilometres
John Bartholomew & Son Ltd

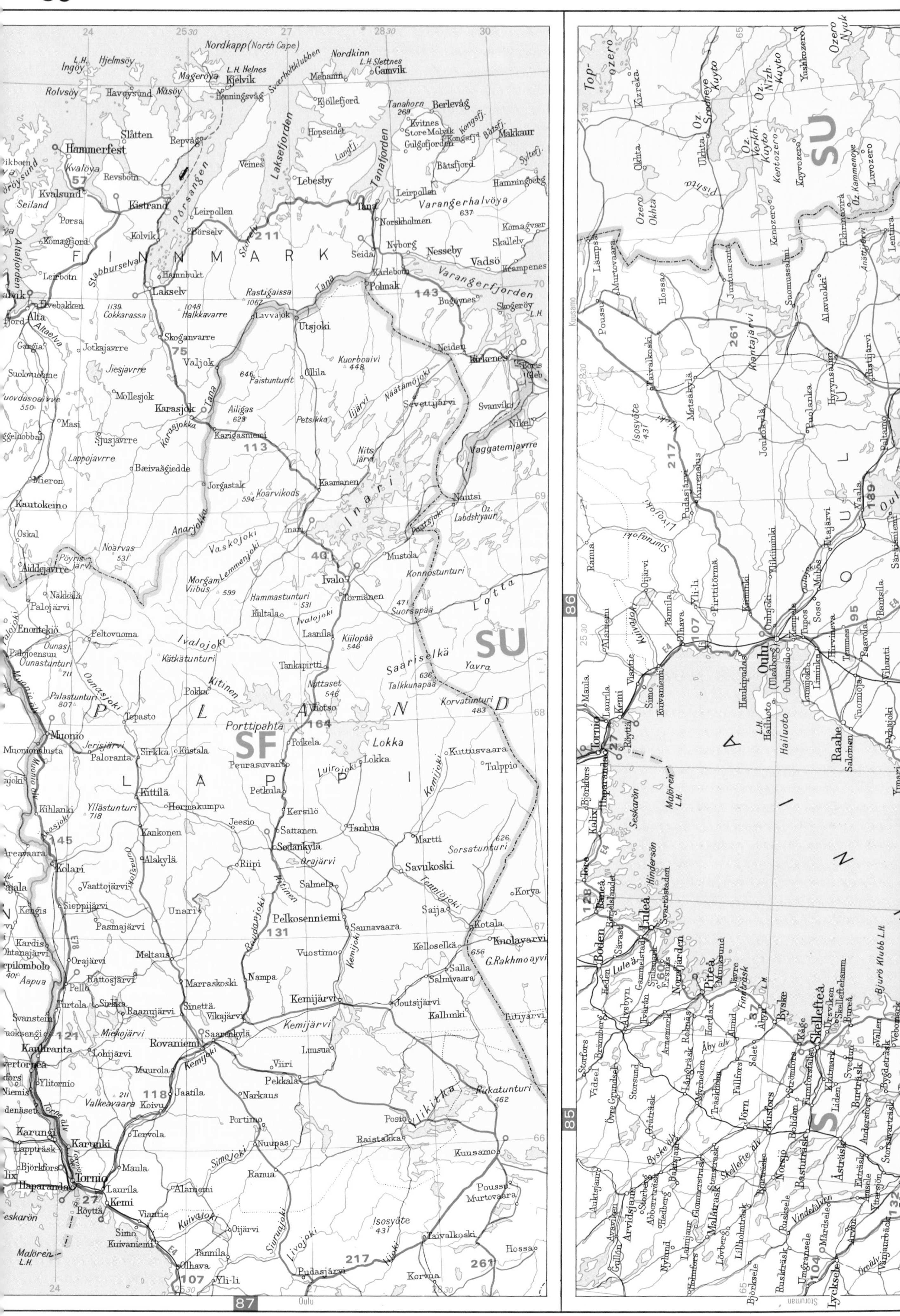
Nordkapp (North Cape)
Nordkinn
Ingöy
Hjelmsöy
Rolvsöy
Havöysund
Måsöy
Magerøya
Kjelvik
Honningsvåg
Mehamn
Gamvik
Kjøllefjord
Tanahorn
Berlevåg
Hammerfest
Kvalöya
Slåtten
Repvåg
Laksefjorden
Tanafjorden
Lebesby
Kistrand
Kvalsund
Seiland
Porsa
Komagfjord
Leirpollen
Börselv
Tana
Varangerhalvöya
Vadsö
Nesseby
Nyborg
Seida
Polmak
Kolvik
Porsangen
FINNMARK
Lakselv
Hammbukt
Leirbotn
Alta
Elvebakken
Varangerfjorden
Bugöynes
Skogeröy
Kirkenes
Neiden
Rastigaissa
Lavvajok
Utsjoki
Cokkarassa
Halkkavarre
Skoganvarre
Valjok
Jotkajavrre
Jiesjavrre
Mollesjok
Karasjok
Masi
Sjusjavrre
Karigasniemi
Lappojavrre
Bævašgiedde
Mieron
Kautokeino
Jorgastak
Koarvikods
Kaamanen
Nantsi
Inari
Oskal
Noarvas
Aiddejavrre
Vaskojoki
Lemmenjoki
Ivalo
Mustola
Konnostunturi
Kuorboaivi
Ollila
Paistunturit
Ailigas
Petsikko
Ijärvi
Näätämöjoki
Sevettijärvi
Svanvik
Nikel
Vaggatemjavrre
Hammastunturi
Törmänen
Suorsapää
Kultala
Ivalojoki
Laanila
Kiilopää
Lotta
SU
Saariselkä
Talkkunapää
Yavra
Korvatunturi
Näkkälä
Palojärvi
Enontekiö
Peltovuoma
Ounasjoki
Ounastunturi
Kätkätunturi
Tankapirtti
Nattaset
Kitinen
Pokka
Palastunturi
Tepasto
Porttipahta
Vuotso
Poikela
Lokka
Muonio
Jerisjärvi
Muonionalusta
Paloranta
Sirkka
Kiistala
Peurasuvanto
Luirojoki
Kuttusvaara
Tulppio
SF
LAPPI
Kittilä
Hormakumpu
Petkula
Kersilö
Kemijoki
Kihlanki
Yllästunturi
Jeesio
Sattanen
Sodankylä
Tanhua
Martti
Sorsatunturi
Kankonen
Alakylä
Riipi
Orajärvi
Savukoski
Kolari
Vaattojärvi
Salmela
Tenniöjoki
Korya
Sieppijärvi
Unari
Saija
Pasmajärvi
Pelkosenniemi
Saunavaara
Kotala
Kuolayarvi
Kelloselkä
Vuostimo
G. Rakhmo ayvi
Meltaus
Marraskoski
Nampa
Salla
Salmivaara
Rattosjärvi
Pello
Turtola
Sirkka
Raanujärvi
Sinettä
Vikajärvi
Kemijärvi
Joutsijärvi
Kallunki
Tuutiyarvi
Miekojärvi
Rovaniemi
Saarenkylä
Luusua
Kaulinranta
Lohijärvi
Muurola
Viiri
Ylitornio
Pekkala
Rukatunturi
Valkeavaara
Koivu
Jaatila
Narkaus
Portimo
Posio
Kitka
Karungi
Karunki
Tervola
Raistakka
Nuupas
Kuusamo
Simojoki
Haparanda
Tornio
Maula
Ranua
Laurila
Alaniemi
Poussu
Murtovaara
Kemi
Röyttä
Viantie
Kuivajoki
Oijärvi
Isosyöte
Taivalkoski
Simo
Kuivaniemi
Siuruanjoki
Livojoki
Iijoki
Tannila
Hossa
Olhava
Pudasjärvi
Maloren L.H.
Yli-Ii
Korvua
Oulu
Kuhmo
Ozero Okhta
Kentozero
Alavuokki
Suomussalmi
Juntusranta
Kiantajärvi
Kurenalus
Ranua
Ii
Haukipudas
Oulu (Uleåborg)
Hailuoto
Kempele
Tyrnävä
Liminka
Raahe
Utajärvi
Muhos
Vaala
Paltamo
Puolanka
Hyrynsalmi
Ristijärvi
Kalix
Töre
Råneå
Luleå
Boden
Piteå
Skellefteå
Byske
Kåge
Burträsk
Jörn
Arvidsjaur
Lycksele
Vindelälven
Skellefte älv
Norsjö
Malå
Bastuträsk
Åsträsk
Storuman
Seskarön
Hindersön
Svartöstaden
Norrfjärden
Åby älv
Arvidsjaur
1:2 500 000

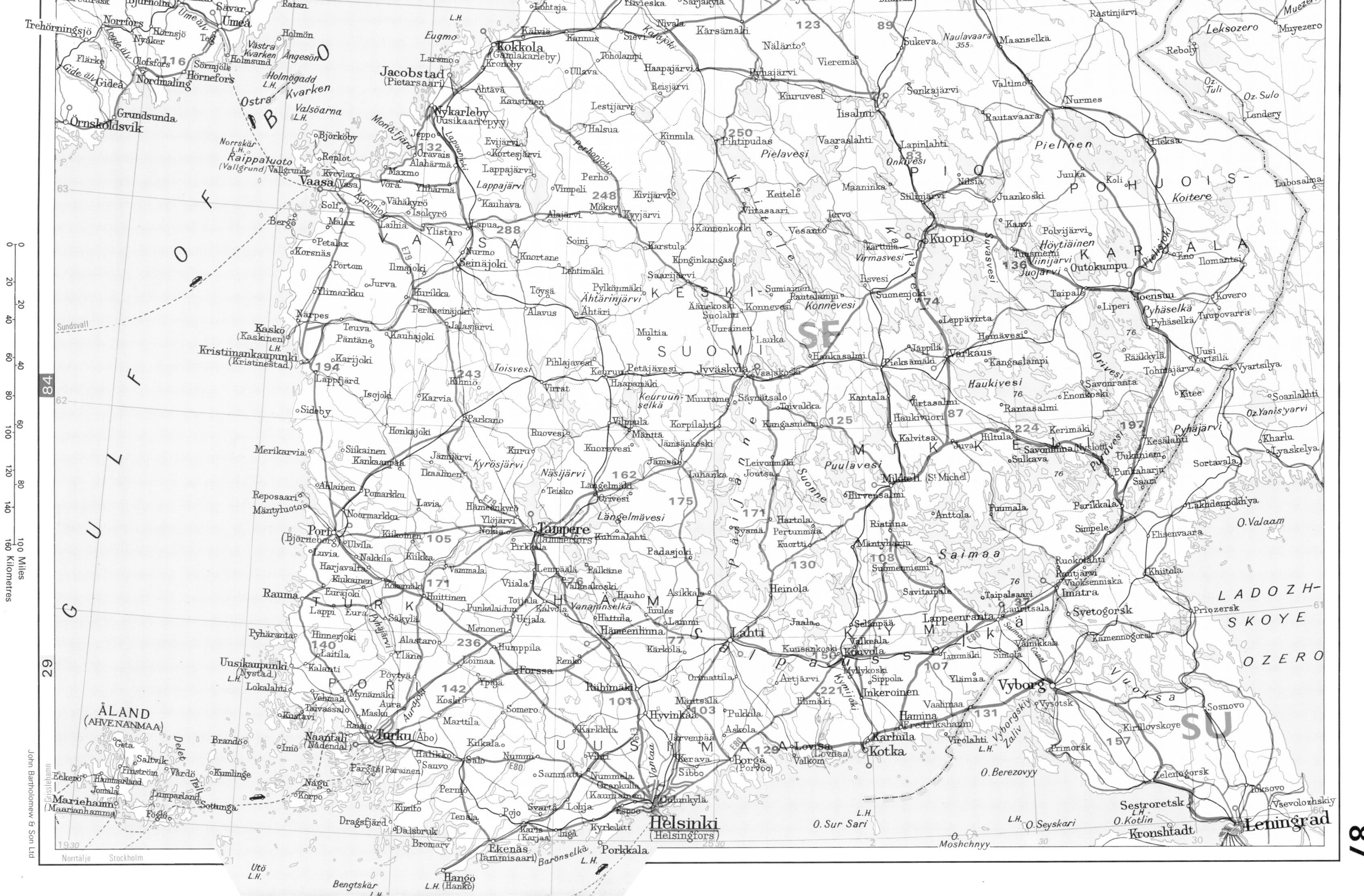

Helsinki (Helsingfors)
Leningrad
Tampere (Tammerfors)
Turku (Åbo)
Vyborg
Kuopio
Jyväskylä
Vaasa (Vasa)
Lahti
Kotka
Pori (Björneborg)
Jacobstad (Pietarsaari)
Kokkola (Gamlakarleby)
ÅLAND (AHVENANMAA)
Mariehamn (Maarianhamina)
GULF OF BOTHNIA
LADOZHSKOYE OZERO
SUOMI
SF
SU
0 20 40 60 80 100 Miles
0 20 40 60 80 100 120 140 160 Kilometres

John Bartholomew & Son Ltd

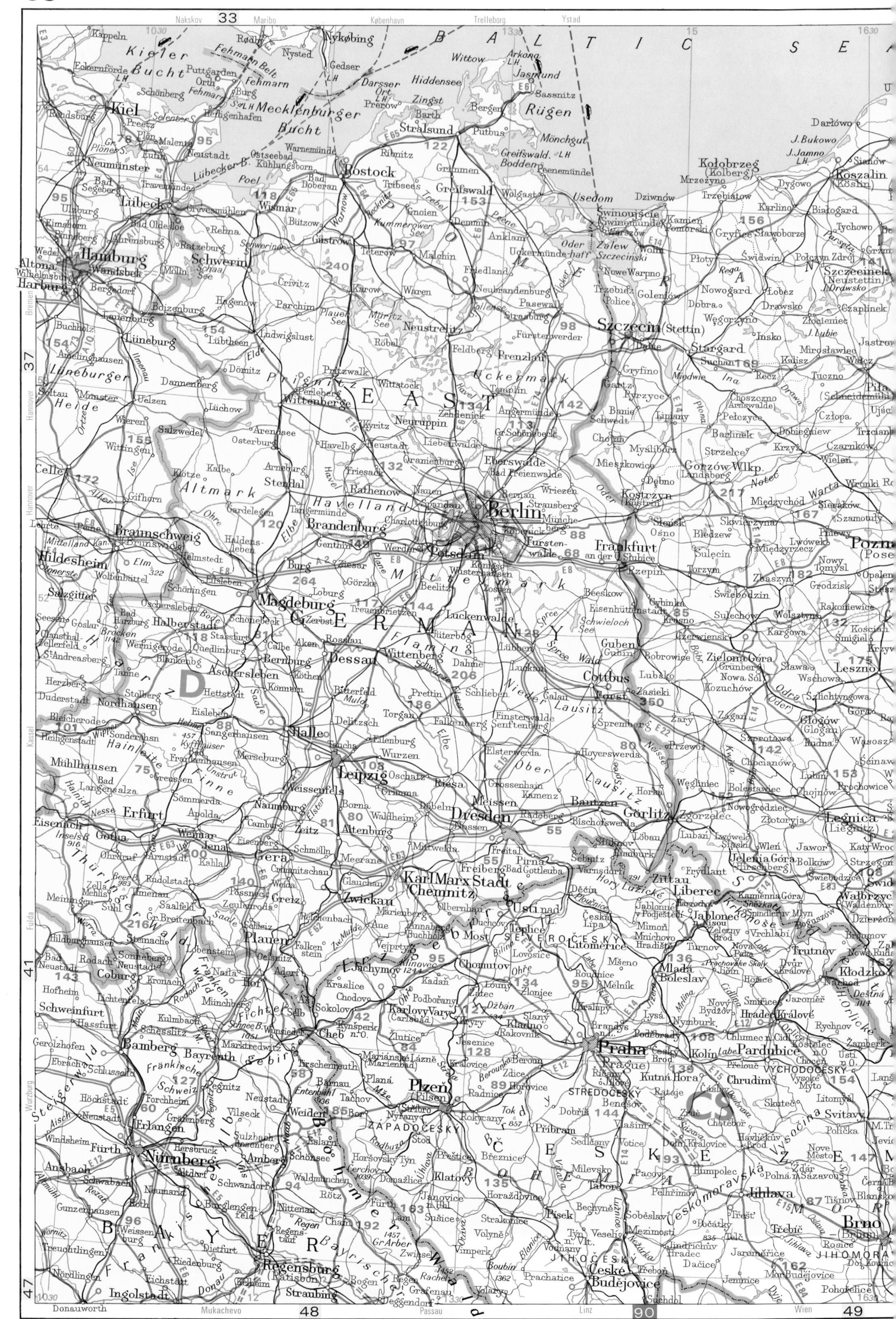
Nakskov
33
Maribo
København
Trelleborg
Ystad
BALTIC SEA
Kieler Bucht
Fehmarn Belt
Fehmarn
Mecklenburger Bucht
Lübecker B.
Rügen
Hiddensee
Usedom
Greifswald. Bodden
Oder Zalew Szczeciński
Kiel
Rendsburg
Neumünster
Lübeck
Hamburg
Altona
Wandsbek
Harburg
Schwerin
Wismar
Rostock
Stralsund
Greifswald
Anklam
Neubrandenburg
Szczecin (Stettin)
Świnoujście
Kołobrzeg (Kolberg)
Koszalin (Köslin)
Stargard
Lüneburg
Lüneburger Heide
Celle
Braunschweig
Hildesheim
Salzgitter
Wolfsburg
Magdeburg
Stendal
Wittenberge
Prignitz
Uckermark
Altmark
Havelland
Brandenburg
Potsdam
Berlin
Frankfurt an der Oder
Słubice
Gorzów Wlkp. (Landsberg)
Kostrzyn (Küstrin)
Zielona Góra (Grünberg)
Leszno
Poznań
Mittel Mark
Fläming
Spree Wald
Cottbus
Guben
Dessau
Halle
Leipzig
Wittenberg
Nieder Lausitz
Ober Lausitz
Dresden
Meissen
Görlitz
Zgorzelec
Legnica (Liegnitz)
Jelenia Góra (Hirschberg)
Wałbrzych
Kłodzko
EAST GERMANY
D
Harz
Nordhausen
Hainleite
Finne
Erfurt
Gotha
Eisenach
Weimar
Jena
Gera
Zwickau
Karl Marx Stadt (Chemnitz)
Plauen
Thüringer Wald
Frankenwald
Erzgebirge
Fichtel gebirge
Coburg
Hof
Bamberg
Bayreuth
Schweinfurt
Fränkische Schweiz
Steigerwald
Erlangen
Fürth
Nürnberg
Ansbach
Amberg
Weiden
Regensburg (Ratisbon)
Straubing
Ingolstadt
Bayerischer Wald
Böhmer Wald
BAYERN
Franken Alb
Cheb
Karlovy Vary (Carlsbad)
Chomutov
Most
Teplice
Ústí nad Labem
Děčín
Liberec
Jablonec
Krkonoše
Mladá Boleslav
Hradec Králové
Pardubice
Praha (Prague)
Kladno
Plzeň (Pilsen)
Příbram
Tábor
Písek
České Budějovice
Jihlava
Brno (Brünn)
ČESKÉ
ČS
SEVEROČESKÝ
STŘEDOČESKÝ
ZÁPADOČESKÝ
JIHOČESKÝ
VÝCHODOČESKÝ
JIHOMORA
Českomoravská Vysočina
Elbe
Oder
Warta
Saale
Mulde
Havel
Spree
Neisse
Vltava
Labe
Donau
37
41
47
48
90
49
Donauworth
Mukachevo
Passau
Linz
Wien
Bremen
Hannover
Kassel
Fulda
Würzburg

1:2,500,000

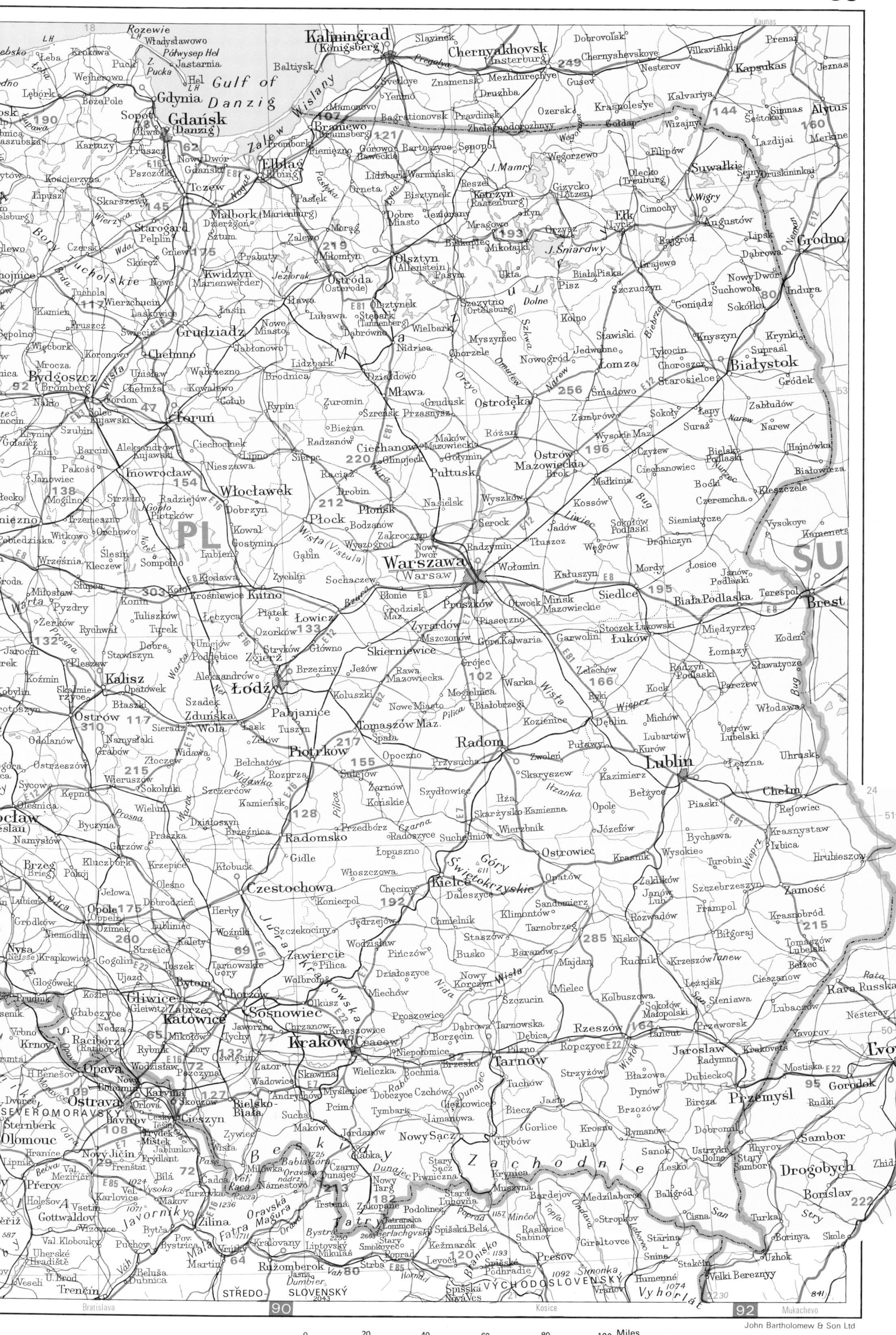

John Bartholomew & Son Ltd

0   20   40   60   80   100 Miles
0 20 40 60 80 100 120 140 160 Kilometres

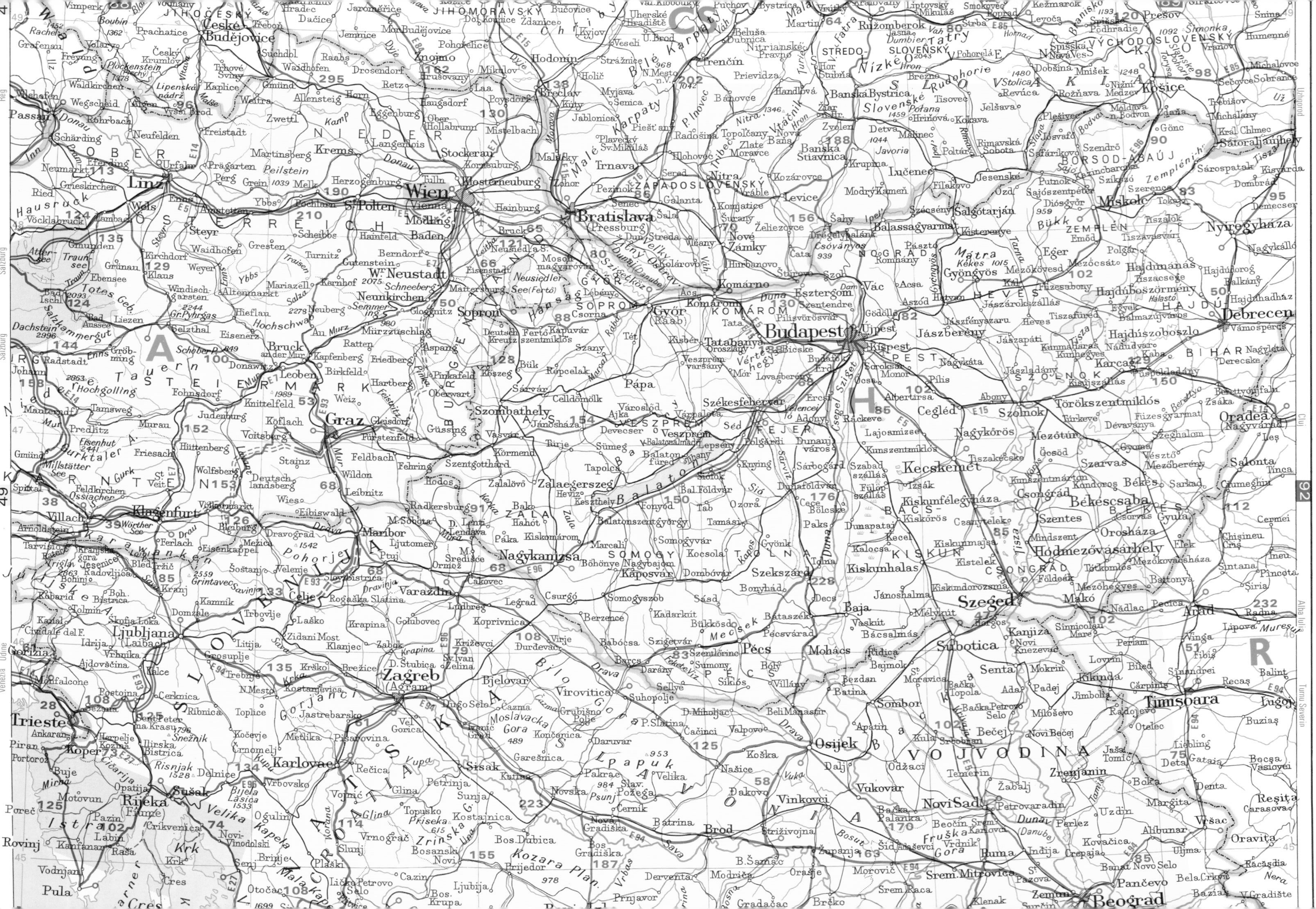
92
Uzhgorod
Cluj
Alba Iulia
Turnu Severin
Salzburg
Reg
Venezia
Udine
CS
A
H
R
JIHOČESKÝ
JIHOMORAVSKÝ
ZÁPADOSLOVENSKÝ
STŘEDOSLOVENSKÝ
VÝCHODOSLOVENSKÝ
BORSOD-ABAÚJ
ZEMPLÉN
NÓGRÁD
HEVES
HAJDÚ
BIHAR
PEST
SZOLNOK
KOMÁROM
GYŐR
SOPRON
VESZPRÉM
FEJÉR
VAS
ZALA
SOMOGY
TOLNA
BARANYA
BÁCS-KISKUN
CSONGRÁD
BÉKÉS
NIEDER
OBER
ÖSTERREICH
STEIERMARK
KÄRNTEN
BURGENLAND
SLOVENIJA
CROATIA
SLAVONIJA
VOJVODINA
České Budějovice
Znojmo
Hodonín
Bratislava
(Pressburg)
Trnava
Nitra
Trenčín
Ružomberok
Banská Bystrica
Košice
Prešov
Linz
Passau
Wien
(Vienna)
St. Pölten
Wr. Neustadt
Graz
Klagenfurt
Villach
Sopron
Győr
(Raab)
Komárno
Esztergom
Budapest
Miskolc
Eger
Debrecen
Nyíregyháza
Szolnok
Kecskemét
Szeged
Szombathely
Zalaegerszeg
Nagykanizsa
Kaposvár
Pécs
Székesfehérvár
Veszprém
Balaton
Oradea
(Nagyvárad)
Arad
Timişoara
Maribor
Ljubljana
(Laibach)
Zagreb
(Agram)
Trieste
Rijeka
Karlovac
Osijek
Subotica
Novi Sad
Beograd
Varaždin
Sisak
Pula
Duna
Dunaj
Donau
Drava
Tisza
Sava
1:2 500 000

1:2 500 000

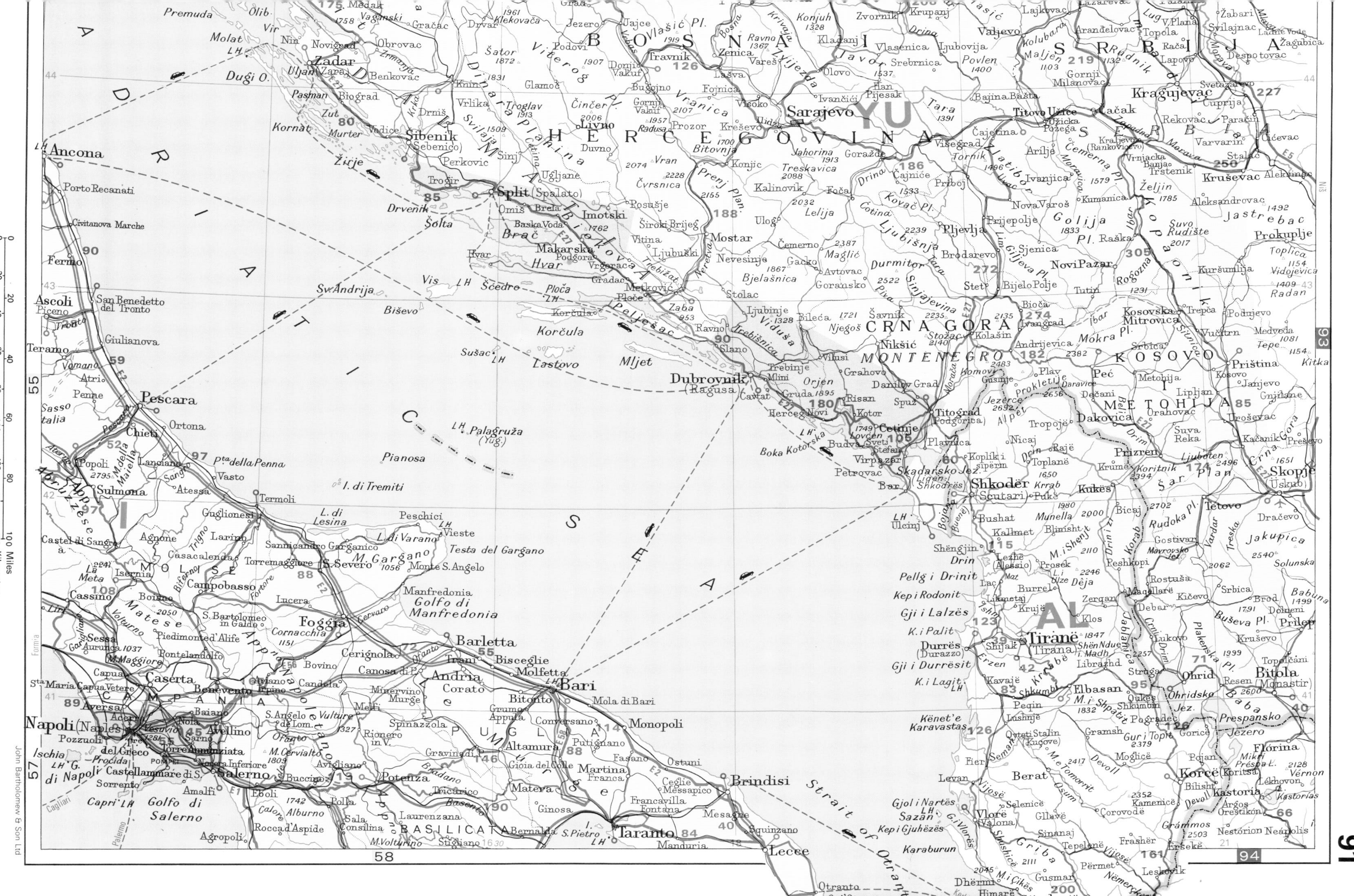
ADRIATIC SEA
Strait of Otranto
YU
AL
I
BOSNA I HERCEGOVINA
SRBIJA
CRNA GORA
MONTENEGRO
KOSOVO I METOHIJA
PUGLIA
CAMPANIA
BASILICATA
MOLISE
Abruzzese
Dinara planina
Biokovo
Sarajevo
Mostar
Split (Spalato)
Šibenik (Sebenico)
Zadar (Zara)
Dubrovnik (Ragusa)
Titograd (Podgorica)
Cetinje
Kotor
Nikšić
Shkodër (Scutari)
Tiranë (Tirana)
Durrës (Durazzo)
Elbasan
Vlorë (Valona)
Berat
Korçë (Koritsa)
Ohrid
Bitola (Monastir)
Prilep
Skopje (Üsküb)
Tetovo
Prizren
Priština
Peć
Kragujevac
Kruševac
Prokuplje
Valjevo
Titovo Užice
Novi Pazar
Ancona
Pescara
Chieti
Sulmona
Ascoli Piceno
Teramo
Termoli
Campobasso
Foggia
Bari
Brindisi
Lecce
Taranto
Potenza
Matera
Salerno
Napoli (Naples)
Caserta
Benevento
Avellino
Barletta
Andria
Manfredonia
Golfo di Manfredonia
Golfo di Salerno
Brač
Hvar
Korčula
Mljet
Lastovo
Vis
Pelješac
Šolta
I. di Tremiti
Pianosa
Palagruža (Yug.)
Capri
Ischia
Otranto
C. d'Otranto
93
94
58
57
55
0 20 40 60 80 100 Miles
0 20 40 60 80 100 120 140 160 Kilometres
John Bartholomew & Son Ltd

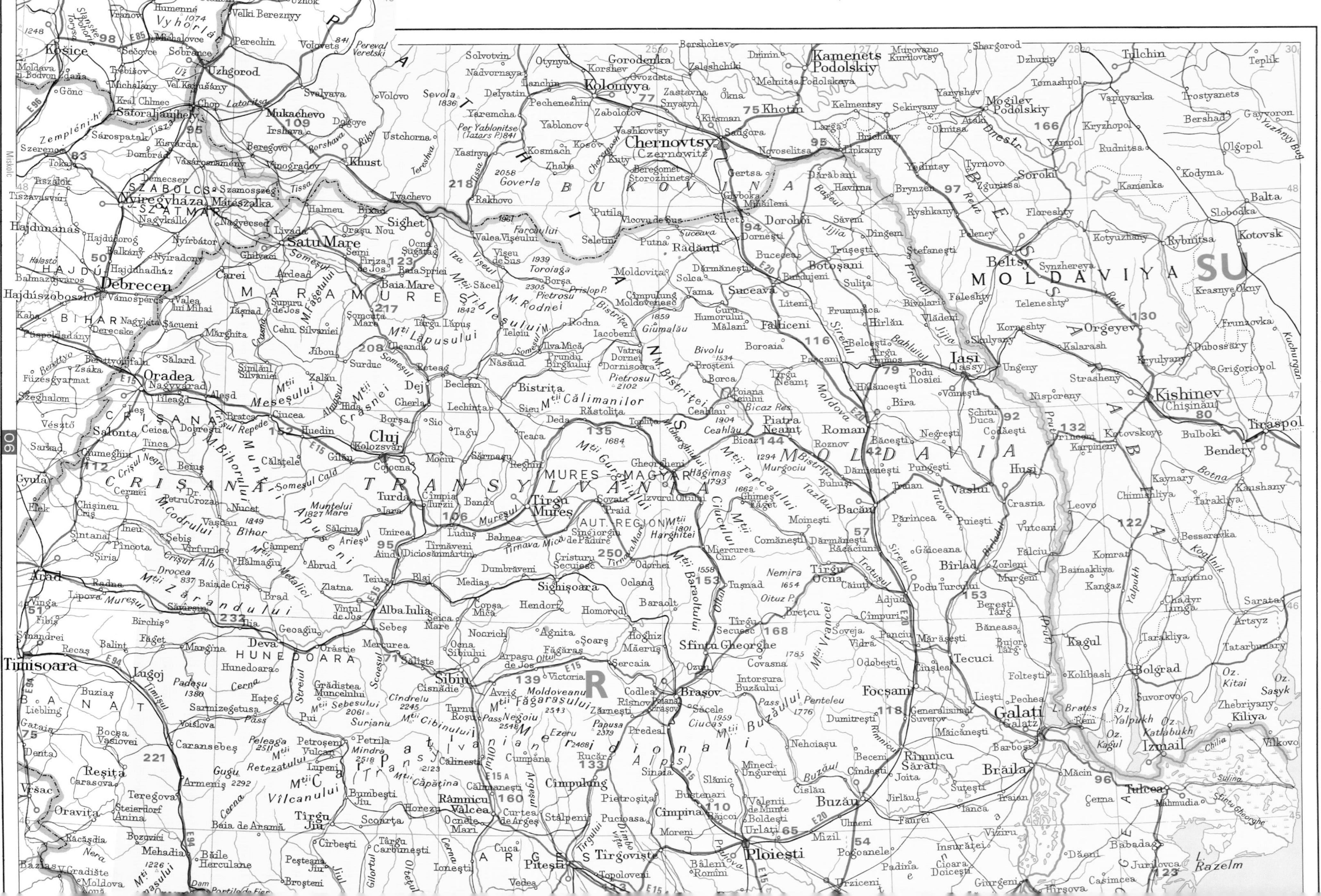

1:2 500 000

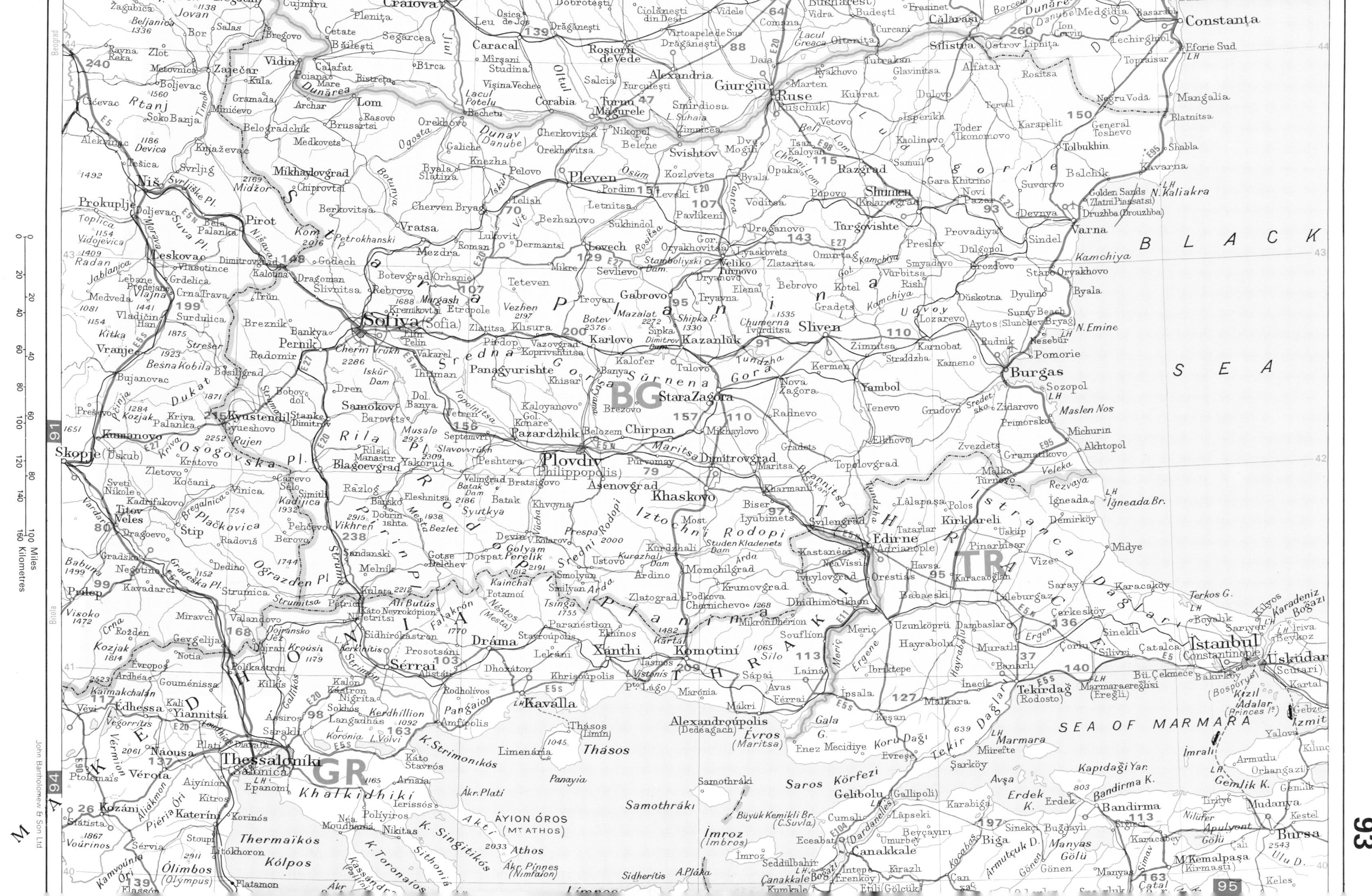

0	20	40	60	80	100 Miles		
0	20	40	60	80	100	120	140	160 Kilometres

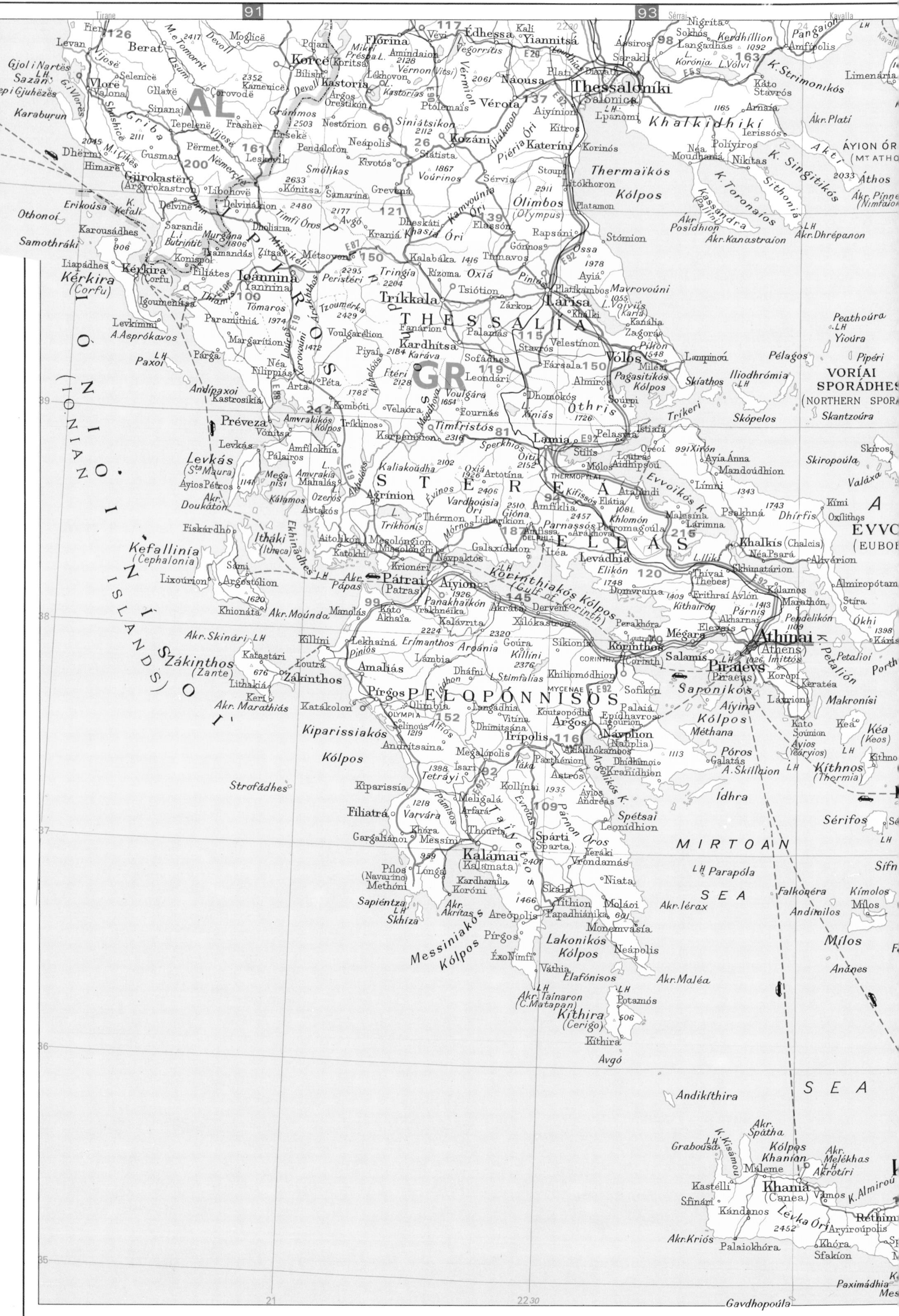
91
93
Tirane
Sérrai
Kavalla
Berat
Vlorë
(Valona)
Karaburun
Sazan
Himarë
Gjirokastër
(Argyrokastron)
Delvinë
Sarandë
Korçë (Koritsa)
Flórina
Édhessa
Yiannitsá
Thessaloníki
(Salonica)
Náousa
Véroia
Kastoria
Kozáni
Grevená
Kateríni
Ólimbos
(Olympus)
Khalkidhikí
Thermaïkós
Kólpos
Ioánnina
Kérkira
(Corfu)
Igoumenítsa
Párga
Paxoí
Préveza
Árta
Levkás
Tríkkala
Kardhítsa
THESSALIA
Lárisa
Vólos
Pagasitikós
Kólpos
Lamía
GR
Agrínion
STEREÁ ELLÁS
Levádhia
Khalkís (Chalcis)
Thívai
(Thebes)
Kefallinía
(Cephalonia)
Argostólion
Itháki
(Ithaca)
Pátrai
(Patras)
Aíyion
Korinthiakós Kólpos
(Gulf of Corinth)
Kórinthos
(Corinth)
Mégara
Athínai
(Athens)
Piraievs
(Piraeus)
Salamís
Saronikós
Kólpos
Zákinthos
(Zante)
Pírgos
Amaliás
PELOPÓNNISOS
Trípolis
Árgos
Návplion
(Nauplia)
Kiparissiakós
Kólpos
Kalámai
(Kalamata)
Spárti
(Sparta)
Messiniakós
Kólpos
Lakonikós
Kólpos
Kíthira
(Cerigo)
Akr. Taínaron
(C. Matapan)
Ídhra
MIRTOAN
SEA
Kéa
(Keos)
Kíthnos
(Thermia)
Sérifos
Mílos
VORÍAI
SPORÁDHES
(NORTHERN SPOR
Skíathos
Skópelos
Skíros
Khaniá
(Canea)
Réthim
Andikíthira
SEA
IÓNIO
(IONIAN
NÍSOI
ISLANDS)

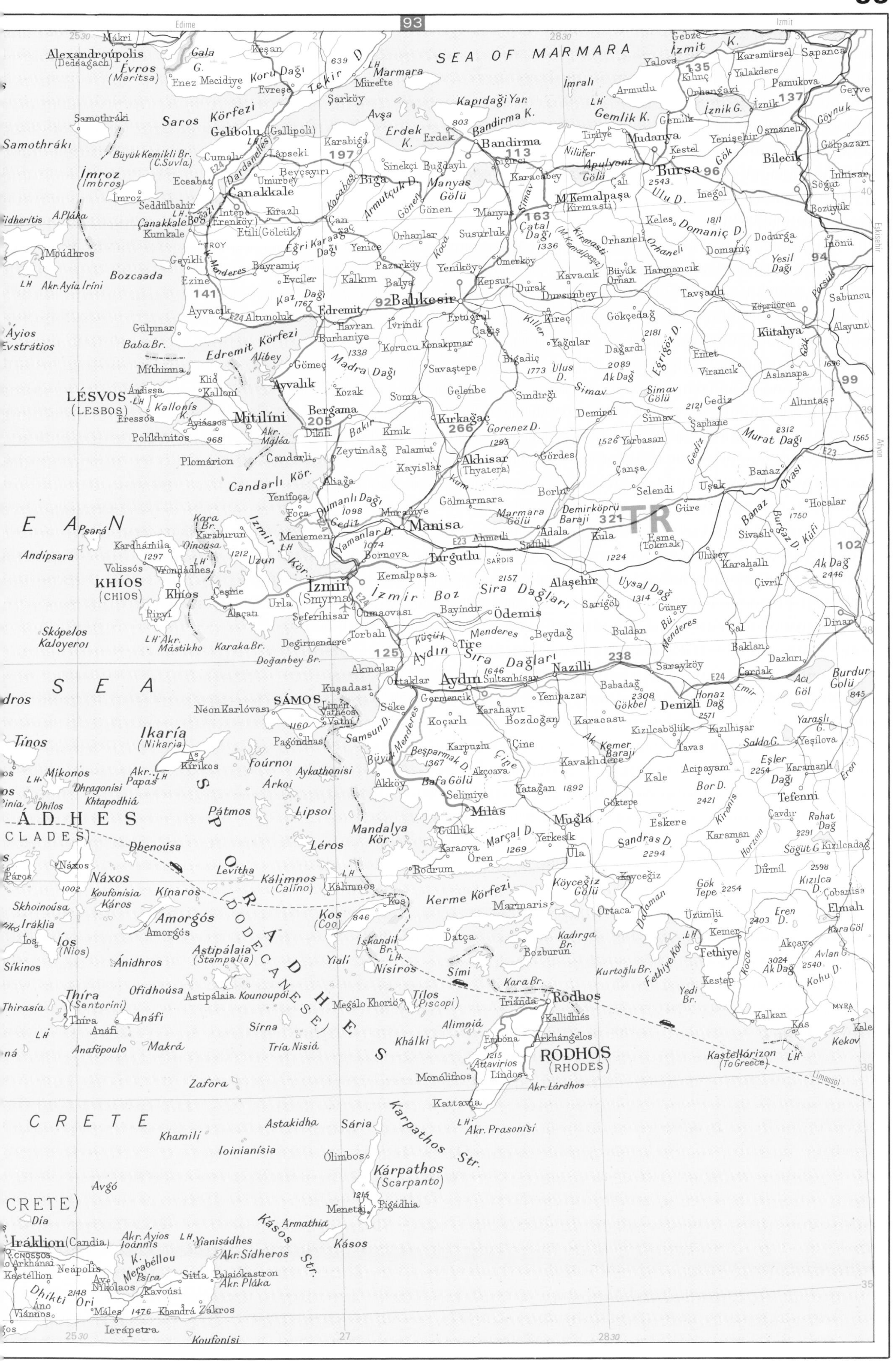

SEA OF MARMARA

Alexandroúpolis (Dedéagach)
Évros (Maritsa)
Gala G.
Keşan
Enez Mecidiye
Koru Dağı
639
Marmara
Mürefte
Şarköy
Kapıdağı Yar.
İzmit K.
İzmit
Yalova
Karamürsel
Sapanca
135
Gebze
Kihnç
Orhangazi
Yalakdere
Pamukova
Gèyve
137
Göynuk
Samothráki
Evreşe
Avşa
Armutlu
İmralı
Samothráki (Imbros)
Saros Körfezi
Gelibolu (Gallipoli)
Karabiğa
Tapseki
197
Sinekçi Buğdaylı
803
Erdek
Erdek
Bandirma K.
Bandirma
113
Tirilye
Armutlu
İznik G.
İznik
Orhaneli
Gemlik K.
Gemlik
Mudanya
Osmaneli
Gölpazarı
İnhisar
Bilecik

İmroz (Imbros)
İdherítis
A.Pláka
İmroz
Cumalı
Eceabat
Beyçayırı
Biga
Gönen
Manyas Gölü
Manyas
163
Çatal Dağı 1336
Nilüfer
Apulyont Gölü
Çalı
M.Kemalpaşa (Kirmasti)
Bursa
96
Kestel
Yenişehir
Söğüt
Bozüyük
40
94

Çanakkale Boğ.
Çanakkale
Seddülbahir
İntepe
Frenköy
Etili (Gölcük)
Çan
Yenice
Orhanlar
Susurluk
Ömerköy
Keles
Domaniç D.
Orhaneli
Orhaneli
Dodurga
Domaniç
İnönü

Kumkale
TROY
Geyikli
Ezine
Bayramiç
Evciler
Kalkım
Balya
Durak
Dursunbey
Büyük Harmancık
Orhan
Yeşil Dağı
1811
Köprüören
Sabuncu

141
Kaz Dağı 1762
Altunoluk
Edremit
Havran
İvrindi
Ertuğrul
Çağış
Balıkesir
Gökçedağ
Tavşanlı
Alayunt

Akr.Ayía İríni
LH
Ayvacık
Burhaniye
Korucu
Kınıkpınar
2181
Dağardı
Kütahya

Áyios Evstrátios
Gülpınar
Baba Br.
Alibey
Gömeç
1338
Madra Dağı
Savaştepe
Bigadiç
1773
Ulus D.
2089
Ak Dağ
Virancık
Emet
Aslanapa
1696
99

LÉSVOS (LESBOS)
Míthimna
Klió
Andissa
Kalloni
Kozak
Soma
Gelenbe
Sındırgı
Simav
Simav Gölü
2121
Şaphane
Gediz
Altıntaş
39

Eressós
Ayiássos
Kallonís
Bergama
205
Kırkağaç
266
Gorenez D.
1293
Gördes
1526
Yarbasan
Çanşa
Selendi
Uşak
Banaz
Murat Dağı 2312
1565

Polikhnitos
968
Akr. Maléa
Dikili
Bakır
Kınık
Akhisar (Thyatera)
Borlu
Demirci
Güre
Banaz Ovası
Burgaz D. Küfi
1750
Hocalar
102

Plomárion
Candarlı
Zeytindağ
Palamut
Kayislar
Demirköprü Baraji
321
TR
Sivaslı
Karahallı
Ak Dağ 2446

E A N
Psará
Kara Br.
Candarlı Kör.
Aliağa
Yenifoça
Dumanlı Dağı
Gediz
Muradiye
Gölmarmara
Marmara Gölü
Adala
Kula
Esme (Tokmak)
Ulubey
Çivril
1224

Andípsara
Kardhámila
Oinoúsa
1297
1212
Foça
Menemen
Yamanlar D. 1074
Manisa
Ahmetli
E23
Spahli
1314
Alaşehir
Uysal Dağ
Güney
Dinar

KHÍOS (CHIOS)
Volissós
Vrondádhes
1098
Bornova
Turgutlu
SARDIS
2157
Sıra Dağları
Sarıgöl
Bü

Khíos
Çeşme
Urla
İzmir (Smyrna)
Kemalpaşa
İzmir Boz
Bayındır
Beydağ
Buldan
Çal

Skópelos Kaloyeroi
Piryí
Alaçata
Seferihisar
Çumaovası
Ödemiş
1646
Sıra Dağları
Nazilli
238
Saraybaşı
E24
Cardak
Burdur Gölü 845

LH Akr. Mástikho
Karaka Br.
Değirmendere
Torbalı
Küçük Menderes
Tire
Aydın
Sultanhisar
Honaz Dağ 2571
Acı Göl
Dazkırı

S E A
Doğanbey Br.
Akmcılar
Ortaklar
Aydın
Babadağ 2308
Gökbel
Denizli
Kızılhisar
Salda G.
Yeşilova
Eren

dros
Kuşadası
Söke
Germencik
Yenipazar
Karacasu
Kızılcabölük
Tavas
Acıpayam
2254
Eşler Dağı
Karamanlı

Tínos
SÁMOS
Liman Vatheos
Samsun D.
Koçarlı
Karahayıt
Bozdoğan
Ak
Kemer Baraji
Kale
Bor D. 2421
2291
Çavdır
Rahat Dağ
Tefenni

İkaría (Nikaria)
Néon Karlóvasi
1160
Vathi
Büyük Menderes
Karpuzlu
Çine
Kavaklıdere
Göktepe
Eskere
Karaman
Söğüt G.
Dirmil
2598

Míkonos
Akr. Papas
Pagóndhas
Fournoi
Besparmak D. 1367
Akçaova
Yatağan 1892
Muğla
Sandras D. 2294
Karaman
2403
Kızılca D.
Çobanisa
Elmalı

Dhragonísi
Kirikos
Árkoi
Akkóy
Bafa Gölü
Selimiye
Mılas
Ula
Gök Tepe 2254
Eren D.
Kara Göl

ÁDHES (CLADES)
Khtapodhiá
Pátmos
Lipsoi
Güllük
Marçal D. 1269
Yerkesik
Kayceğiz
Köyceğiz Gölü
Üzümlü
Kemer
Akçay
3024 Ak Dağ 2540
Kohu D.

Dhenoúsa
Náxos
Levitha
Léros
Mandalya Kör.
Karaova
Ören
Bodrum
Kerme Körfezi
Marmaris
Ortaca
Dalaman
Fethiye
Kesteb

Koufonisia
Káros
Kínaros
Kálimnos (Calino)
Kálimnos
Kos
Datça
Bozburun
Kurtoğlu Br.
Fethiye Kör.
Yedi Br.
Kalkan
Kaş
MYRA

Amorgós
Amorgós
Astipálaia (Stampalia)
Kos (Coo) 846
İskandil Br.
Yiali
Nísiros
Simi
Kara Br.
Triánda
Ródhos
Kallidhiés
Kastellórizon (To Greece)
Kale
Kekov
36

Thíra (Santorini)
Anáfi
Astipálaia
Kounoupói
Megálo Khorió
Tílos (Piscopi)
Alimniá
Arkhángelos
RÓDHOS (RHODES)

Thirasía
Thíra
Anáfi
Sírna
Tría Nisiá
Khálki
Emboná
Attavirios 1215
Líndos

Anafópoulo
Makrá
Zafora
Monólithos
Akr. Lárdhos

CRETE
Khamilí
Astakidha
Sária
Karpathos Str.
Kattavia
Akr. Prasonísi

Avgó
İoinianísia
Ólimbos
Kárpathos (Scarpanto)

Ierápetra
Kásos Str.
Menetái
Pigádhia

Iráklion (Candia)
KNOSSOS
Neápolis
Psíra
Sitía
Yianisádhes
Kásos
35

0   20   40   60   80   100 Miles
0 20 40 60 80 100 120 140 160 Kilometres

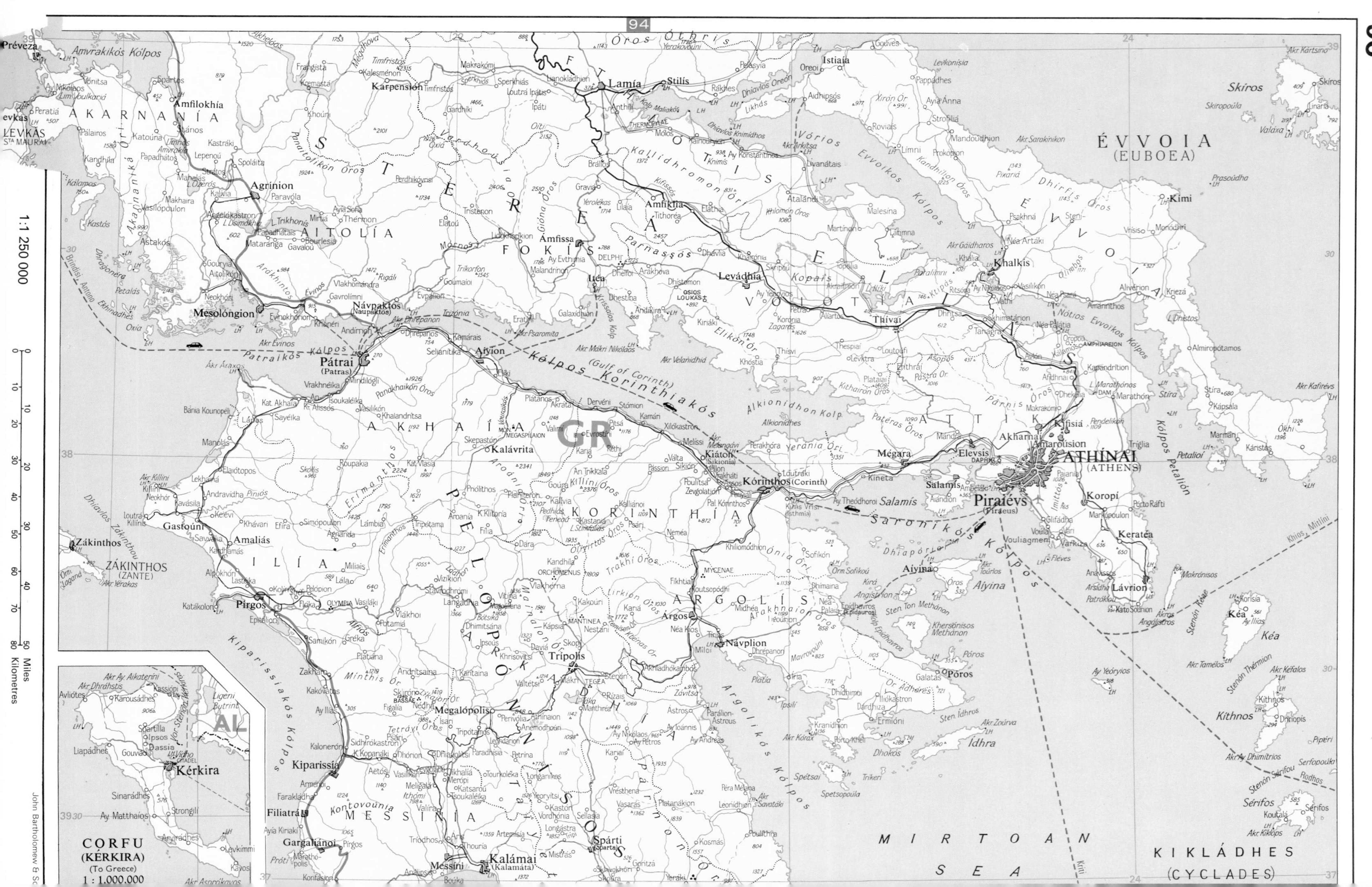

1:1 250 000

0	10	20	30	40	50 Miles
0	10 20	30 40	50	60 70	80 Kilometres

John Bartholomew & Son

INDEX

References

Most names in the index are followed by a page number and an arrow. The arrow refers to one of six squares, as shown in the diagram, in which the name will be found. Names followed by an equals sign (=) and a further name are cross references to correct entries.

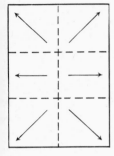

Abbreviations

Ákr. = Ákra, Akrotírion
Anc. = Ancient
Ay. = Ayía, Áyioi, Áyion, Áyios
B. = Bahía, Baia, Bay, Bjerge
Bgem. = Barragem
Bos. = Bosanski
Br. = Burnu, Burun
C. = Cabo, Cap, Cape
Chan. = Channel
D. = Dağ, Dağı, Dağlar, Dağları, Donja, Donji
Dist. = District
E. = East
Emb. = Embalse
F. = Firth
Fj. = Fjord
G. = Göl, Golfe, Golfo, Gölü
Geb. = Gebirge
Gr. = Grosse
Hd. = Head
I. = Île, Island, Isle, Isole
Is. = Islands
J. = Jezioro
K. = Kep
Kólp. = Kólpos
L. = Lac, Lago, Lake, Lìmni, Loch, Lough
L.H. = Light House
M. = Monte, Murska, Mursko
Mt. = Mont, Mount, Mountain
Mte. = Monte
Mtes. = Montes
Mti. = Monti, Munti
Mtii. = Muntii
Mts. = Monts, Mountains
N. = Noord, Nord, Nordre, Nörre, Nørre, North
N.D. = Notre Dame
Nat. = National
O. = Oost, Ost, Ostrov, Ostrova, Ostrówa
Ö. = Öster, Östra, Östre
Ø. = Øster, Østre
Ór. = Óri, Óros
Oz. = Ozero
P. = Pass
Pen. = Peninsula, Penìsola
Pl. = Planina
Plat. = Plateau
Pnte. = Pointe
Prom. = Promontory
Prov. = Province
Pt. = Point
Pta. = Ponta, Punta
Pto. = Portillo, Puerto
R. = Río, River
Reg. = Region
Res. = Reservoir
S. = See, Sör, Sør, South, Syd
Sa. = Sierra
Sd. = Sound
Sr. = Sönder, Sønder, Søndre
St. = Saint, Stara
Sta. = Santa
Ste. = Sainte
Str. = Strait
Strs. = Straits
Sv. = Sveti
U.K. = United Kingdom
V. = Väster, Vatn, Vest, Vester, Vestre
Vel. = Velika, Velká, Velké
W. = West, Wester
Z. = Zalew, Zatoka

Aachen 39↖
Aalen 40↘
Aalst 38↗
Aalten 35→
Äänekoski 87←
Aarau 47←
Aarburg 47←
Aardenburg 35↙
Aarschot 39↖
Abadin 63↗
Abano Terme 54↗
Abarán 74↗
Abárzuza 65→
Abbekås 30↘
Abbeville 38←
Abbey Town 20→
Abbeydorney 25↙
Abbeyfeale 25↙
Abbeyleix 25←
Abbiategrasso 53↗
Abbots' Bromley 22↗
Abbotsbury 22↘
Abejar 65→
Abelnes 27↘
Abelvær 84←
Abenójar 69↘
Åbenrå 33↙
Abensberg 41↙
Aberaeron 22↖
Aberarder 18→
Aberchirder 19←
Aberdåre 22→
Aberdaron 22↖
Aberdeen 19←
Aberdour 19↙
Aberdovey 22↖
Aberfeldy 18↘
Aberford 21↙
Aberfoyle 18↘
Abergavenny 22→
Abergele 20↘
Abergwili 22←
Aberlady 19↙
Aberlour 19←
Abernethy 19↙
Aberporth 22↖
Abersoch 22↖
Abertillery 22→
Abertridwr 22↗
Aberystwyth 22↖
Abetone 54←
Abingdon 23←
Abington 20↗
Abisko 85→
Abla 74←
Ablainen 87↘
Ablis 45↖
Ablitas 65→
Åbo = Turku
Abondance 46↘
Abony 90↖
Åbosjö 84↖
Aboyne 19←
Abram 21↙
Abrantes 68←
Abrets, les 52↗
Abridge 23→
Abriès 53←
Abrunhosa Velha 68↗
Abruzzo e Molise, prov. 57←
Åby 29↙
Åbybro 32←
Åbyn 86↘
Acate 60↗
Acceglio 53←
Accrington 21↙
Acebo 68→
Acehuche 69←
Acerenza 58↘
Acerra 57↙
Aceuchal 69↙
Acharacle 18←
Achenkirch 48←
Achensee 13→
Achern 40↙
Achfary 18↗
Achill 24↘
Achiltibuie 18↗
Achim 36↗
Achnasheen 18→
Achnashellach 18→
Achterwehr 33→
Aci Castello 60↗
Acipayam 95→
Acireale 60↗
Acle 23↗
Acquacadda 61↙
Acquacalda 60↖
Acquapendente 56↗
Acquasanta Terme 57↖
Acquasparta 56↗
Acri 59←
Ács 90←
Acton Turville 22→
Ada 90↗
Ådal = Viker
Adala 95→
Ådalsliden 84↖
Adamello, mt. 47↘
Adamstown 25↖
Adamuz 73↗
Adanero 64↘
Adare 25↙
Adasevci 90↗
Adderbury 23←
Addingham 21←
Adelboden 47↙
Adelöv 30↗
Ädelsfors 31↖
Adelsheim 40↘
Adelso 29→
Ademúz 71←
Adhámi 96→
Adjud 92↗
Adlington 21↙
Admont 49←
Adneram 27←
Adolfsberg 29←
Adolfsström 85↙
Adony 90←
Adorf 41←
Adra 74↙
Adradas 65↘
Adrano 60→
Ádria 54→
Adrianople = Edirne
Adrigole 25↘
Aduanas San Antonio 15↘
Adwick le Street 21↙
Adzaneta 71→
Æbeltoft 32→
Ærø, island 33←
Ærøskøbing 33←
Aesch 47←
Aeschi 12→
Aëtós, *Greece* 96↙
Aëtós, *Greece* 96↘
Affoltern am Albis 47←
Afídhnai 96↖
Afjord 84←
Aflenz 49←
Afráti 96↖
Aga 26↘
Agalianí 96↘
Agay 53↙
Agde 67↗
Agen 50→
Ager 67←
Agger 32↙
Aggius 61↖
Aghaboe 25←
Agivey 24↖
Aglish 25→
Ágnanda 96↘
Agnandi 96←
Agnone 57←
Agognate 53↗
Agoncillo 65→
Agordo 48↙
Agost 71↙
Agramunt 67←
Agrapidhokhóri 96↘
Agrate 53↗
Agreda 65→
Agria 60↗
Agrídhion 96↘
Agrigento 60→
Agrínion 96↙
Agropoli 58↘
Agrós 96↘
Aguadulce 73←
Aguarón 66↙
Aguaviva 66↘
Agudo 69↘
Agueda 68↖
Aguero 66→
Aguilafuente 65↙
Aguilar de Campoo 65↖
Aguilar de la Frontera 73→
Aguilar del Río Alhama 65→
Aguilas 74→
Agunnaryd 30→
Ahascragh 25←
Ahaus 36←
Aheim 26↙
Ahillones 73↖
Ahlen 36↙
Ahmetli 95→
Ahoghill 24↖
Ahrdor 39→
Ahrensburg 33→
Ähtäri 87←
Ähtävä 87↙
Ahun 45↙
Åhus 30↘
Aiándion 96↗
Aibar 66←
Aich 40↘
Aicha 41↘
Aichach 48↖
Aiddejavrre 86←
Aidhipsós 96←
Aigen 41↘
Aigle 46↘
Aigle, l' 43→
Aignan 50↘
Aigre 44↘
Aiguablava 67→
Aiguafreda 15↖
Aiguebelle 52↗
Aigueperse 45↘
Aigues-Mortes 52↙
Aiguillon 50→
Aigurande 45↙
Ailefroide 52→
Aillant-sur-Tholon 45↗
Aillevillers 46↗
Ailly-sur-Noye 38←
Aime 52↗
Aïnhoa 65↗
Ainsa 66→
Airdrie 18↘
Aire, *France* 38↖
Aire, *France* 50↘
Airolo 47↙
Airth 18↘
Airvault 44→
Aissey 46→
Aiterhofen 41↘
Aith 19↗
Aitolikón 96↙
Aiud 92↘
Aix-d'Angillon, les 45←
Aix-les-Bains 52↗
Aix-en-Othe 45↗
Aix-en-Provence 52↘
Aíyina 96↗
Aiyínion 94↗
Aíyion 96↙
Áiyos Míron 95↙
Aizenay 44←
Ajaccio 62←
Ajka 90←
Ajofrin 69→
Åkarp 30↙
Akçay 95↘
Aken 37↘
Åker, *Sweden* 29←
Åker, *Sweden* 30↗
Akernes 27→
Åkersberg 29→
Åkersjön 84←
Åkersstyckebruk 29←
Akersund 26↘
Akharnai 96↖
Akhílion 96←
Akhisar 95↗
Akhladherí 96↖
Akhladhiní 96↘
Akhladhókambos 96→
Akhtopol 93↗
Akincilar 95←
Åkirkeby 31↙
Akköy 95←
Akksjöseter 26↖
Akksojösetrene 26↖
Akmangit 92↗
Åkovos 96→
Akraífnion 96←
Akráta 96←
Åkrehamn 27↙
Aksta 26↙
Al 26→
Alaçati 95←
Alæfors 30↖
Alaejos 64↘
Alagna-Valsésia 47↙
Alagón 66←
Alahärmä 87↙
Alajar 72→
Alajärvi 87←
Alakylä 86←
Alameda 73→
Alamedilla 74←
Alamillo 69↘
Alanäs 84↖
Åland 29←
Åland, island 87↘
Alandroal 68↘
Ålandsbro 84↗

Alanieni 86→
Alanis 73↖
Alar del Rey 65←
Alaraz 69↖
Alarcón 70→
Alaró 75←
Alaşehir 95→
Alássio 53←
Alastaro 87↘
Alatri 57←
Alavuokki 86↗
Alavus 87←
Alayor 75→
Alba 53←
Alba de Tormes 64↘
Alba Iulia 92↘
Albacete 70↘
Albaida 71↙
Albal 71←
Albaladejo 70↘
Albalate del Arzobispo 66↘
Alban 51↘
Albánchez 74→
Albano Laziale 56→
Albarracín 71↖
Albares de la Ribera 64←
Albatana 71↙
Albatera 75↖
Albenga 53→
Albens 52↗
Albergaria-a-Velha 68↖
Alberique 71↙
Alberobello 58→
Albersdorf 33↘
Albersloh 36←
Albert 38←
Albertville 52↗
Albi 51↙
Albires 64→
Albisola Marina 53→
Albocácer 71↗
Alböke 31←
Albólote 73→
Albondón 74↙
Alborea 71←
Ålborg 32←
Albox 74→
Albrighton 22↗
Albufeira 72←
Albufereta 15↙
Albuñol 74↙
Albuñuelas 73↘
Alburquerque 68↘
Alby, *Öland, Sweden* 31←
Alby, *Sweden* 84↗
Alcabideche 68↙
Alcácer do Sal 68↙
Alcáçovas 68↙
Alcadozo 70↘
Alcains 68→
Alcalá de Chivert 71↗
Alcalá de Guadaira 73←
Alcalá de Henares 70↖
Alcalá de Júcar 71↙
Alcalá de la Selva 71↖
Alcalá de los Gazules 73↙
Alcala del Rio 73←
Alcalá dell Valle 73↙
Alcalá la Real 73→
Alcamo 60↙
Alcampet 66→
Alcanar 71↗
Alcanede 68←
Alcanena 68←
Alcanhões 68←
Alcañices 64↙
Alcañiz 66↘
Alcántara 68→
Alcantarilla 74→
Alcantud 70↗
Alcaracejos 69↘
Alcaraz 70↘
Alcaria Ruiva 72←
Alcarrás 66↘
Alcaudete 73→
Alcaudete de la Jara 69→
Alcázar de San Juan 70→
Alceda 65↖
Alcester 22↗
Alcira 71↙
Alcoba 69→
Alcobaca 68←
Alcobendas 70↖
Alcócer 70↗
Alcochete 68↙
Alcolea 73→
Alcolea de Calatrava 69↘
Alcolea de Cinca 66→
Alcolea del Pinar 65↘
Alcolea del Rio 73←
Alconchel 68↘
Alcora 71→
Alcorisa 66↘
Alcoutim 72→
Alcover 67↙
Alcoy 71↙
Alcubierre 66→
Alcubillas 70↘
Alcublas 71←
Alcudia 75←
Alcudia de Carlet 71↙
Alcudia de Crespins 71↙
Alcudia de Guadix 74←
Alcuéscar 69↙
Aldbourne 23←
Aldbrough 21→
Aldea 66↘
Aldea de Trujillo 69←
Aldea del Cano 69←
Aldea del Rey 69↘
Aldea del Rey Nino 69↗
Aldeacentenera 69←
Aldeadavila de la Ribera 64↙
Aldealpozo 65→
Aldeamayor de San Martin 64↘
Aldeanueva de Barbarroya 69→
Aldeanueva de Ebro 65→
Aldeanueva de la Vera 69←
Aldeaquemada 70↙
Aldeavieja 69↗
Aldeburgh 23↗
Aldeia Nova de São Bento 72→
Alderley Edge 21↙
Aldermaston 23←
Aldersbach 41↘
Aldershot 23←
Aldridge 22↗
Aldsworth 23←
Alegrete 68→
Aleksandrovac 91↖
Aleksandrów Kujawski 89←
Aleksinac 91↖
Ålem 31←
Alençon 44↗
Alenquer 68↙
Aleria 62←
Alès, *France* 52←
Ales, *Sardinia* 61←
Aleşd 92↙
Alessandria 53→
Alessio = Lezhë
Ålestrup 32↘
Ålesund 84↘
Aletschhorn, mt. 47↙
Alexandria, *Romania* 93←
Alexandria, *Scotland* 18↘
Alexandroúpolis 93→
Alf 39→
Alfajarin 66→
Alfambra 71↖
Alfândéga da Fe 63↘
Alfara 66↘
Alfarnate 73→
Alfaro 65→
Alfarras 66→
Alfatar 93↖
Alfedena 57←
Alfeizerao 68←
Alfeld 36→
Alfioúsa 96↘
Alfonsine 54→
Alford, *England* 21↘
Alford, *Scotland* 19←
Ålfoten 26↙
Alfoz 63↗
Alfreton 21↙
Alfriston 23↘
Alfta 84↗
Algaida 75↙
Algar, *Spain* 73↙
Algar, *Spain* 75←
Ålgarås 28↘
Ålgard 27↘
Algarinejo 73→
Algeciras 73↙
Algemesí 71↙
Algete 70↖
Alghero 61↖
Ålghult 31←
Alginet 71←
Algodonales 73↙
Algora 65↘
Algorta 65↖
Algoz 72←
Ålgsjön 84↖
Alguazas 74↗
Algueña 71↙
Alhama de Almeria 74↙
Alhama de Aragón 65↘
Alhama de Granada 73→
Alhama de Murcia 74→
Alhambra 70↘
Alhandra 68↙
Alhaurín de la Torre 73↘
Alhaurín el Grande 73↘
Alhendin 73→
Alhus 26↙
Ali Terme 60↖
Alia, *Italy* 60←
Alia, *Spain* 69→
Aliaga, *Spain* 66↘
Aliağa, *Turkey* 95←
Alíartos 96←
Alibunar 90↗
Alicante 71↙
Alija de los Melones 64→
Alijo 63↘
Alimena 60→
Alingsås 30↖
Aliseda 69←
Alise-Ste-Reine 46←
Alistáti 93↘
Alivérion 96↖
Aljezur 72←
Aljucén 69↙
Aljustrel 72←
Alkmaar 35↖
Allanche 51↗
Allariz 63→
Alleen 27→
Alleghe 48↙
Allègre 51↗
Allendale Town 21↖
Allendorf 40↖
Allenstein = Olsztyn
Allentsteig 49↖
Allepuz 71↖
Allevard 52↗
Allgunnen 31←
Allhallows 23→
Allingåbro 32→
Allinge 31↙
Allo 65→
Alloa 18↘
Allonby 20→
Allos 52→
Alloway 20↗
Alloza 66↘
Allstakan 28←
Allstedt 37↙
Alltwalis 22←
Alm 13↖
Almacellas 66→
Almada 68↙
Almadén 69↘
Almadén de la Plata 73←
Almadenejos 69↘
Almagro 70↙
Almajano 65→
Almansa 71↙
Almanza 64→
Almaraz 69←
Almargen 73←
Almarza 65→
Almazán 65↘
Almazora 71→
Almazul 65↘
Ålmeboda 31←
Almedinilla 73→
Almeida, *Portugal* 68↗
Almeida, *Spain* 64↙
Almeirim 68↙
Almelo 35→
Almenar 66←
Almenar de Soria 65↘
Almenara 71→
Almendral 68↘
Almendralejo 69↙
Almeria 74↙
Älmhult 30→
Almiropótamos 96↖
Almirós 94↗
Almklov 26↙
Almodóvar 72←
Almodóvar del Campo 69↘
Almodóvar del Pinar 70→
Almodóvar del Río 73←
Almogia 73↘
Almoharin 69↙
Almonaster la Real 72→
Almondsbury 22→
Almonte 72→
Almoradi 75↖
Almorox 69↗
Almoster 68←
Almudébar 66→
Almundsryd 30→
Almuñecar 73↘
Almunge 29→
Almuradiel 70↙
Almvik 31↖
Alness 18→
Alnö 84↗
Alnmouth 21↖
Alnwick 21↖
Alonistaína 96→
Alora 73↘
Alos 67↖
Alosno 72→
Alozaina 73↙
Alp 67←
Alpalhão 68→
Alpbach 48←
Alpe-d'Huez, l' 52↗
Alpen, mts. 47→
Alpera 71↙
Alpes, mts. 52→
Alphen am der Rijn 35←
Alpi, mts. 47↙
Alpiarça 68←
Alpirsbach 47↖
Alpokhóri 96↘
Alquézar 66→
Alrewas 23↖
Als, island 33←
Alsace, reg. 46↗
Alsager 21↙
Alsasua 65↗
Alsdorf 39↗
Alsfeld 40↗
Ålshult 30→
Alsleben 37↙
Alstahaug 85↙
Alster 28→
Alston 21←
Alt Landsberg 37→
Alt St Johann 12↗
Alta 85↗
Altamura 58→
Altarejos 70→
Altavilla 58↙
Altdorf, *Switzerland* 47←
Altdorf, *W. Germany* 41↙
Altea 71↘
Altena 40↖
Altenahr 39→
Altenau 37↙
Altenberg 39↗
Altenberge 36←
Altenburg 41↖
Altenfelden 49↖
Altenhagen 37↗
Altenhunden 40↖
Altenkirchen 39↗
Altenmarkt 49←
Altenschlirf 40→
Altensteig 40↙
Alter do Chão 68↘
Altguish Inn 18→
Altheim 48↗
Altıntaş 95↗
Altkirch 46→
Altnaharra 18↗
Alto Adige = Trentino-Alto Adige
Alton 23←
Altrincham 21↙
Altruppin 37↖
Altshausen 47↗
Altstätten 47→
Altuna 29←
Altunoluk 95↖
Altura 71←
Ålund 86↘
Alunda 29→
Alustante 66↙
Alva 18↘
Alvaiázere 68←
Alvalade 72←
Alvaneu 47↘
Älvängen 30↖
Alvares 68→
Alvdal 26↖
Alvdalen 84→
Alverca 68↙
Alvesta 30→
Alveston 22→
Älvho 84↗
Alvignac 51←
Alvik, *Norway* 26↙
Alvik, *Norway* 26↘
Alvito 68↘
Alvkarleby 29↗
Älvkarleö 29↗
Alvor 72←
Alvöy 26↘
Ålvros, *Sweden* 84↗
Ålvros, *Sweden* 84→
Ålvsbacka 28→
Alvsered 30←
Alvsjökytan 28→
Alyth 19↙
Alytus 89↗
Alzano Lombardia 54↖
Alzey 40←
Alzonne 67↖
Amadora 68↙
Åmal 28↙
Amalfi 57↙
Amance 46↗
Amándola 57↖
Amantea 59←
Amarante 63↙
Amareleja 72↗
Amares 63↙
Amárinthos 96↖
Amaroúsion 96↖
Amaseno 57←
Amatrice 57↖
Ambazac 45↙
Ambeláki 96↗
Ambelókambos 96↘
Amberg 41↙
Ambergate 21↙
Ambérieu-en-Bugey 46↙
Ambert 51↗
Ambjörby 28↗
Amble 21↖
Ambleside 20→
Ambleteuse 38↖
Ambleve 39←
Amboise 44→
Ambrières 44↗
Åmdal 27→
Amden 47→
Ameixial 72←
Åmelfot 26↙
Amélia 56↗
Amélie-les-Bains-Palada 67→
Amendolara 58→
Amelinghausen 37↖
Amer 67→
Amerongen 35→
Amersfoort 35→
Amersham 23←
Amesbury 23←
Amfia 96↘
Amfíklia 96←
Amfilokhia 96↙
Amfípolis 93↘
Åmfissa 96←
Amaliás 96↘
Amiens 38←
Amieva 64↗
Amigdhaliá 96←
Amigdhaliés 96↘
Åmli 27→
Amlwch 20↘
Ammanford 22←
Ammarnäs 85↙
Åmmeberg 28↘
Ammerschwihr 46↗
Ammerwald 13←
Amo 26↗
Amorbach 40→
Amorebieta 65↗
Amorgós 95↙
Åmot, *Norway* 26↗
Åmot, *Norway* 27↖
Åmot, *Norway* 27←
Åmot, *Sweden* 29↖
Åmot, *Sweden* 84→
Amotfors 28←
Åmotsdal 27←
Ampezzo 48↘
Amphiareion 96↖
Amplepuis 46↙
Ampthill 23←
Amposta 66↘
Ampudia 64→
Ampuero 65↖
Amriswil 47→
Amsele 86↘
Amsterdam 35←
Amstetten 49↖
Amulree 18↘
Amurrio 65↖
Amusco 65←
An Uaimh 24↗
Ana Sira 27↘
Anadia 68↖
Anáfi 95↙
Anagni 57←
Análipsis, *Greece* 96↙
Análipsis, *Greece* 96↘
Analpládhes 96↘
Anascaul 25↘
Anávissos 96↗
Anavrití, *Greece* 96↙
Anavrití, *Greece* 96→
Ançã 68↖
Ancaster 21↘
Ancenis 44←
Anchuras 69→
Ancião 68←
Ancona 55↙
Ancora 63←
Ancy-le-Franc 46↖
Andalsnes 84↘
Andau 49→
Andebu 27↖
Andeer 47↘
Andelot 46↖
Andelys, les 38↙
Andenes 85↖
Andenne 39←
Andermatt 47↙
Andernach 39→
Andernos 50←
Andersfors 86↘
Anderslöv 30↘
Anderstorp 30→
Andíkira 96←
Andírrion 96↙
Åndissa 95↖
Andoain 65↗
Andora 10↘
Andorra 66↘
Andorra la Vella 67←
Andosilla 65→
Andover 23←
Andoversford 22→
Andraitx 75↙
Andravídha 96↘
Andria 58←
Andrijevica 91↖
Andrítsaina 96↘
Andróni 96↘
Androniánoi 96↖
Andrychów 89↙
Andselv 85↗
Andújar 73↗
Aneby 30↗
Anemodhoúri 96→
Anemokhór 96↘
Anet 38↙
Anga 31↗
Ånge 84↗
Angebo 84↗
Angeja 68↖
Ängelholm, *Sweden* 29↙
Ängelholm, *Sweden* 30↙
Angelókastron, *Greece* 96↙
Angelókastron, *Greece* 96→
Ängelsberg 29←
Angelsfors 29↖
Angelstad 30→
Angera 13↙
Angermünde 88↗
Angers 44→
Ångersjö 84↗
Angerville 45↖
Ängesâ 85↘
Angístri 96→
Angle 22←
Anglés 67→
Anglesola 67←
Anglure 45↗
Angoulême 50↗
Angoumois, reg. 50↗
Angsnäs 29↖
Angsö 29←
Angûés 66→
Anguiano 65→
Aniche 38→
Anifí 96→
Änimskog 28↙
Anizy le-Château 38↘
Anjou, reg. 44←
Ankaran 55↖
Ankarede 84←
Ankarsrum 31↖
Ankarsund 85↙
Anklam 88↗
Anlaby 21↘
Ånn 84←
Anna 71↙
Annaberg-Buchholz 41↗
Annabichl, airport 49↙
Annaburg 37↘
Annalong 24↗
Annan 20↗
Annecy 46↘
Annemasse 46↘
Annestown 25→
Annonay 52↖
Annot 53↙
Annslung 30↗
Annweiler 40↙
Ano-Akhaia 96↙
Ano-Alissós 96↙

Ano Arkhánai 95↙
Ano-Kerásovon 96↙
Ano Klitoría 96↘
Ano Korakiána 96↘
Ano-Liósia 96↖
Ano-Makrinou 96↙
Ano-Palaioxári 96↙
Áno-Próstovas 96↙
Ano-Sélitsa 96→
Ano-Tríkkala 96→
Ano Váthia 96↖
Ano Virón 96↘
Añover de Tajo 70←
Anóyia 96→
Anröchte 36↘
Ans 32↘
Ansbach 40↘
Ansó 66↗
Anstruther 19↙
Anten 30↖
Antequera 73→
Anthíl 96←
Antibes 53↙
Antigüedad 65←
Antoñana 65→
Antrain 43↖
Antrim 24↖
Antrodoco 56↗
Anttola 87↗
Antwerpen 39↖
Anundsjö 84↖
Anwick 21↘
Anzánigo 66→
Anzio 56→
Aoiz 66↖
Aosta 53↖
Apames, anc. site 95→
Apatin 90→
Apeldoorn 35→
Apen 36↖
Apolda 41↖
Apollonía 94→
Äppelbo 28↗
Appelhülsen 36←
Appenino, mts. 54←
Appenweier 40↙
Appenzell 47→
Appingedam 35↗
Appleby 21←
Applecross 18←
Appledore, *England* 22←
Appledore, *England* 23→
Apricena 58↙
Aprilia 56→
Apt 52↘
Apulia = Puglia
Aracena 72→
Araches 12↘
Arad 90↗
Aragoncillo 65↘
Arakhnánon 96→
Arákhova, *Greece* 96↘
Arákhova, *Greece* 96←
Arákhova, *Greece* 96↙
Araksbo 27→
Aram 84↘
Aranda de Duero 65↙
Aranda de Moncayo 65↘
Aranjuez 70←
Aránzazu, Convento de 65↗
Arápis 96↗
Aras de Alpuente 71←
Arbatax 61→
Arbeca 67↙
Arbéost 66↗
Arbesbach 49↖
Arboga 29←
Arbois 46←
Arbon 47→
Arboréa 61←
Arbrå 84↗
Arbroath 19↙
Arbucias 67→
Arc-en-Barrois 46↖
Arcachon 50←
Arce 57←
Arcen 35→
Arceniega 65↖
Arcévia 55↙
Archar 93↙
Archena 74↗
Archidona 73→
Archiestown 19←
Archivel 74↗
Arcidosso 56↖
Arcis 46↖
Arco 54↖
Arco de Baulhe 63↘
Arcos 65←
Arcos de Jalón 65↘
Arcos de la Frontera 73↙
Arcos de las Salmas 71←
Arcos de Valdevez 63←
Ardagh 25↙
Årdal, *Norway* 27↙
Årdal, *Norway* 27→
Årdal, *Norway* 84↘
Ardala 29↙
Ardales 73↙
Årdalstangen 26←
Ardara 24←
Ardavasar 18←
Ardbeg 18↙
Ardead 92↙
Ardee 24↗
Arden 32←
Ardennes, mts. 39←
Ardentes 45←
Ardentinny 18↘
Ardeonaig 18↘
Ardes 51↗
Ardfert 25↙
Ardfinnan 25←
Ardgay 18↗
Ardglass 24↗
Ardgour 18→
Ardhea 93↘
Ardino 93→
Ardisa 66→
Ardleigh 23→
Ardlui 18↘
Ardminish 20↖
Ardmore 25→
Ardnagashel 25↘
Ardres 38↖
Ardrishaig 18↘
Ardrossan 20↗
Åre 84←
Areavaara 85→
Aremark 28←
Aremarksjöen 28←
Arenas de San Juan 70←
Arenas de San Pedro 69→
Arenas del Rey 73↘
Arendal 27→
Arendonk 39↖
Arendsee 37←
Arénys de Mar 67↘
Arénys de Munt 67→
Arenzano 53→
Arépolís 94↘
Ares 63↖
Arévalo 64↘
Arezzo 54↘
Arfará 96↘
Argamasilla de Alba 70↘
Argamasilla de Calatrava 69↘
Arganda 70↖
Arganil 68→
Argegno 13↘
Argelès-Gazost 66↗
Argelès-sur-Mer 67→
Argent 45←
Argenta 54→
Argentan 43←
Argentat 51↖
Argente 66↙
Argentera 53←
Argentiera 61↖
Argentière 13↘
Argentière, l' 52→
Argenton 45↙
Argenton-Château 44→
Argentona 67↘
Árgos 96→
Argos Orestikón 94↖
Argostólian 94←
Arguedas 65→
Arguellite 70↘
Arguis 66→
Argyrokastron = Gjirokastër
Arhult 31←
Århus 32→
Ariano Irpino 58↙
Ariano nel Polèsi 54→
Arija 65↖
Arild 30←
Arinagour 18↙
Aríni 96↘
Ariño 66↘
Áris 96↘
Arisaig 18←
Aristoménis 96↘
Aritzo 61→
Arive 66↖
Ariza 65↘
Årjang 28←
Arjeplog 85↘
Arjona 73→
Arjonilla 73→
Arkelstorp 30↘
Arkhángelos 95↘
Arkhondokhóri 96↙
Arkhánion 96←
Arklow 25↖
Arkösund 29↙
Ärla 29←
Arlanc 51↗
Arlanda 29→
Arlanzón 65←
Arles, *France* 52↙
Arles, *Spain* 67→
Arlon 39←
Arlöv 30↙
Arluno 53↗
Arma di Tággia 10↘
Armacão de Pera 72←
Armadale 18↘
Armagh 24↗
Armamar 63↘
Armeniş 92↘
Arménoi 96↘
Armentera 67→
Armentières 38↗
Armilla 73→
Armot 26↘
Armoy 24↘
Arna 26↘
Arnafjord 26↘
Arnaía 94↗
Årnäs 28↘
Arnay-le-Duc 46←
Arnborg 32↘
Arnedillo 65→
Arnedo 65→
Arnemark 86↘
Årnes 28←
Arnhem 35→
Arnisdale 18→
Arnprior 18↘
Arnsberg 40↖
Arnside 21←
Arnstadt 41↖
Arnswalde = Choszczno
Aroanía 96↘
Aroche 72→
Arolla 46↘
Arolsen 40↖
Arona 53↗
Åros 27↖
Arosa 47→
Arøsund 33←
Arouca 63↙
Arpajon 38↙
Arpaşu de Jos 92→
Arques 38↖
Arques-la-Bataille 38←
Arquillos 73↗
Arraiolos 68↘
Arras 38→
Årre 33↙
Arreau 66↗
Arrens 66↗
Arrifana 68↗
Arrigorriaga 65↗
Arrild 33↙
Arrington 23↖
Arriondas 64↗
Arroba 69↘
Arrochar 18↘
Arromanches 43↖
Arronches 68↘
Arróyabe 65↗
Arroyo de la Luz 69←
Arroyo de San Serván 69↙
Arroyomolinos 69↙
Arroyomolinos de León 72↗
Arruda dos Vinhos 68↙
Ars, *France* 44↙
Års, *Denmark* 32↙
Ars, *Belgium* 39↙
Arsídha 96↗
Arsiè 54↗
Arsiero 54↗
Arsinói 96↘
Arskogen 84↗
Årstad 30←
Arsunda 29↖
Árta, *Greece* 94↖
Artá, *Spain* 75→
Artajona 65→
Artana 71→
Arteaga 65↗
Arteijo 63↖
Artemisía 96→
Arten 48↙
Artenay 45↖
Artern 37↙
Artés 67←
Artesa de Segre 67←
Arth 47←
Arthurstown 33↗
Artix 66↗
Artjärve 87→
Artotína 96↙
Arudy 66↗
Arundel 23↙
Årup, *Denmark* 32↙
Årup, *Denmark* 33←
Arva 24→
Arvån, *Norway* 84↖
Arvån, *Sweden* 86↘
Arvika 28←
Åryd, *Sweden* 30→
Åryd, *Sweden* 30↘
Aryirá 96↙
Aryirádhes 96↘
Arzachena 61↗
Arzacq-Arraziquet 66↗
Arzberg 41←
Arzignano 54↗
Arzúa 63↖
As, *Belgium* 39↖
Aš, *Czechoslovakia* 41←
Ås, *Norway* 27↖
Ås, *Norway* 84→
Ås, *Sweden* 30→
Aså, *Denmark* 32←
Åsa, *Sweden* 30↖
Åsa, *Sweden* 30→
Åsarna 84→
Åsarp 30↗
Asarum 30↘
Åsbro 28↘
Asby 31↖
Ascha 41↘
Aschach 49↖
Aschaffenburg 40→
Aschau 48→
Aschbach 49↖
Ascheberg 36↙
Aschendorf 36↖
Aschersleben 37↙
Ascó 66↘
Ascoli Piceno 57↖
Ascona 47↙
Ascot 23←
Aseda 31←
Åsele 84↖
Åsen 84→
Åsenhöga 30↗
Åsensbruk 28↙
Åseral 27→
Asfeld 38↘
Åsgårdstrand 27↖
Åshammar 29↖
Ashbourne, *England* 21↙
Ashbourne, *Rep. of Ireland* 24↗
Ashburton 22↙
Ashbury 23←
Ashby de la Zouch 23↖
Ashford 23→
Ashington 21↖
Ashkirk 20↗
Ashton under Lyne 21↙
Asiago 54↗
Asikkala 87→
Asin 66←
Ask, *Norway* 26↘
Ask, *Sweden* 30↘
Askeaton 25↙
Asker 27↖
Askern 21↙
Askersund 28↘
Askilje 84↖
Askim 30↖
Åskloster 30←
Askola 87→
Askyoll 26↙
Aslanapa 95↗
Aslested 27←
Åsljunga 30→
Asmansbo 29↖
Asmarka 26↗
Ásola 54↖
Asolo 54↗
Aspang 49→
Aspås 84←
Aspatria 20→
Aspe 71↙
Aspeå 84↖
Asperg 40↙
Aspet 67↖
Aspres-sur-Buëch 52→
Asprópirgos 96↖
Assen, *Netherlands* 35↗
Assen, *Norway* 84←
Ássiros, *Greece* 94↗
Assisi 56↗
Assling 48↖
Ássos 96→
Assumar 68↘
Assy 12↘
Åsta 28↖
Astaffort 50↘
Astakós 96↙
Asten 35↘
Asti 53→
Astillero 65↖
Astipálaia 95↙
Aston Clinton 23←
Astorga 64←
Åstorp 30↙
Åsträsk 86↘
Ástros 96→
Astudillo 65←
Aszód 90↖
Ataki 92↖
Atalaia 68↙
Atalándi 96←
Atalho 68↙
Ataquines 64↘
Atea 65↘
Ateca 65↘
Atessa 57←
Ath 38↗
Athboy 24↗
Athenry 25←
Athens = Athínai
Atherstone 23↖
Athíkia 96→
Athínai 96↗
Athínaion 96→
Athleague 24→
Athlone 24→
Athy 25↖
Atienza 65↘
Atina 57←
Atlingbo 31↗
Atna 26↖
Atnbrua 26↖
Atrå 27←
Ätran 30←
Atrani 11↘
Åtransk 86↘
Atri 57↖
Attendorn 40↖
Attigny 39↙
Attleborough 23↗
Attlebridge 23↗
Attmar 84↗
Attnang-Pucheim 48↗
Atvidaberg 29↙
Atzendorf 37←
Au 47→
Aua 40↗
Aub 40→
Aubagne 52↘
Aubenas 52←
Auberive 46↖
Aubiet 51↙
Aubigny-sur-Nère 45←
Aubin 51←
Auboue 39↙
Aubrac 51→
Aubusson 45↙
Auch 50↘
Auchenblae 19←
Auchencairn 20→
Auchinleck 20↗
Auchterarder 19↙
Auchtermuchty 19↙
Auchtertyre 18→
Audenge 50←
Audierne 42↙
Audincourt 46→
Audlem 21↙
Audnedal 27→
Audruicq 38↖
Audun-le-Roman 39↙
Aue 41↗
Auerbach, *E. Germany* 41←
Auerbach, *W. Germany* 41←
Auffach 13→
Augher 24→
Aughnacloy 24↗
Aughrim, *Rep. of Ireland* 25↖
Aughrim, *Rep. of Ireland* 25←
Augsburg 47↗
Augusta 60↗
Augustenborg 33←
Augustów 89↗
Auktsjaur 85↘
Auldearn 18→
Aulla 54←
Aulléne 62←
Aulnat, airport 51↗
Aulnay 44↘
Aultbea 18↗
Aultnamain Inn 18↗
Ault-Onival 38←
Aulus-les-Bains 67↖
Auma 41↖
Aumale 38←
Aumont 51→
Aunay-sur-Odon 43←
Auneau 45↖
Auneuil 38↙
Auning 32→
Aunis, reg. 44↙
Auñón 70↗
Aups 52↘
Aura 87↘
Auray 42↘
Aurdal 26↗
Aure 84↙
Aurich 36↖
Aurignac 67↖
Aurillac 51→
Aurland 26→
Aurlandsvargen = Aurland
Auron 53←
Auronzo 48↘
Aurskog 28←
Ausejo 65→
Ausmetz 39↙
Aust 22→
Austbö 27←
Austbygda 27←
Austefjord 26↙
Austenå 27→
Austerlitz = Slavkov
Austmannli 27←
Austmarka 28←
Austråt 84↙
Austre Moland 27→
Auterive 67↖
Authon-du-Perche 44↗
Autol 65→
Autreville 46↗
Autun 45→
Auxerre 45↗
Auxi-le-Château 38↙
Auxonne 46←
Auvergne, mts. 51↗
Auzances 45↙
Auzon 51↗
Avafors 85↘
Availles-Limouzine 44↘
Avaldsnes 27↙
Avallon 45→
Avants, les 46↘
Avaträsk 84↖
Avaudden 85↘
Avaviken 85↘
Avebury 22→
Aveiras de Cima 68↙
Aveiro 68↖
Avelgem 38↗
Avellino 58↘
Åvendal 27↘
Aversa 57↙
Avesnes 38→
Avesnes-le-Comte 38←
Avesta 29↖
Avezzano 57←
Aviano 48↘
Aviemore 18→
Avigliana 53↖
Avigliano 58↘
Avignon 52↙
Avila 69↗
Avilés 64↗
Avión 63←
Aviz 68↘
Avize 38↘
Avliótes 96↘
Avlón, *Greece* 96↖
Avlón, *Greece* 96↘
Avlonárion 96↖
Avlum 32↘
Avoca 25↖
Avoch 18→
Avola 60↗
Avonmouth 22→
Avord 45←
Avoriaz 12↘
Avramió 96↘
Avranches 43←
Avratsberg 28↗
Avrig 92→
Avtovac 91←
Ax-les-Thermes 67↖
Axams 13←
Axat 67↗
Axbridge 22→
Axel 35↙
Axelfors 30↗
Axmarsbruk 29↖
Axminster 22↘

Brora 18↗
Brørup 33↙
Brosärp 30↘
Broşteni, *Romania* 92←
Broşteni, *Romania* 92↘
Bröstrud 26→
Brotas 68↘
Brotherton 21↙
Broto 66→
Bröttum 26↗
Brou, *France* 45↖
Brou, *France* 46↙
Brough 21←
Broughshane 24↖
Broughton, *England* 20→
Broughton, *England* 23↖
Broughton, *Scotland* 20↗
Broughty Ferry 19↙
Brouwershaven 35←
Brovst 32↙
Broxburn 19↙
Brozas 68→
Brozzo 54↖
Bruay, *France* 38←
Bruay, *France* 38→
Bruchhausen-Vilsen 36→
Bruchsal 40↙
Bruck, *Austria* 49←
Bruck, *Austria* 48→
Brück, *E. Germany* 37→
Bruck, *W. Germany* 41↙
Brückenau 40→
Bruel 33↗
Bruff 25←
Bruflat 26↗
Brugg 47←
Brugge 38↗
Brulon 44↗
Brumath 39↘
Brummen 35→
Brumunddal 28↖
Brunate 13↘
Brunau 37←
Bruneck 48↙
Brünen 36↙
Brunete 69↗
Brunflo 84↗
Brunico 48↙
Brunkeberg 27←
Brünn, *Czechoslovakia* = Brno
Brunn, *E. Germany* 49↗
Brunnen 47←
Brunsberg 28→
Brunsbüttelkoog 33↘
Bruntál 89↙
Bruree 25←
Brusago 48↙
Brusali 27↘
Brusand 27↘
Brusane 55→
Brusio 47↘
Brusovaca 55↗
Brussel = Bruxelles
Brusson 53↖
Bruton 22→
Bruvik 26↘
Bruxelles 38↗
Bruyères 46↗
Bruzaholm 31↖
Bryggesåk 27→
Bryggja 26↙
Bryn Mawr 22→
Brynamman 22←
Bryne 27↘
Brynilen 85↗
Brynzen 92↖
Brza Palanka 92↘
Brzeg 89←
Brzesko 89↘
Brzeziny 89←
Brzeźnica 89←
Bü Çekmece 93↗
Bua 30←
Buarcos 68←
Buavåg 27↙
Bubwith 21↘
Buccino 58↘
Bucecea 92←
Bucelas 68↙
Bucharest = Bucureşti
Buchau 47↗
Buchboden 47→
Büchen, *W. Germany* 37↖
Buchen, *W. Germany* 40↘
Buchholz 36↗
Büchlberg 41↘
Buchloe 47↗
Buchs 47→
Buchy 38←
Buckden, *England* 21←
Buckden, *England* 23↖
Bückeburg 36→
Buckfastleigh 22↙
Buckhaven and Methil 19↙
Buckie 19←
Buckingham 23←
Bucks Green 23←
Buck's Mills 22↙
Bučovice 89↙
Bucureşti 93←
Bud 84↘
Budafok 90←
Budapest 90←
Budduso 61↗
Bude 22↙
Büderich 36↙
Budeşti 93←
Budia 70↗
Büdingen 40←
Budleigh Salterton 22↘
Búdrio 54→
Budva 91→
Budyne 41→
Bue 27↘
Buenache de Alarcón 70→
Buenavista de Valdavia 64→
Buendía 70↗
Bugarra 71←
Buğdaylı 95↖
Bugeat 51↖
Bugle 22↙
Bugojno 91←
Bugöynes 86↖
Bugue, le 51←
Bühl 40↙
Buhuşi 92→
Builth Wells 22↗
Buitenpost 35↗
Buitrago del Lozoya 65↙
Bujalance 73→
Bujan 63↖
Bujanovac 93↙
Bujaraloz 66↘
Buje 55↖
Bük 49→
Buksnes 85←
Bülach 47←
Bulboki 92↖
Buldan 95→
Bulkava 87↗
Bulken 26↘
Bullange 39→
Bullas 74↗
Bulle 46↘
Bulltofta 30↙
Bulwell 21↙
Bumbeşti Jiu 92↘
Bunbeg 24←
Bunclody 25↖
Buncrana 24←
Bunde, *W. Germany* 36↖
Bünde, *W. Germany* 36→
Bundoran 24→
Bunessan 18↙
Bungay 23↗
Bunic 55→
Bunnahabhainn 18↙
Buñol 71←
Buñola 75←
Buntingford 23←
Buñuel 66←
Buonconvento 56↖
Burbage 23←
Burdujeni 92←
Bureå 86↘
Büren 36↘
Burford 23↘
Burg, *E. Germany* 37←
Burg, *W. Germany* 33→
Burgas 93↖
Burgau 47↗
Burgdorf, *Switzerland* 47←
Burgdorf, *W. Germany* 36→
Burgebrach 40→
Bürgel 41↖
Burgenland, prov. 49→
Bürgenstock 47←
Burggrub 41←
Burgh, *Netherlands* 35←
Burgh, *Scotland* 20↗
Burgh-le-Marsh 21↘
Burgh-on-Bain 21↘
Burghausen 48↗
Burghead 19←
Burgjoss 40→
Burgkunstadt 41←
Burglengenfeld 41↙
Burgo 73↙
Burgos 65←
Burgsalach 41↙
Burgsinn 40→
Burgstädt 41↗
Burgsteinfurt 36←
Burguete 66↖
Burgui 66↖
Burguillos del Cerro 69↙
Burgundy = Bourgogne
Burhaniye 95↖
Burhave 36↗
Burjasot 71←
Burkal 33↙
Burkhardtsdorf 41↗
Burkyrke 29→
Burnfoot 24←
Burnham Market 23↗
Burnham-on-Crouch 23→
Burnham-on-Sea 22→
Burniston 21→
Burnley 21↙
Burntisland 19↙
Burón 64↗
Burravoe 19↗
Burrel 91↗
Burriana 71→
Burry Port 22←
Bursa 95↗
Burseryd 30→
Bürstadt 40←
Burton Agnes 21→
Burton upon Trent 23↖
Burtonport 24←
Burträsk 86↘
Burwash 23↘
Burwell, *England* 21↘
Burwell, *England* 23↗
Burwick 19↖
Bury 21↙
Bury St Edmunds 23↗
Busalla 53→
Busby Stoop 21←
Busca 53←
Bushat 91↗
Bushey 23←
Bushmills 24↖
Busko 89↘
Busnesgrend 27↖
Busquistar 74↙
Bussang 46↗
Busseto 54←
Bussoleno 53↖
Bussum 35→
Busto Arsizio 53↗
Büsum 33↘
Butera 60→
Buttermere 20→
Buttevant 25→
Buttlar 40↗
Buttle 31↗
Buttstädt 41↖
Butzbach 40←
Bützow 33↗
Buvassbrenna 26→
Buvika 84→
Buxtehude 36↗
Buxton 21↙
Buxy 46↙
Buzançais 45←
Buzancy 39↙
Buzău 92→
Buzet 55↖
Byala 93←
Byala Slatina 93↙
Byczyna 89←
Bydgoszcz 89↖
Byfield 23↖
Bygdeå 87↙
Bygdeträsk 86↘
Bygdin 26←
Bygdsiljum 86↘
Bygget 30→
Bygstad 26↙
Bykle 27←
Bylandsfjord 27→
Bylchau 20↘
Byremo 27→
Byringe 29←
Byrkjedal 27↘
Byrkjelo 26↙
Byrness 21↖
Byske 86↘
Bystrzyca Kłodzka 88↘
Bytča 89↙
Bytom 89↙
Bytów 89↖
Byvalla 29↖
Byxelkrok 31↖

Cabañaquinta 64↗
Cabanes 71→
Cabar 55↗
Cabeça Gorda 72←
Cabeceiras de Basto 63↘
Cabeço de Vide 68↘
Cabeza de Buey 69↘
Cabeza la Vaca 72↗
Cabezas del Villar 69↗
Cabezas Rubias 72→
Cabezón de Liebana, *Spain* 64↗
Cabezón de Liebana, *Spain* 65↖
Cabezuela del Valle 69←
Cabourg 43↖
Cabra 73→
Cabra del Santo Cristo 74←
Cabrach 19←
Cábras 61←
Cabrerets 51←
Cabrillas 69↖
Cacabelos 63→
Čačak 91↖
Cáccamo 60←
Cáceres 69←
Cachopa 72←
Čačinci 90→
Cadalen 51↙
Cadalso de los Vidrios 69↗
Cadaqués 67→
Cadaval 68←
Cadca 89↙
Cadelbosco 54←
Cadenábbia 13↘
Cadenet 52↘
Cádiar 74↙
Cadillac 50→
Cádiz 72↘
Cadnam 23↙
Caen 43↖
Caenby Corner 21↘
Caerleon 22→
Caernarvon 20↘
Caerphilly 22→
Caersws 22↗
Caerwent 22→
Çağaş 95↗
Cagli 54↘
Cágliari 61↙
Cagnano Varano 58↙
Cagnes-sur-Mer 53↙
Caherdaniel 25↘
Cahir 25←
Cahirciveen 25↘
Cahors 51←
Caiazzo 57↙
Cairndow 18↘
Cairngorms, mts. 18→
Cairnryan 20↖
Cairnsmore of Carsphairn 20↗
Cáiro Montenotte 53→
Caister 23↗
Caistor 21↘
Căiuţi 92→
Cajarc 51←
Čajetina 91↖
Čajniče 91↖
Cakovec 49↘
Çal 95→
Cala 72→
Cala Ratjada 75→
Calabria, prov. 59←
Calaceite 66↘
Calacuccia 62←
Calaf 67←
Calafat 93↙
Calafell 67↙
Calahonda 14↙
Calahorra 65→
Calais 38↖
Calamocha 66↙
Calamonte 69↙
Calañas 72→
Calanda 66↘
Calangiánus 61↗
Călăraşi 93↖
Calasetta 61↙
Calasparra 74↗
Calatafimi 60↙
Calatañazor 65↘
Calatayud 65↘
Călăţele 92↙
Calatorao 66↙
Calbe 37←
Calcinelli 55↙
Caldas da Rainha 68←
Caldas das Taipas 63↙
Caldas de Bohi 67←
Caldas de Gerez 63↙
Caldas de Malavella 67→
Caldas de Mombuy 67←
Caldas de Reyes 63←
Calderari 60→
Caldicot 22→
Caledon 24↗
Calella, *Spain* 15↖
Calella, *Spain* 15↖
Calenzana 62←
Calenzano 54↙
Calera de León 72↗
Calera y Chozas 69→
Caleruega 65←
Caletta 61↗
Caltavuturo 60←
Calgary 18↙
Çalı 95↗
Cálig 71↗
Călimăneşti 92→
Călineşti 92→
Calitri 58↘
Callac 42←
Callan 25←
Callander 18↘
Callanish 18↖
Callington 22↙
Callosa de Ensarria 71↘
Callosa de Segura 75↖
Calmbach 40↙
Calne 22→
Calonge 67→
Calore 58↙
Caloy, le 50↘
Calpe 71↘
Caltabellotta 60→
Caltagirone 60↗
Caltanissetta 60→
Caltojar 65↘
Calvarrasa de Abajo 64↘
Calver 21↙
Calvi 62↖
Calvine 18→
Calvörde 37←
Calw 40↙
Calzada de Valdunciel 64↘
Calzadilla de los Barros 69↙
Camaiore 54↙
Camáldoli 54↘
Camarasa 67←
Camarena 69→
Camargo 65↖
Camarillas 71↖
Camariñas 63↖
Camaross 25↖
Camarzana de Tera 64←
Camas 73←
Camasnacroise 18↘
Cambados 63←
Camberg 40←
Camberley 23←
Cambil 73→
Cambo 66↖
Camborne 22↙
Cambrai 38→
Cambridge, *England* 22→
Cambridge, *England* 23↗
Cambrils 67↙
Camburg 41↖
Cambuslang 18↘
Camelford 22↙
Camerino 56↗
Camerota 58↘
Camigliatello Silano 59←
Caminha 63←
Caminreal 66↙
Cammarata 60→
Camogli 53→
Camp 25↘
Campanario 69↙
Campania 57↙
Campanet 75←
Campaspero 65↙
Campbeltown 20↖
Campdevánol 67←
Campello 71↘
Câmpeni 92↘
Campiglia Marittima 56↖
Campiglia Soana 53↖
Campile 25↖
Campillo de Altobuey 71←
Campillo de Arenas 73→
Campillo de Llerena 69↙
Campillos 73←
Campione d'Italia 13↙
Campo, *Spain* 66→
Campo, *Portugal* 68↘
Campo de Criptana 70→
Campo de San Pedro 65↙
Campo Maior 68↘
Campo Real 70↖
Campo Tures 48←
Campobasso 58↙
Campobello di Licata 60→
Campomarino 58↗
Campomoro 62←
Camporeale 60↙
Camporrobles 71←
Campos del Puerto 75↙
Camposampiero 54↗
Campotosto 57↖
Camprodón 67→
Camuñas 70←
Çan 95↖
Cañada del Hoyo 70→
Cañada Rosal 73←
Cañadajuncosa 70→
Canadel, le 52↘
Canakkale 95↖
Canale 53←
Canals 71↙
Cañamares 70↗
Cañamero 69←
Canaston Bridge 22←
Cañaveral 69←
Cañaveras 70↗
Canazei 48↙
Cancale 42→
Cancárix 71↙
Cancon 50→
Candarli 95↖
Candas 64↗
Candasnos 66↘
Candé 44←
Candela 58↙
Candeleda 69→
Candia = Iráklion
Cándia Lomellina 53↗
Candin 63→
Canea = Khaniá
Canelli 53→
Canena 73↗
Canes de Senhorim 68↗
Canet de Mar 67→
Canet lo Roig 71↗
Canet-Plage 67↗
Cañete 71←
Cañete de las Torres 73→
Cañete la Real 73↙
Canfranc 66↗
Cangas 63←
Cangas de Nárcea 64↖
Cangas de Onís 64↗
Canha 68↙
Canhestros 72↖
Canicattì 60→
Caniles 74←
Canillo 67←
Canino 56↗
Cañizal 64↘
Canjáyar 74←
Cánnero Riviera 13↙
Cannes 53↙
Cannet, le 10↙
Canneto 54↙
Canneto sull' Oglio 54↖
Cannich 18→
Cannóbio 47↘
Cannock 22↗
Canonbie 20↗
Canosa di Puglia 58→
Cantabrica, Cordillera, mts. 64↖
Cantalapiedra 64↘
Cantalpino 64↘
Cantanhede 68↖
Cantavieja 71↖
Canterbury 23→
Cantillana 73←
Cantalejo 65↙
Cantoria 74→
Cantù 53↗
Canvey 23→
Cany-Barville 43↗
Cáorle 55↖
Cap d'Ail 53↙
Cap-Ferret 50←
Cap Martin 10↙
Caparroso 65→
Capbreton 50↙
Capdella 67←
Capdenac 51←
Capdepera 75→
Capel 23←
Capel Curig 20↘
Capellades 67↙
Capelle-en-Thierache, la 38→
Capendu 67↗
Capestang 67↗
Capistrello 57←
Capo Rizzuto 59←
Capoterra 61↙
Cappamore 25←
Cappercleuch 20↗
Cappoquin 25→
Capráia 62↗
Caprarola 56→

Châtillon-sur-Chalaronne 46↙
Châtillon-sur-Indre 45←
Châtillon-sur-Loire 45←
Châtre, la 45↙
Chatteris 23↗
Chatton 21↖
Chaudes-Aigues 51→
Chaudrey 46↖
Chauffailles 46↙
Chaulnes 38→
Chaumont, *France* 45←
Chaumont, *France* 46↖
Chaumont-en-Vexin 38↙
Chauny 38→
Chauvigny 44↘
Chaux-de-Fonds, la 46→
Chaves 63↘
Chazelles-sur-Lyon 52↖
Cheadle, *England* 21↙
Cheadle, *England* 21↙
Cheb 41←
Chęciny 89↙
Cheddar 22→
Cheekpoint 25↖
Chef-Boutonne 44↘
Cheles 68↘
Chelford 21↙
Chelles 38↙
Chełm 89→
Chełmno 89↖
Chelmsford 23→
Chełmża 89↖
Cheltenham 22→
Chelva 71←
Chemillé 44→
Chemnitz = Karl-Marx-Stadt
Chenérailles 45↙
Chenonceaux 45←
Chepstow 22→
Chera 71←
Cherbourg 43↖
Cherkovitsa 93←
Chernovtsy 92←
Chernyakhovsk 89↗
Chernyshevskoye 89↗
Cherta 66↘
Chertsey 23←
Cherven Bryag 93←
Chesham 23←
Cheshunt 23←
Chesne, le 39↙
Cheste 71←
Chester 20↘
Chester-le-Street 21←
Chesterfield 21↙
Cheval-Blanc 39↘
Chevreuse 38↙
Chexbres 46↘
Cheylard, le 52←
Chiaramonte Gulfi 60↗
Chiaravalle Centrale 59→
Chiari 54↖
Chianciano Terme 56↗
Chiasso 47↙
Chiàvari 53→
Chiavenna 47↘
Chichester 23↙
Chiclana 70↘
Chiclana de la Frontera 73↙
Chieri 53↖
Chiesa 47↘
Chieti 57←
Chilham 23→
Chillón 69↘
Chimay 38→
Chimeneas 73→
Chimishliya 92↗
Chinale 53←
Chinchilla de Monte Aragón 71↙
Chinchón 70←
Chinon 44→
Chióggia 54↗
Chipiona 72↘
Chippenham 22→
Chipping Campden 23←
Chipping Norton 23←
Chipping Sodbury 22→
Chiprana 66↘
Chirivel 74→
Chirk 20↘
Chirnside 19↙
Chirpan 93→
Chişinău = Kishinev
Chişineu Criş 90↗
Chiusa 48↙
Chiusa di Pésio 53←
Chiusi 56↗
Chiva 71←
Chivasso 53↖
Chlumec nad Cidlinou 88↘
Chmielnik 89↘
Chodzież 88↗
Chojna 88↗
Chojnice 89↖
Chojnów 88→
Cholderton 23←
Cholet 44←
Chollerford 21↖
Chomérac 52←
Chomutov 41→
Chop 90↖
Chop Gate 21←
Chorges 52→
Chorley 21↙
Choroszcz 89↗
Chorzele 89↗
Chorzów 89↙
Choszczno 88↗
Christchurch 23↙
Christiansfeld 33↙
Chrudim 88↘
Chrzanow 89↙
Chudleigh 22↙
Chulilla 71←
Chulmleigh 22↙
Chur 47→
Church Stoke 22↗
Church Stretton 22↗
Churchill 22→
Cianciana 60→
Ćićevac 91↖
Cidones 65→
Ciéchanow 89↖
Ciechocinek 89←
Cięjkowice 89↘
Ciempozuelos 70←
Cieszanow 89↘
Cieszyn 89↙
Cieza 71↙
Cifuentes 65↘
Cigales 64→
Cigliano 53↖
Čilipi 16↙
Cillas 65↘
Cilleros 68→
Cîmpia Turzii 92↘
Cîmpina 92→
Cîmpulung 92→
Cîmpulung Moldovenesc 92←
Cinctorres 71↗
Cinderford 22→
Çine 95→
Ciney 39←
Cíngoli 55↙
Cinquefrondi 59→
Cintegabelle 67↖
Cintruénigo 65→
Cioara Doiceşti 92↗
Ciotat, la 52↘
Ciperez 64↙
Ciruqui 65→
Cirat 71←
Cîrbeşti 92↘
Cirencester 22→
Cirey-sur-Vezouz 39↘
Ciria 65↘
Ciriè 53↖
Ciro 59←
Cislău 92→
Cisneros 64→
Cistá 41→
Cisterna di Latina 56→
Cistierna 64↗
Citta del Vaticano 56→
Cita della Pieve 56↗
Città di Castello 54↘
Cittadella 54↗
Cittaducale 56↗
Cittanova 59→
Ciucea 92↙
Ciudad Encantada 70→
Ciudad Real 69↘
Ciudad Rodrigo 69↖
Ciudadela 75→
Ciulniţa 93↖
Ciumeghiu 90↗
Cividale 48↘
Cívita Castellana 56→
Civitanova Marche 55↙
Civitavécchia 56→
Civitella del Tronto 57↖
Civitella Pagánico 56↖
Civray 44↘
Çivril 95→
Clachan, *Scotland* 18←
Clachan, *Scotland* 18↘
Clackmannan 18↘
Clacton on Sea 23→
Cladich 18↘
Clady 24←
Clairvaux-les-Lacs 46↙
Clamecy 45→
Clane 25↖
Claonaig 18↘
Clapham 21←
Clara 25←
Clare 23↗
Clarecastle 25↙
Claregalway 25↙
Claremorris 24↘
Clarholz 36←
Clarinbridge 25↙
Clashmore 25→
Claudy 24←
Claughton 21←
Clausthal Zellerfeld 37↙
Claviere 53←
Clawton 22↙
Clay Cross 21↙
Claydon 23↗
Claye-Souilly 38↘
Clayette, la 46↙
Cleethorpes 21↘
Cleggan 24↘
Cleobury Mortimer 22↗
Cleobury North 22↗
Clères 38←
Clermont 38↙
Clermont-en-Argonne 39↙
Clermont-Ferrand 51↗
Clermont-l'Hérault 51↘
Clervaux 39←
Cléry 45↖
Cles 48↙
Clevedon 22→
Cleveland Tontine 21←
Cleveleys 20→
Cley 23↗
Clifden 24↘
Cliffony 24→
Clifton 21←
Clisson 44←
Clitheroe 21←
Clogh 24↖
Clogh Mills 24↖
Cloghan 25←
Clogheen 25←
Clogher 24→
Clonakilty 25↘
Clonaslea 25←
Clonbur 24↘
Clonco Bridge 25←
Clondalkin 25↖
Clonea 25→
Clonee 24↗
Clones 24→
Clonfert 25←
Clonmacnois 25←
Clonmel 25←
Clonroche 25↖
Clontibret 24↗
Cloonfad 24↘
Clophill 23←
Cloppenburg 36←
Clough 24↗
Clova 19←
Clovelly 22↙
Clovenfords 20↗
Clowne 21↙
Clows Top 22↗
Cloyes 45↖
Cloyne 25→
Cluanie 18→
Cluj 92↙
Clun 22↗
Clunes 18→
Cluny 46↙
Clusaz, le 12↙
Cluse, la 46↙
Cluses 46↘
Clusone 47↘
Clutton 22→
Clydebank 18↘
Clyro 22↗
Cnossos, anc. site 95↙
Coachford 25↘
Coagh 24↖
Coalisland 24↖
Coalville 23↖
Coatbridge 18↘
Çobanisa 95→
Cobh 25→
Coburg 41←
Coca 64↘
Cocentaina 71↙
Cochem 39→
Cochstedt 37←
Cock Bridge 19←
Cockburnspath 19↙
Cockerham 20→
Cockermouth 20→
Codigoro 54→
Codlea 92→
Codogno 53↗
Codróipo 54↖
Coesfeld 36←
Coevorden 35↗
Cofrentes 71←
Cogealac 92↗
Cogeces del Monte 65↙
Coggeshall 23→
Cognac 50↗
Cogne 53↖
Cognin 52↗
Cogolin 52↘
Cogollos 65←
Cogolludo 65↙
Cogollus Vega 73→
Coimbra 68←
Coin 73↘
Cointrin, airport 46↘
Colares 68↙
Colbitz 37←
Colchester 23→
Cold Ashton 22→
Coldingham 19↙
Colditz 41↗
Coldrano 47↘
Coldstream 21↖
Coleford 22→
Coleraine 24↖
Coleshill 23↖
Coligny 46↙
Colindres 65↖
Colintraive 18↘
Collado-Villalba 69↗
Collagna 54←
Colle di Val d'Elsa 54↙
Colle Isarco 48←
Colle Salvetti 54↙
Colleferro 56→
Collesano 60←
Collin 20↗
Collingbourne Ducis 23←
Collingham, *England* 21←
Collingham, *England* 21↘
Cóllio 47↘
Collioure 15↗
Collobrières 52↘
Collon 24↗
Collooney 24→
Colmar 46↗
Colmenar 73↘
Colmenar de Oreja 70←
Colmenar Viejo 70↖
Colmonell 20↖
Colne 21←
Cologna Veneta 54↗
Cologne 51↙
Colombier 46→
Colombey-les-Belles 46↗
Colomera 73→
Colonna, la 54↙
Colsterworth 23↖
Coltishall 23↗
Colunga 64↗
Colwell 21↖
Colyton 22↘
Comácchio 54→
Comăneşti 92→
Comarruga 15←
Combe Martin 22←
Comber 24↖
Combourg 43↖
Combronde 45↘
Comeglians 48↘
Comillas 65↖
Comiso 60↗
Commercy 39↙
Como 47↙
Cómpeta 73↘
Compiègne 38↘
Comporta 68↙
Comps 52↘
Comrie 18↘
Concarneau 42↙
Conches 43→
Condat-en-Féniers 51↗
Condé 38→
Condé-sur-Noireau 43←
Condeixa-a-Nova 68←
Condom 50↘
Condrieu 52↖
Conegliano 54↗
Conflans 38↙
Conflans-en-Jarnisy 39↙
Confolens 44↘
Cong 24↘
Congesbury 22→
Congleton 21↙
Conil 73↙
Coningsby 21↘
Conisbrough 21↙
Coniston 20→
Coniston Water 20→
Connel 18↘
Connerre 44↗
Conon Bridge 18→
Conques, *France* 51→
Conques, *Spain* 67←
Conques, *France* 67↗
Conquet, le 42←
Conquista 69↘
Conselve 54↗
Consenvoye 39↙
Consett 21←
Constance, Lake = Boden-See
Constância 68←
Constanţa 93↖
Constanti 67↙
Constantina 73←
Constantinople = İstanbul
Consuegra 70←
Contamines, les 46↘
Contin 18→
Contis-Plage 50←
Contres 45←
Contrexéville 46↗
Conty 38←
Convento de la Rábida 72→
Conversano 58←
Conway 20↘
Conwil Elvet 22←
Cookstown 24↖
Coolham 23↙
Coombe Bissett 22↘
Cootehill 24→
Copenhagen = København
Copertino 58↗
Copparo 54→
Copplestone 22↙
Copşa Mică 92→
Corabia 93←
Corato 58←
Corbeil 38↙
Corbie 38←
Corbigny 45→
Corbridge 21↖
Corby 23↖
Corby Glen 23↖
Corconte 65↖
Corcubión 63↖
Cordes 51←
Córdoba 73→
Cordovilla de Lácara 69↙
Corella 65→
Coreses 64↘
Corfe Castle 22↘
Corfu = Kérkira
Corgo 63↗
Coria 69←
Coria del Rio 73←
Corigliano 59←
Corinth = Kórinthos
Coripe 73↙
Coristanco 63↖
Cork 25→
Corleone 60←
Corleto Perticara 58→
Çorlu 93↗
Cormano 53↗
Cormeilles 43↗
Corna 47↘
Cornago 65→
Cornellana 64↖
Cornhill 19←
Cornhill-on-Tweed 21↖
Corniglio 54←
Cornimont 46↗
Corniolo 54↘
Cornudella 67↙
Corofin 25↙
Çorovodë 91↗
Corpach 18→
Corps 52→
Corral de Almaguer 70→
Corral de Calatrava 69↘
Corrales 64↘
Corre 46↗
Corredoiras 63↗
Corréggio 54←
Corrèze 51↖
Corrie 20↖
Corse, I. 62←
Corsica = Corse
Corsico 53↗
Corte 62←
Cortegada 63←
Cortegana 72→
Cortemilia 53→
Cortés 66←
Cortes de Aragón 66↙
Cortes de Baza 74←
Cortes de la Frontera 73↙
Cortes de Pallas 71←
Cortina d'Ampezzo 48↙
Corton 23↗
Cortona 54↘
Coruche 68↙
Corunna = La Coruña
Corvara in Badia 48↙
Corwen 20↘
Coryton 23→
Cosenza 59←
Cosham 23↙
Coshieville 18↘
Cosne 45→
Cosne-d'Allier 45↘
Cossonay 46↘
Costa da Caparica 68↙
Costa Nova 68↖
Costelloe 25↙
Costeşti 92→
Costigliole Saluzzo 53←
Costock 23↖
Coswig, *E. Germany* 37→
Coswig, *E. Germany* 41↗
Côte d'Azur, airport 53↙
Côte-St-André, la 52↗
Cottanello 56↗
Cottbus 88→
Couches 46←
Couço 68↙
Coucy-le-Château-Auffrique 38↘
Couhé-Vérac 44↘
Couiza 67↗
Coulange-la-Vineuse 45→
Coulanges-sur-Yonne 45→
Coulommiers 38↘
Coulonges 44↙
Coulport 18↘
Coupar Angus 19↙
Cour-Cheverny 45←
Courcelles-Chaussy 39↘
Courcheval 52↗
Courmayeur 53↖
Courpière 51↗
Cours 46↙
Coursan 67↗
Courseulles 43↖
Courson 45→
Courtacon 38↘
Courtenay 45↗
Courtine, la 51↖
Courtisols 39↙
Courtmacsherry 25→
Courtown Harbour 25↖
Courville 45↖
Coutainville 43←
Coutances 43←
Coutras 50→
Couvet 46→
Couvin 39←
Covadonga 64↗
Covarrubias 65←
Cove 18↘
Coventry 23↖
Covilha 68↗
Cowbit 23↖
Cowbridge 22→
Cowdenbeath 19↙
Cowes 23↙
Cowfold 23↙
Coylton 20↗
Coylumbridge 18→
Cózar 70↘
Cozes 50↗
Crabbs Cross 22↗
Crackington Haven 22↙
Cracow = Kraków
Craigellachie 19←
Craighouse 18↙
Craignure 18↙
Crail 19↙
Crailsheim 40↘
Craiova 93↙
Cranagh 24←
Cranborne 22↘
Cranbrook 23→
Cranleigh 23←
Crans 46↘
Cranshaws 19↙
Craon 44↖
Craonne 38↘
Craponne-sur-Arzon 52↖
Crarae 18↘
Crasna 92↗
Crathie 19←
Crato 68→
Craughwell 25←

Fjaler 84↘
Fjalkinge 30↘
Fjällbacka 28↙
Fjällnäs 84→
Fjällsjö 84↖
Fjärås 30↖
Fjell 26↘
Fjellerup 32→
Fjellheim 26↗
Fjellstrud 28←
Fjerritslev 32↙
Fjöle 28←
Fjone 27←
Fjugesta 28→
Flå, *Norway* 26↗
Flå, *Norway* 84→
Fladungen 40→
Flagavik 30↗
Flaikhórion 96→
Flaine 12↘
Flakstad 85←
Flåm 26→
Flammersfeld 39↗
Flanders = Flandre
Flandre, reg. 38↖
Flärke 87↙
Flassón 94↗
Flatabo 26↘
Flatåker 26→
Flateland 27←
Flatford Mill 23→
Flatöydegarden 26↗
Flatråker 27↙
Flattnitz 49←
Flawil 47→
Flèche, la 44→
Fleda 29↙
Fleet 23←
Fleetmark 37←
Fleetwood 20→
Flekke 26↙
Flekkefjord 27↘
Flen 29↙
Flensburg 33↙
Flers 43←
Flesburg 27↖
Flesland 26↘
Fleurance 50↘
Fleurier 46→
Fleury 43↖
Fleury-sur-Andelle 38↙
Flieden 40→
Flikka 27↘
Flims 47→
Flimwell 23→
Flinsberg 40↗
Flint 20↘
Fliseryd 31←
Flix 66↘
Flixecourt 38←
Flize 39←
Floby 30↗
Flockton 21↙
Floda, *Sweden* 28↗
Floda, *Sweden* 30↖
Flodden Field 21↖
Flöha 41↗
Flóka 96↘
Flora 26↙
Florac 51→
Florenville 39←
Floreshty 92↖
Florídia 60↗
Flórina 94↖
Florvag 26↘
Flotte, la 44↙
Flottsund 29→
Flouch 21↙
Fluberg 26↗
Flumet 46↘
Fluminimaggiore 61↙
Flums 47→
Foča 91←
Fochabers 19←
Focșani 92↗
Fogdö 29←
Foggia 58↙
Föglö 87↘
Fohnsdorf 49←
Foiano di Chiana 54↘
Foix 67↖
Foix, reg. 67↖
Fokstua 26↖
Foldereid 84←
Folelli 62←
Folgoet, le 42←
Folie, la 46←
Foligno 56↗
Folkestone 23→
Folkingham 21↘
Follandsvangen 26↖
Folldal 84↘
Follingbo 31↗
Föllinge 84↖
Foltești 92↗
Fon 27↖
Fond de France 52↗
Fondi 57↙
Fondo 48↙
Fonelas 74←
Fonfría 64↙
Fonn 26↙
Fonnes 26↘
Fonni 61→
Fons 52↙
Fonsagrada 63↗
Font-Romeu 67←
Fontaine 45↖
Fontainebleau 45↖
Fontane 53←
Fontanélice 54→
Fonte Blanda 56↖
Fontenay-le-Compte 44↙
Fontenay-Trésigny 38↘
Fontevrault-l'Abbaye 44→
Fontfroide 67↗
Fontiveros 64↘
Fonyód 90→
Fonz 66→
Fonzaso 48↙
Föppolo 47↘
Föra 31←
Forbach, *France* 39↘
Forbach, *W. Germany* 40↙
Forcall 66↘
Forcalquier 52↘
Forcarey 63←
Forchheim 41←
Ford 18↘
Förde, *Norway* 26↙
Förde, *Norway* 27↙
Fördesfjord 27↙
Fordham 23↗
Fordingbridge 22↘
Fordon 89↖
Fordoun 19←
Fore 85←
Forfar 19↙
Forges-les-Eaux 38←
Forio 57↙
Forli 54→
Formby 20↘
Formerie 38←
Formia 57↙
Formígine 54←
Formofoss 84←
Fornåsa 29↙
Fornebu 27↖
Fornelli 61↖
Fornélls 75→
Forno Alpi Gráie 53↖
Forno d'Allione 47↘
Forno di Zoldo 48↙
Fornos 63↘
Fornos de Algodres 68↗
Fornovo di Taro 54←
Förre 27↙
Forres 19←
Fors, *Sweden* 29↖
Fors, *Sweden* 84↗
Forsa, *Sweden* 84↗
Forsa, *Norway* 85←
Forsand 27↘
Forsba 29↖
Forsbacka 29↖
Forserum 30↗
Forshaga 28→
Forsheda 30→
Forshem 28↘
Förland 27→
Formentera, island 74↘
Forsinard 18↙
Förslövsholm 30←
Forsmark 29↗
Forsmo 84↖
Forsnäs 85↘
Forssa 87→
Forst 88→
Fort Augustus 18→
Fort George 18→
Fort l'Ecluse 46↘
Fort-Mahon 38←
Fort William 18→
Fortanete 71↖
Forte dei Marmi 54↙
Forth 18↘
Forth, river 18↘
Fortingall 18↘
Fortrose 18→
Fortun 84↘
Fortuna 74↗
Fortuneswell 22↘
Foruby 30→
Forsvik 28↘
Fos 67↖
Fosnes 84←
Fossano 53←
Fossheim 26↗
Fossli 26→
Fossombrone 54↘
Fougères 44↖
Four Crosses 20↘
Fourchambault 45→
Fourmies 38→
Fournels 51→
Fours 45→
Fousseret, le 67↖
Fowey 22↙
Foxen 28←
Foxford 24↘
Foyers 18→
Foynes 25↙
Foz 63↗
Foz do Arelho 68←
Foz do Giraldo 68→
Frabosa Soprana 53←
Fraddon 22↙
Frafjord 27↘
Fraga 66↘
Frailes 73→
Fraize 46↗
Framlingham 23↗
Främlingshem 29↖
Frammersbach 40→
Frampol 89↘
França 63→
Francardo 62←
Francavilla 58↗
Francavilla al Mare 57↖
Frändefors 28↙
Franeker 35↗
Frangísta 96↙
Frangy 46↘
Frankenberg, *W. Germany* 40↖
Frankenberg, *E. Germany* 41↗
Frankenthal 40←
Frankfurt 40←
Frankfurt an der Oder 88→
Frant 23→
Frantiskovy Lazne 88↙
Fraserburgh 19←
Frashër 94↖
Frasne 46→
Frastanz 47→
Frauenfeld 47←
Frauenkirchen 49→
Frechen 39↗
Frechilla 64→
Freckenhorst 36←
Fredensborg 32↗
Fredericia 33←
Frederikshavn 32←
Frederikssoord 35↗
Frederikssund 32↗
Frederiksværk 32↗
Fredrika 84↖
Fredriksberg 28→
Fredrikshamn = Hamina
Fredrikstad 28←
Fredros 28←
Fregenal de la Sierra 72↗
Fregene 56→
Frei 84↘
Freiberg 41↗
Freiburg, *W. Germany* 36↗
Freiburg, *W. Germany* 47↖
Freila 74←
Freilassing 48→
Freilingen 39→
Freising 48↖
Freistadt 49↖
Freital 41↗
Freixianda 68←
Freixo de Espada á Cinta 63↘
Fréjus 53↙
Frekhaug 26↘
Fremdingen 40↘
Frenchpark 24→
Freren 36←
Freshford 25←
Freshwater 23↙
Fresnay 44↗
Fresne-St-Mamès 46→
Fresnes-en-Woëvre 39↙
Fresno Alhándiga 69↖
Fresvik 26→
Frétigney 46→
Frettes 46←
Freudenstadt 47↖
Frévent 38←
Freyburg 41↖
Freyenstein 37↗
Freystadt 41↙
Freyung 41↘
Fribourg 46→
Fridafors 30→
Fridaythorpe 21→
Fridingen 47↖
Friedberg, *W. Germany* 40←
Friedberg, *W. Germany* 48↖
Friedberg, *Austria* 49→
Friedland, *W. Germany* 36↘
Friedland, *E. Germany* 37↗
Friedrichroda 40↗
Friedrichshafen 47→
Friedrichskoog 33↘
Friedrichstedt 33↘
Friedrichsthal 39↘
Friesach 49←
Friesack 37→
Friesoythe 36↖
Frigiliana 73↘
Friisbua, *Norway* 26↖
Friisbua, *Norway* 26←
Frillesås 30↖
Frinton 23→
Friockheim 19↙
Friol 63↗
Fristad 30↗
Frithville 21↘
Fritsla 30↖
Fritton 23↗
Fritzlar 40↗
Froan 84↙
Frodsham 21↙
Frogel 31↗
Frogn 27↖
Frohburg 41↖
Froland 27←
Frombork 89↖
Frome 22→
Fromista 65←
Fronteira 68↘
Frontenay 44↘
Frontenhausen 41↙
Frontignan 51↘
Fronton 51↙
Frösaråsen 28↗
Fröseke 31←
Frosinone 57←
Fröskog 28↙
Froslida 30←
Frosolone 57←
Frosta 84←
Frösthult 29←
Frøstrup 32↙
Frostviken 84←
Frosunda 29→
Frousína 96→
Frovi 29←
Fröystul 27←
Fruges 38←
Frumușica 92←
Frutigen 47↙
Frydek Mistek 89↙
Frydenlund 26↗
Fryvollen 26↖
Fucécchio 54↙
Fucine 48↙
Fuencaliente 69↘
Fuencarra 70↖
Fuendejalón 65→
Fuendetodos 66↙
Fuengirola 73↘
Fuenlabrada de los Montes 69↘
Fuenmayor 65→
Fuensanta de Martos 73→
Fuente Alamo 71↙
Fuente-Alamo de Murcia 74→
Fuente de Cantos 72↗
Fuente de Pedro Naharro 70→
Fuente del Arco 73↖
Fuente del Maestre 69↙
Fuente el Fresno 70←
Fuente Encarroz 71↘
Fuente la Higuera 71↙
Fuente-Obejuna 69↙
Fuente Palmera 73←
Fuentealbilla 71←
Fuentecén 65↙
Fuenteguinaldo 69↖
Fuentelapeña 64↘
Fuentepinilla 65↘
Fuenterrabia 65↗
Fuentes Claras 66↙
Fuentes de Andalucia 73←
Fuentes de Ayódar 71←
Fuentes de Ebro 66↘
Fuentes de León 72↗
Fuentes de Nava 64→
Fuentes de Oñoro 68↗
Fuente de Piedra 73→
Fuentes de Valdepero 64→
Fuentesaúco 64↘
Fuerte del Rey 73→
Fügen 13→
Fuglebjerg 33↖
Fuhrberg 36→
Fulda 40→
Fulpmes 48←
Fulunäs 84→
Fumay 39←
Fumel 51←
Funäsdalen 84→
Funbo 29→
Fundao 68→
Fundulea 93←
Funen = Fyn
Funusund 29→
Funzie 19↗
Furadouro 63↙
Furnace 18↘
Fürnitz 48↘
Fürstenau 36←
Fürstenberg 37↗
Fürstenfeld 49→
Fürstenfeldbruck 48↖
Fürstenwalde 88→
Furstenwerder 37↗
Fürth, *W. Germany* 40←
Fürth, *W. Germany* 41↙
Furth im Wald 41↘
Furtwangen 47↖
Furulund 30↙
Fusa 26↘
Fusch 48→
Fusio 47↙
Füssen 47→
Füstenberg 40↖
Fustiñana 66←
Füzesabony 90↖
Fuzeta 72←
Fyn, island 33←
Fyresdal 27←
Fyvie 19←

Gaal 49←
Gabarret 50↘
Gabia la Grande 73→
Gabicce Mare 11↗
Gabrovo 93←
Gacé 43→
Gacilly, la 42↘
Gacko 91←
Gäddede 84←
Gadebusch 33→
Gadmen 47↙
Gádor 74↙
Găești 92→
Gaeta 57↙
Gaggenau 40↙
Gagnef 28↗
Gaildorf 40↘
Gailey 22↗
Gaillac 51↙
Gaillon 38↙
Gainford 21←
Gainsborough 21↘
Gairloch 18↗
Gairlochy 18→
Gajary 49↗
Gålå 26↖
Galanito 85↗
Galanta 90←
Galapagar 69↗
Galar 65↗
Galaroza 72→
Galashiels 21↖
Galatás, *Greece* 96↗
Galatás, *Greece* 96→
Galați 92↗
Galatina 58↗
Galátone 58↗
Galatz = Galați
Galaxídhion 96←
Galbally 25←
Galdhöppigen, mt. 26←
Galera 74←
Galeria 62←
Galisteo 69←
Gallarate 53↗
Gallardon 45↖
Gällared 30←
Gállego 66→
Galleno 54↙
Galliate 53↗
Gallinge 30↖
Gallípoli, *Italy* 58↗
Gallipoli = Gelibolu
Gällivare 85→
Gällö 84↗
Gällstad 30↗
Gallur 66←
Galson 18↖
Galston 20↗
Galten 85↗
Galterud 28↖
Galtström 84↗
Galtür 47→
Galve de Sorbe 65↙
Galveias 68↘
Gálvez 69→
Galway 25↙
Gamaches 38←
Gámbara 54↖
Gambárie 59↘
Gamla Uppsala 29→
Gamleby 31↖
Gamlingay 23↖
Gammelstad 86↘
Gammertingen 47↗
Gamvik 86↖
Gan 66↗
Ganacker 41↘
Ganddal 27↘
Ganderkesee 36↗
Gandesa 66↘
Gandia 71↘
Ganges 51↘
Gangi 60←
Gangkofen 48↗
Ganllwyd 22↖
Gannat 45↘
Gänserndorf 49↗
Gap 52→
Gara Khitrino 93↖
Garboldisham 23↗
Garclaz 69←
Garda 54↖
Garda, Lago di, lake 54↖
Gårdby 31←
Gardelegen 37←
Gardenstown 19←
Gardermoen 28↖
Gardhíki 96↙
Garding 33↘
Gardone Riviera 54↖
Gardsjö 29↖
Gardsjönas 85↙
Gärdslösa 31←
Gårdstånga 30↘
Garelochhead 18↘
Garéssio 53←
Gargaliánoi 96↘
Gargazo 62↙
Gargellen 47→
Gargia 86↗
Gargilesse 45↙
Gargnano 54↖
Gargnäs 85↘
Garlasco 53↗
Garlieston 20→
Garlin 66↗
Garmisch-Partenkirchen 48←
Garmo 26→
Garmouth 19←
Garnlakarleby = Kokkola
Garonne, river 50→
Garpenberg 29↖
Garphyttan 28→
Garpnytan 28→
Garraf 67↙
Garrafe de Torio 64→
Garray 65→
Garrel 36↖
Garrison 24→
Garrovillas 69←
Garrucha 74→
Garsås 28↗
Gärsnäs 30↘
Garstang 21←
Garstedt 33→
Garsten 49↖
Garth 22↗
Garthmyl 22↗
Garthus 26↗
Gartocharn 18↘
Gartow 37↖
Gartz 88↗
Garvagh 24↖
Garve 18→
Garwolin 89→
Garynahine 18↖
Gåsborn 28→
Gascogne, reg. 50↘
Gascueña 70↗
Gasny 38↙
Gassin 10←
Gastoúni 96↘
Gastoúr 96↘

Gastropol 63↗
Gata 69↖
Gata de Gorgas 71↘
Gatehouse-of-Fleet 20↗
Gateshead 21↖
Gatten 32↙
Gatteo a Mare 11↗
Gattinara 53↗
Gatwick, airport 23←
Gau-Algesheim 40←
Gaucin 73↙
Gaukheihytta 27→
Gaular 84↘
Gaupen 26↗
Gaupne 26←
Gausdal 26↖
Gautefall 27←
Gauting 48↖
Gavá 67↙
Gavaloú 96↙
Gavardo 54↖
Gavarnie 66↗
Gavi 53→
Gavião 68→
Gavirate 47↙
Gävle 29↖
Gavray 43←
Gavrion 94→
Gavolimni 96↙
Gavunda 28↗
Gäxsjö 84↖
Gaydon 23↖
Gayton 23↗
Gbely 49↗
Gdańsk 89↖
Gdynia 89↖
Gea de Albarracin 71↖
Geashill 25←
Geaune 50↘
Gedern 40→
Gediz 95↗
Gedser 33↖
Geel 39↖
Geertruidenberg 35←
Geesthact 37↖
Gefrees 41←
Gehren 41↖
Geijersholm 28→
Geilenkirchen 39↖
Geilo 26→
Geiranger 26←
Geiselhöring 41↙
Geisenfeld 41↙
Geisingen 47↖
Geisling 41↙
Geislingen 40↘
Geiteryggen 26→
Geithain 41↗
Geithus 27↖
Gela 60→
Geldermalsen 35→
Geldern 39↗
Geldrop 35↘
Gelibolu 95↖
Gelnhausen 40→
Gelsa 66↘
Gelsdorf 39→
Gelsenkirchen 36↙
Gelves 73←
Gembloux 39←
Gémenos 52↘
Gemert 35→
Gemlik 95↗
Gemona 48↘
Gémozac 50↗
Gemünd 39→
Gemünden, *W. Germany* 39→
Gemünden, *W. Germany* 40→
Genarp 30↘
Genazzano 56→
Gencay 44↘
General Toshevo 93↖
Generalisimul Suverov 92↗
Generalski Stol 55↗
Genevad 30←
Geneva = Genève
Geneva, Lake of = Léman, Lac
Genève 46↘
Gengenbach 47↖
Gennep 35→
Gennes 44→
Génova 53→
Gent 38↗
Genthin 37→
Georgsheil 36↖
Gera 41↖
Geraardsbergen 38↗
Gérardmer 46↗
Geras 49↗
Gerbéviller 46↗
Gerbini 60↗
Gerbstedt 37↙
Gerena 73←
Gérgal 74←
Gerlos 48←
Germade 63↗
Germain-du-Bois 46↙
Germay 46↖
Germencik 95←
Germersheim 40↙
Gernrode 37↙
Gernsbach 40↙
Gernsheim 40←
Gerola Alta 47↘
Gerolstein 39→
Gerolzhofen 40→
Gerona 67→
Gerovo 55↗
Gerri 67←
Gersau 47←
Gersfeld 40→
Gerstetten 40↘
Gertsa 92←
Gerusshamn 26↘
Gesäter 28↙
Geseke 36↘
Gessertshausen 47↗
Gesunda 28↗
Geta 87↘
Getafe 70↖
Getinge 30←
Gets, les 46↘
Gettorf 33→
Gevelsberg 40↖
Gevgelija 93↘
Gex 46↘
Geyiklı 95↖
Geyve 95↗
Gföhl 49↖
Ghedi 54↖
Gheorgheni 92→
Gherla 92↙
Ghilarza 61←
Ghimeș Făget 92→
Ghisonaccia 62←
Ghisoni 62←
Giarre 60↖
Giat 51↗
Gibellina 60↙
Gibostad 85↗
Gibraleón 72→
Gibraltar 73↙
Gideå 87↙
Giebelstadt 40→
Gieboldehausen 37↙
Gien 45←
Giengen 40↘
Giens 52↘
Giermundshamn 26↘
Giessen 40↖
Gieten 35↗
Giffnock 18↘
Gifford 19↙
Gifhorn 37←
Gigaro 10←
Giglio Porto 56↖
Gignac 51↘
Gijón 64↗
Giláu 92↙
Gildeskal 85←
Gilleleje 32↗
Gillhov 84↗
Gillingham, *England* 22↘
Gillingham, *England* 23→
Gillstad 28↙
Gilserberg 40↖
Gilsland 21↖
Gilwern 22→
Gimo 29↗
Gimont 51↙
Gimsöy 85←
Ginzling 48←
Ginzo de Limia 63→
Gioia del Colle 58→
Gióia Táuro 59↘
Giornico 47↙
Giraltovce 89↘
Girecourt-sur-Durbion 46↗
Girifalco 59←
Giromagny 46→
Gironella 67←
Gironville-sous-les-Côtes 39↙
Girstad 29↙
Girvan 20↗
Gisburn 21←
Gislaved 30→
Gislinge 32↗
Gisors 38↙
Gissi 57←
Giubasco 13↘
Giulianova 57↖
Giurgeni 92↗
Giurgiu 93←
Gíve 32↘
Givet 39←
Givors 52↖
Givry 46←
Givry-en-Argonne 39↙
Gizeux 44→
Gizycko 89↗
Gjelten 26↖
Gjendesheim 26←
Gjengedal 26↙
Gjerde 26←
Gjerstad 27↗
Gjersvik 84←
Gjirokastër 94↖
Gjövdal 27→
Gjovik 26↗
Gjuvvasshytta 26←
Gladbeck 36↙
Gladenbach 40↖
Glamis 19↙
Glamoc 91←
Glamsbjerg 33←
Glandore 25↘
Glandorf 36←
Glanshammar 29←
Glanworth 25→
Glarus 47←
Glasbury 22→
Glasgow 18↘
Glashiltte 41↗
Glaslough 24↗
Glastonbury 22→
Glatz = Kłodzko
Glauchau 41↖
Glava 28←
Glavaglasbruck 28←
Gleidorf 40↖
Gleisdorf 49→
Gleiwitz = Gliwice
Glemmen 28←
Glen 24←
Glenamoy 24↘
Glenarm 24↖
Glenavy 24↖
Glenbarr 20↖
Glenbeigh 25↘
Glenbervie 19←
Glenbrittle 18←
Glencaple 20↗
Glencar 25↘
Glencarse 19↙
Glencoe 18↘
Glencolumbkille 24←
Glendalough 25↖
Glenelg 18→
Glenfarg 19↙
Glenfarne 24→
Glenfinnan 18→
Glengarriff 25↘
Glenisla 19←
Glenluce 20↖
Glenmaye 20→
Glenrothes 19↙
Glenties 24←
Gletness 19↗
Gletsch 47↙
Glifádha 96↗
Glimåkra 30→
Glin 25↙
Glina 90↘
Glinojeck 89←
Glinton 23↖
Glissjöberg 84→
Glitterheim 26←
Glittertind, mt. 26←
Gliwice 89↙
Gllavë 91↗
Glogau = Głogów
Gloggnitz 49→
Głogów 88→
Glogówek 89↙
Glomfjord 85↙
Glomma, river 84→
Glommen 30↘
Glommerstrask 85↘
Gloppen 26↙
Glossop 21↙
Glostrup 32↗
Gloucester 22→
Głowno 89←
Głubczyce 89↙
Glücksburg 33↙
Glückstadt 33↘
Glumslöv 30↙
Glyboka 92←
Glyfada 96↗
Glyn-Ceiriog 20↘
Glyn Neath 22←
Glyncorrwg 22←
Gmund, *W. Germany* 48←
Gmünd, *Austria* 48↘
Gmund, *Austria* 49↖
Gmunden 48↗
Gnarp 84↗
Gnesta 29↘
Gniew 89↖
Gniezno 89←
Gnissau 33→
Gnjilane 91↖
Gnoien 33↗
Gnosjö 30↗
Goathland 21→
Gobowen 20↘
Goch 35→
Godalming 23←
Godegard 28↘
Godelheim 36↘
Goderville 43↗
Godmanchester 23↖
Gödöllö 90↖
Godovič 49↙
Godstone 23←
Goes 35↙
Goirle 35↘
Góis 68→
Góito 54↖
Goizueta 65↗
Gokels 33↘
Gökhem 30↗
Göktepe 95→
Gol 26↗
Golada 63→
Gölcük = Etili
Gołdap 89↗
Goldberg 37↗
Goldegg 13↖
Goldelund 33↙
Golden Sands 93↖
Goldenstedt 36→
Goldrein 48↙
Goleen 25↘
Golegã 68←
Goleniów 88↗
Golfe-Juan 53↙
Golfo Aranci 61↗
Gollin 37↗
Golling 48→
Gölmarmara 95→
Gölpazarı 95↗
Golpejas 64↘
Gols 49↗
Golspie 18↗
Gołuli 89↖
Golzow 37→
Gómara 65↘
Gömeç 95↖
Gomecello 64↘
Gomes Aires 72←
Gommern 37←
Gondomar, *Spain* 63←
Gondomar, *Portugal* 63↙
Gönen 95↖
Goniądz 89↖
Gónnos 94↗
Goodwick 22←
Goole 21↘
Goor 35→
Göppingen 40↘
Goppollen 26↖
Gor 74←
Góra 88→
Góra Kalwaria 89→
Goražde 91←
Gordes, *France* 52↘
Gördes, *Turkey* 95↗
Gordola 47↙
Gordon 21↖
Gordoncillo 64→
Gordonstown 19←
Goresbridge 25↖
Gorey 25↖
Gorgonzola 53↗
Gorice 91↗
Gorinchem 35←
Goring 23←
Göringen 84↗
Goritzá 96→
Gorleston 23↗
Gørlev 33←
Gorlice 89↘
Görlitz 88→
Gornje Jelenje 55↗
Gornji Milanovac 91↖
Gornji Vakuf 91←
Gorodenka 92←
Gorodok 89↘
Górowo Iławeckie 89↖
Gorran Haven 22↙
Gorron 44↖
Gorssel 35→
Gort 25↙
Gortahork 24←
Gorteen 24→
Gortin 24←
Gorzów Wielkopolski 88→
Gosberton 21↘
Gosforth 20→
Goslar 37←
Gospic 55→
Gosport 23↙
Gossensass 48↙
Gössnitz 41↖
Gostivar 91↗
Gostyn 89←
Gostynin 89←
Göta 30↖
Göteborg 30↖
Götene 28↘
Gotha 40↗
Gothem 31↗
Gotland, island 31↗
Götlunda 29←
Göttingen, *W. Germany* 36↘
Göttingen, *W. Germany* 40↖
Gottne 85↖
Gottskär 30↖
Gotse Delchev 93↘
Götzendorf 49↗
Götzis 47→
Gouda 35←
Goudhurst 23→
Gouffre-de-Padirac 51←
Goumaíoi 96↙
Gouménissa 93↘
Goúra 96→
Gourdon, *Scotland* 19←
Gourdon, *France* 51←
Gourdon, *France* 53↙
Gourin 42←
Gournay-en-Bray 38↙
Gourock 18↘
Gouryiá 96↙
Gouveia 68↗
Goúves 96←
Gouviá 96↘
Gouzon 45↙
Gowran 25↖
Graal-Muritz 33↗
Gråbo 30↖
Grabow 37↖
Gračac 91↙
Gračanica 90→
Graçay 45←
Gradac 91←
Graddis 85↙
Gradefes 64→
Gradets, *Bulgaria* 93←
Gradets, *Bulgaria* 93→
Grado, *Italy* 55↖
Grado, *Spain* 64↖
Gradsko 93↘
Grafenau 41↘
Gräfenhainichen 37↘
Gräfentonna 40↗
Gräfinau 41↖
Grafsnas 30↖
Gragnano 57↙
Grahovo 91←
Graiguenamanagh 25↖
Grain 23→
Grajewo 89↗
Gram 33↙
Gramat 51←
Gramatikovo 93↗
Gramisdale 18←
Grammichele 60↗
Grampians, mts. 18→
Grampound 22↙
Gramsh 91↗
Gran 26↗
Gran Paradiso, mt. 53↖
Granåbron 28↗
Granada 73→
Granadella 66↘
Granadilla 69↖
Granard 24→
Granátula 70↙
Grand Bornand, le 12↘
Grand Bourg, le 45↙
Grand-Champ 42↘
Grand' Combe, la 52←
Grand Couronne 38↙
Grand-Fougeray, le 44↖
Grandas de Salime 63↗
Grandcamp 43↖
Grândola 72↖
Grandpré 39↙
Grandrieu 51→
Grandvillars 46→
Grane 85↙
Graned 84↖
Granén 66→
Grangärde 28↗
Grange 24→
Grange over Sands 20→
Grangemouth 18↘
Grangesberge 28→
Granheim, *Norway* 26→
Granheim, *Norway* 27←
Graninge 84↗
Granja, *Portugal* 63↙
Granja, *Portugal* 68↘
Granja de Moreruela 64→
Granja de Torrehermosa 69↙
Grankulla 87→
Grankullavik 31↖
Granli 28↖
Granlunda 84←
Granna 30↗
Granollers 67→
Gransee 37↗
Gransherad 27↖
Grantham 21↘
Grantown-on-Spey 19←
Granträsk 84↖
Grantshouse 19↙
Granville 43←
Granvin 26↘
Gråsala 29↖
Gräsberg 29↖
Grasdorf 37←
Gräsmark 28→
Grasmere 20→
Gräsö 29↗
Grassau 48→
Grasse 53↙
Grassington 21←
Gråsten 33↙
Grästorp 28↙
Graubünden, prov. 47↘
Grau-du-Roi, le 52↙
Graulhet 51↙
Graulinster 39→
Graus 66→
Grava 28→
Gravabotn 26→
Grávalos 65→
Gravarne 28↙
Gravberget 28↖
Grave 35→
Grave, la 52↗
Gravedona 47↘
Gravelines 38↖
Gravellona 47↙
Gravendal 28→
Gravenhage, 's = Den Haag
Gravesend 23→
Graviá 96←
Gravina 58→
Gravvik 84←
Gray 46←
Grays Thurrock 23→
Graz 49←
Grazalema 73↙
Grdelica 93↙
Greaker 28←
Great Ayton 21←
Great Baddow 23→
Great Bernera 18↖
Great Casterton 23↖
Great Chesterford 23↗
Great Cumbrae 18↘
Great Dalby 23↖
Great Driffield 21→
Great Dunmow 23→
Great Hockham 23↗
Great Limber 21↘
Great Malvern 22↗
Great Missenden 23←
Great Oakley 23→
Great Ponton 21↘
Great Shefford 23←
Great Shelford 23↗
Great Torrington 22↙
Great Waltham 23→
Great Witley 22↗
Great Yarmouth 23↗
Grebbestad 28↙
Grebenstein 36↘
Grebo 29↙
Green Hammerton 21←
Greenhead 21↖
Greenlaw 21↖
Greenloaning 18↘
Greenock 18↘
Greenodd 20→
Greenore 24↗

Greenwich 23←
Grefrath 39↗
Gréggio 53↗
Greifenberg 48↖
Greifenburg 48↘
Greifswald 88↗
Grein 49↖
Greiz 41↖
Gréka 96↘
Grenå 32→
Grenade, *France* 51↙
Grenade, *France* 50↘
Grenchen 46→
Grenoble 52↗
Gréoux-les-Bains 52↘
Gressenich 39↗
Gressoney la Trinité 47↙
Greta Bridge 21←
Gretna 20↗
Gretna Green 20↗
Gretz 38↘
Grevback 28↘
Greve 54↘
Greven 36←
Grevená 94↖
Grevenbroich 39↗
Grevenmacher 39→
Grevesmühlen 33→
Greyabbey 24↖
Greystoke 20→
Greystones 25↖
Gries 48←
Gries im Sellrain 48←
Griesbach 48↗
Griesheim 40←
Grieskirchen 48↗
Grignan 52←
Grignols 50→
Grigoriopol 92↖
Grijpskerk 35↗
Grimaldi 53↙
Grimma 41↗
Grimmen 33↗
Grimmenstein 49→
Grimo 26↘
Grimsby 21↘
Grimsdalshytta 26↖
Grimstad 27→
Grindaheim 26→
Grinde 27↙
Grindelwald 47↙
Grinneröd 28↙
Griñón 69→
Grinstad 28↙
Grinsted 32↘
Gripport 46↗
Gripsholm 29←
Grisignano 54↗
Grisolles 51↙
Grisslehamn 29→
Grizebeck 20→
Grjotli 84↘
Gröbers 37↘
Gröbzig 37↙
Gródek 89↗
Grodków 89↙
Grodno 89↗
Grodzisk 88→
Groenlo 35→
Grogport 20↖
Groitzsch 41↖
Grójec 89→
Gronau, *W. Germany* 36←
Gronau, *W. Germany* 36→
Grönbua 26←
Grong 84←
Gronheim 36←
Grönhögen 31↙
Groningen, *Netherlands* 35↗
Gröningen, *E. Germany* 37←
Grono 47↘
Grönöy 85↙
Grönskåra 31←
Groothusen 36↖
Gropello 53↗
Grorud 28←
Gross Beeren 37→
Gross Gerau 40←
Gross Glockner, mt. 48→
Gross Oesingen 37←
Gross Reken 36↙
Gross Schönebeck 88↗
Gross Siegharts 49↖
Gross Sittensen 36↗
Gross Umstadt 40←
Gross-Warasdorf 49→
Grossalmerode 40↗
Grossau 49↖
Grossbodungen 37↙
Grossen Buseck 40↖
Grossenbrode 33→
Grossengstingen 47↗
Grossenhain 41↗
Grossenlüder 40↗
Grosseto 56↖
Grossgerungs 49↖
Grosshabersdorf 40↘
Grosshöchstetten 47←
Grossmugl 49↗
Grosspetersdorf 49→
Grostol 27→
Grotli 26←
Grötöy 85←
Grottáglie 58↗
Grottaminarda 58↙
Grottammare 57↖
Grotteland 27↘
Grotteria 59→
Grova 27←
Grove 63←
Grövelsjön 84↙
Grua 26↗
Grubenwald 12←
Gruda 91←
Grudovo 93↖
Grudusk 89↖
Grudziadz 89↖
Gruinard 18↗
Gruissan 67↗
Gruline 18↙
Grumello 54↖
Grums 28→
Grünbach 49→
Grünberg, *W. Germany* 40↖
Grünberg = Zielona Góra
Grundsund 28↙
Grundsunda 87↙
Grundtjärn 84↖
Grungedal 27←
Grünstadt 40←
Gruvberget 29↖
Gruyères 46↘
Grybów 89↘
Grycksbo 29↖
Gryfice 88↗
Gryfino 88↗
Gryon 46↘
Gryt 31↖
Grythyttan 28→
Grytten 84↘
Gryttjam 29↗
Grzmiąca 88↗
Gschnitz 48←
Gschwend 40↘
Gstaad 46↘
Gstadt 49↖
Gsteig 46↘
Guadahortuna 74←
Guadalajara 70↗
Guadalaviar 71↖
Guadalcanal 73↖
Guadalcazar 73←
Guadalquivir, river 73←
Guadalupe 69←
Guadamur 69→
Guadarrama 69↗
Guadarrama, Sierra de, mts. 70↖
Guadasuar 71↙
Guadiana, river 69↙
Guadix 74→
Gualdo Tadino 54↘
Guarda 68↗
Guardamar de Segura 75↖
Guardiagrele 57←
Guardiola 67←
Guardo 64↗
Guareña 69↙
Guarromán 73↗
Guastalla 54←
Gúbbio 54↘
Guben 88→
Gubin = Guben
Guddal 26↙
Gudensberg 40↗
Gudhem 30↗
Gudhjem 31↙
Gudvangen 26→
Guebwiller 46↗
Guecho 65↖
Güejar Sierra 74←
Guéméné Penfao 44↖
Guémené-sur-Scorff 42↘
Güeñes 65↖
Guer 42↘
Guerche, la 44↖
Guerche-sur-l'Aubois, la 45→
Guéret 45↙
Guernica y Luno 65↗
Guetaria 14↗
Guéthary 65↗
Gueugnon 45↘
Gühlen Glienicke 37↗
Guichen 42↘
Guido de Granadilla 69←
Guignes-Rabutin 38↘
Guijuelo 69↖
Guildford 23←
Guillaumes 53←
Guillena 73←
Guimarães 63↙
Guincho 14←
Guines 38↖
Guingamp 42→
Guisborough 21←
Guiscard 38→
Guise 38→
Guist 23↗
Guitiriz 63↗
Guîtres 50→
Gujan 50←
Guldborg 33↖
Gulen 26↘
Gulgofjorden 86↖
Gulin 55↘
Gullabo 31←
Gullane 19↙
Gullbrå 26↘
Gullhollmen 28↙
Gullön 85↘
Gullspång 28↘
Gullträsk 85↘
Güllük 95←
Gülpınar 95↖
Gulsele 84↖
Gulsvik 26↗
Gumboda 86↘
Gumiel de Hizán 65←
Gummersbach 40↖
Gundelfingen 40↘
Gunderup 32←
Gunnarn 84↖
Gunnarskog 28←
Gunnebo 31↖
Gunnilbo 29←
Guntin 63↗
Guntramsdorf 49↗
Günzburg 47↗
Gunzenhausen 40↘
Güre 95→
Guru Humorului 92←
Gusev 89↗
Gusmar 94↖
Guspini 61←
Gustafs 29↖
Gustav Adolf, *Sweden* 28→
Gustav Adolf, *Sweden* 30↗
Gustavsberg 29→
Gustavsfors 28←
Güsten 37↙
Gustrow 33↗
Gusum 29↙
Gutau 49↖
Gutcher 19↗
Gütersloh 36→
Guyenne, reg. 51↖
Guyhirne 23↗
Gvarv 27↖
Gvozdets 92←
Gwbert-on-Sea 22↖
Gweedore 24←
Gweek 22↙
Gwyddelwern 20↘
Gy 46←
Gya 27↘
Gyland 27↘
Gyljen 85↘
Gyoma 90↘
Gyöngyös 90↖
Gyönk 90→
Győr 90←
Gypsera 12←
Gysinge 29↖
Gyttorp 28→
Gyula 90↗

Haag 48↖
Haaksbergen 36←
Haan 39↗
Haapajärvi 87←
Haapamäki 87←
Haapavesi 87←
Haar 48↖
Haarlem 35←
Habay-la-Neuve 39←
Hablingbo 31→
Habo 30↗
Haby 28↙
Hachenburg 39↗
Hackas 84↗
Hacketstown 25↖
Hadamar 40←
Haddington 19↙
Hade 29↖
Haderslev 33↙
Hadleigh 23↗
Hadlow 23→
Hadsel 85←
Hadsten 32→
Hadsund 32→
Hægebostad 27→
Hægeland 27→
Haga 29→
Hagby 31←
Hagenow 37↖
Hagetmau 50↘
Hagfors 28→
Häggenås 84↖
Häggsjön 84←
Haglebu 26↗
Hagley 22↗
Haguenau 39↘
Hagyfalu 49→
Hahellerhytta 27←
Hahot 49↘
Haiger 40↖
Haigerloch 47↖
Hailsham 23↘
Hailuoto 86→
Hainburg 49↗
Hainchen 40↖
Hainfeld 49↗
Hajdúböszörmény 90↖
Hajduhadház 92↙
Hajdúszoboszlo 90↖
Hají 96↘
Hajnówka 89→
Hakadal 26↗
Håkånes 27←
Håkantorp 28↙
Hakkas 85→
Håksberg 29↖
Halahult 30↘
Håland 27↙
Hălăuceşti 92←
Halberstadt 37←
Halberton 22↘
Halden 28←
Haldensleben 37←
Hale Street 23→
Halesowen 22↗
Halesworth 23↗
Halfway House 25→
Halhjem 26↘
Hâlia 28↗
Halicarnassus, anc. site 95←
Halifax 21↙
Halikko 87↘
Halkirk 19↖
Hälla 84↖
Hålland 84←
Halle, *W. Germany* 36→
Halle, *E. Germany* 37↘
Halle, *Belgium* 38↗
Hällefors, *Sweden* 28→
Hällefors, *Sweden* 29←
Hallein 48→
Hallen 84→
Hallencourt 38←
Halles, les 52↖
Hällesjö 84↗
Hällestad 29↙
Hällevadsholm 28↙
Hällevik 30↘
Hallevikstrand 28↙
Hallingsjö 30↖
Hallingskeid 26→
Hällnäs, *Sweden* 29↗
Hällnäs, *Sweden* 86↘
Hallsberg 29↙
Hållsta 29←
Hallstahammar 29←
Hallstatt 13↗
Hallstavik 29→
Halluin 38↗
Hallviken 84↖
Hällybrunn 29←
Halma 39←
Halmagiu 92↘
Halmeu 92↙
Halne 26→
Hals 32←
Halsa 84↘
Hälsingborg 30↙
Halstead 23→
Halsua 87←
Halstenbek 33→
Haltdalen 84→
Haltern 36↙
Haltwhistle 21↖
Halver 40↖
Halwell 22↙
Ham, *Scotland* 19↗
Ham, *France* 38→
Hamar 28↖
Hamaröy 85←
Hamburg 37↖
Hamburgsund 28↙
Hamdorf 33↘
Hämeenkyrö 87↘
Hämeenlinna 87→
Hameln 36→
Hamilton 18↘
Hamina 87→
Hamm 36↙
Hammar 28↘
Hammarland 87↘
Hammel 32→
Hammenhög 30↘
Hammerdal 84↖
Hammerfest 86↖
Hammersåk 27↙
Hammershøj 32→
Hammaro 28→
Hamnavoe 19↗
Hamnbukt 86↖
Hamneda 30→
Hamneidet 85↗
Hamoir 39←
Håmojokk 85→
Hamra, *Sweden* 31→
Hamra, *Sweden* 84↗
Hamrånge 29↖
Hamrångefjorden 29↖
Hamremoen 26↗
Hamstreet 23→
Hån, *Sweden* 28←
Hån, *Sweden* 84→
Han-sur-Nied 39↘
Hanau 40←
Handcross 23←
Handen 29→
Handest 32→
Handlová 90←
Hanestad 26↖
Hangö 87↘
Hankasalmi 87←
Hanko, *Finland* = Hangö
Hankö, *Norway* 27↖
Hanley 21↙
Hannover 36→
Hannut 39↖
Hansbu 26→
Hansjö 84↗
Hansted 32↙
Haparanda 86→
Happisburgh 23↗
Harads 85↘
Haram 84↘
Harbo 29←
Harburg, *W. Germany* 36↗
Harburg, *W. Germany* 40↘
Hårby 33←
Hardangerfjorden 27↙
Hardegsen 36↘
Hardelot 38↖
Hardenborg 35↗
Harderwijk 35→
Hardheim 40→
Hardwicke 22→
Hareid 84↘
Haren 35↗
Harestad 30↖
Harestua 26↗
Harewood 21←
Harfleur 43↗
Harg 29↗
Harjavalta 87↘
Harkmark 27→
Harlech 20↘
Harleston 23↗
Harlingen 35↗
Harlosa 30↘
Harlow 23→
Härlunda 30→
Harmancık 95↗
Harmånger 84↗
Härnösand 84↗
Haro 65→
Haroldswick 19↗
Harpefoss 26↖
Harpenden 23←
Harpley 23↗
Harplinge 30←
Harrachov 88↘
Harrachsthal 49↖
Harran 84←
Harre 32↘
Harrogate 21←
Harrow 23←
Harrsjön 84↖
Härryda 30↖
Harsefeld 36↗
Harsewinkel 36←
Harsleben 37↙
Harsprånget 85→
Harstad 85←
Hartberg 49→
Hartfield 23→
Harthill 18↘
Hartland 22↙
Hartland Quay 22↙
Hartlepool 21←
Hartley Wintney 23←
Hartmannsdorf 41↗
Hartmannshain 40→
Hartola 87→
Harwell 23←
Harwich 23→
Harz, mts. 37↙
Harzgerode 37↙
Haseby 30→
Häselgehr 47→
Haselünne 36←
Håsjö 84↗
Haskvarna 30↗
Haslach 47↖
Hasle 31↙
Haslemere 23←
Haslev 33↖
Haslingden 21↙
Hasparren 66↖
Hassel 36→
Hassela 84↗
Hässelby 29→
Hasselfelde 37↙
Hasselfors 28↘
Hasselt 39↖
Hassfurt 40→
Hasslach 41←
Hassle 28↘
Hässleholm 30↘
Hästbo 29↖
Hästholmen 28↘
Hastière-Lavaux 39←
Hastings 23↘
Hästveda 30→
Hasvik 85↗
Haţeg 92↘
Hatfield, *England* 21↙
Hatfield, *England* 23←
Hatherleigh 22↙
Hathersage 21↙
Hattem 35→
Hattfjelldal 85↙
Hattingen 39↗
Hatton, *Scotland* 19←
Hatton, *England* 21↙
Hattula 87→
Hatvan 90↖
Haubourdin 38↗
Haugastol 26→
Hauge, *Norway* 27↘
Hauge, *Norway* 84↘
Haugesund 27↙
Haughom 27↘
Haugsdorf 49↗
Haugseter 26↖
Hauho 87→
Haukedal 26↙
Haukeland 26↘
Haukeligrend 27←
Haukeliseter 27←
Haukivuori 87←
Haunersdorf 41↘
Haurida 30↗
Hausach 47↖
Häusern 47←
Hausmening 49↖
Haute Thorame 52→
Hauteville-Lompnès 46↙
Hauzenberg 41↘
Havant 23↙
Havdhem 31→
Håvedalen 28↙
Havelberg 37→
Haverfordwest 22←
Haverhill 23↗
Havero 84↗
Håvilsrud 28←
Havixbeck 36←
Hävla 29↙
Havlíčkův Brod 88↘
Havnsø 32→
Havoysund 86↖
Havran 95↖
Havre, le 43↙
Havsnäs 84↖
Havstenssund 28↙
Hawarden 20↘
Hawes 21←

Hawes Water 21←
Hawick 21↖
Hawkhurst 23→
Hawkshead 20→
Haworth 21↙
Hay-on-Wye 22↗
Hayange 39↙
Haydon Bridge 21↖
Haye-Descartes, la 44→
Haye-du-Puits, la 43↖
Hayfield 21↙
Hayle 22↘
Hayrabolu 93↗
Haywards Heath 23↙
Hazel Grove 21↙
Hazebrouck 38↖
Heacham 21↘
Headcorn 23→
Headford 24↘
Headless Cross 22↗
Heanor 21↙
Héas 66↗
Heath End 23←
Heathrow, airport 23←
Hebburn 21↖
Hebden Bridge 21↙
Heberg 30←
Heby 29←
Hechingen 47↖
Hecho 66↗
Hechtel 39↖
Hechthausen 36↗
Heckington 21↘
Hedal 26↗
Hedared 30↖
Hedås 28→
Heddal 27↖
Hedderen 27→
Heddon-on-the-Wall 21↖
Hédé, *France* 43↖
Hede, *Sweden* 84→
Hedehusene 32↗
Hedemora 29↖
Heden, *Sweden* 28→
Heden, *Sweden* 85↘
Hedenäset 86↙
Hedensted 32→
Hedersleben 37↙
Hedesunda 29↖
Hedeviken 84→
Hedon 21↘
Hee 32↘
Heerde 35→
Heerenveen 35↗
Heessen 36↙
Heeze 35↘
Hegge 26↖
Heggenes 26↗
Heggheim 26↙
Heidal 84↘
Heide 33↘
Heidelberg 40↙
Heiden 47→
Heidenau 41↗
Heidenheim 40↘
Heidenreichstein 49↖
Heiderscheid 39←
Heikendorf 33→
Heiland 27←
Heilbronn 40↘
Heiligenberg 47↗
Heiligenblut 48→
Heiligenhafen 33→
Heiligenkreuz 49→
Heiligenstadt 40↗
Heilsbronn 40↘
Heimdal, *Norway* 26↘
Heimdal, *Norway* 84←
Heinola 87→
Heinsberg 39↖
Heist 38↗
Hejde 31↗
Hejlsminde 33↙
Helbra 37↙
Heldrungen 41↖
Helechal 69↙
Helensburgh 18↘
Helgeroa 27↖
Helgöy 85↗
Helgum 84↖
Héliopolis 10←
Hell 84←
Hella 26↙
Hellandsbygda 27↙
Hellehalsen 26→
Helleland 27↘
Hellemobotn 85←
Hellesylt 26←
Hellevik 26↙
Hellevoetsluis 35←
Hellifield 21←
Helligskogen 85↗
Hellin 71↙
Helmbrechts 41←
Helmond 35↘
Helmsdale 19↖
Helmsley 21←
Helmstedt 37←
Helsinge 32↗
Helsingfors = Helsinki
Helsingør 32↗
Helsinki 87→
Helstad 26←
Helston 22↙
Hem 27↖
Hemel Hempstead 23←
Hemer 36↙
Heming 39↘
Hemmendorf 36→
Hemne 84↙
Hemnesberget 85↙
Hempstead 23→
Hemse 31→
Hemsedal 26→
Hemslingen 36↗
Hemsö 84↗
Hemsworth 21↙
Hen 26↗
Henan 28↙
Hendaye 65↗
Hendon 23←
Henfield 23↙
Hengelo 36←
Hénin-Liétard 38→
Henley in Arden 23↖
Henley on Thames 23←
Henlow 23←
Hennan 84↗
Henndorf 48→
Hennebont 42↘
Hennef 39↗
Hennigsdorf 37→
Henningsvær 85←
Hennöy 26↙
Hennstedt 33→
Henri-Chapelle 39↖
Henrichemont 45←
Henstridge 22↘
Heppenheim 40←
Herad 27↘
Heradsbygd 28↖
Herand 26↘
Herbault 45←
Herbern 36↙
Herbertingen 47↗
Herbiers, les 44←
Herborn 40↖
Herby 89↙
Herceg Novi 91←
Hercegovina = Bosna i Hercegovina
Herchen 39↗
Herdal 26←
Herdecke 36↙
Hereford 22→
Herefoss 27→
Herencia 70←
Herentals 39↖
Hérépian 51↘
Herfølge 33↖
Herford 36→
Herforst 39→
Héricourt 46→
Heriot 19↙
Herisau 47→.
Hérisson 45↙
Herk-de-Stad 39↖
Hermagor 48↘
Hermannsburg 37←
Hermansverk 26←
Herment 51↗
Hermeskeil 39→
Hernani 65↗
Herne 36↙
Hernes 28↖
Herning 32↘
Heroy 84↘
Herräng 29↗
Herrenalb 40↙
Herrenberg 40↙
Herrera 73←
Herrera de Alcántara 68→
Herrera de Pisuerga 65←
Herrera de los Navarros 66↙
Herrera del Duque 69↘
Herreruela 68→
Herrhamra 29↘
Herritslev 33↖
Herrljunga 30↗
Herrsching 48↖
Hersbruck 41↙
Herscheid 40↖
Herselt 39↖
Herstal 39↖
Herten 36↙
Hertford 23←
Hertogenbosch, 's 35→
Hervas 69↖
Herzberg, *W. Germany* 37↙
Herzberg, *E. Germany* 37↗
Herzberg, *E. Germany* 37↘
Herzlake 36←
Herzogenaurath 40→
Herzogenbuchsee 47←
Herzogenburg 49↗
Hesdin 38←
Hesel 36↖
Heskestad 27↘
Hesselager 33←
Hessen 37←
Hessen, land 40↖
Hessisch Lichtenau 40↗
Hessisch Oldendorf 36→
Hessen, prov. 40↖
Hessle 21↘
Hestra 30↗
Heswall 20↘
Hetton-le-Hole 21←
Hettstedt 37↙
Heusweiler 39↘
Heves 90↖
Heviz 90→
Hexham 21↖
Heyrieux 52↗
Heysham 20→
Heytesbury 22→
Heywood 21↙
Hibaldstow 21↘
Hida 92↙
Hieflau 49←
Hiendelaencina 65↙
High Hesket 20→
High Wycombe 23←
Higham Ferrers 23↖
Highbridge 22→
Highclere 23←
Highworth 23←
Higuera de Arjona 73→
Higuera de Vargas 68↘
Higuera de la Serena 69↙
Higuera de la Sierra 72→
Higuera la Real 72↗
Higueruela 71↙
Hijar 66↘
Hildburghausen 40→
Hilden 39↗
Hilders 40↗
Hildesheim 36→
Hille 36→
Hillegom 35←
Hillerstorp 30↗
Hillesheim 39→
Hillesöy 85↗
Hillestad 27↖
Hillington 23↗
Hillsborough 24↗
Hillswick 19↗
Hilltown 24↗
Hilpoltstein 41↙
Hilton 22↗
Hiltula 87↗
Hilversum 35→
Himanka 87←
Himarë 94↖
Himley 22↗
Hinckley 23↖
Hindas 30↖
Hindelang 47→
Hinderwell 21→
Hindhead 23←
Hindley 21↙
Hindon 22→
Hindseter 26↖
Hingham 23↗
Hinnerjoki 87↘
Hinneryd 30→
Hinojales 72↗
Hinojos 72→
Hinojosa del Duque 69↘
Hinojosas 69↘
Hinstock 22↗
Hinterglemm 13↖
Hintermoos 13↖
Hinterrhein 47↘
Hinterthal 48→
Hintertux 48←
Hirläu 92←
Hirmoen 26↖
Hirsau 40↙
Hirschaid 41←
Hirschau 41←
Hirschbach 49↖
Hirschberg = Jelenia Góra
Hirschburg 41←
Hirschhorn 40←
Hirson 38→
Hirşova 92↗
Hirtenberg 49↗
Hirtshals 32←
Hirvensalmi 87→
Hirvineva 86→
Hirwaun 22←
Hishult 30→
Hisingen 30↖
Hita 65↙
Hitchin 23←
Hitra 84↙
Hittarp 30↙
Hitzacker 37↖
Hjallerup 32←
Hjarnarp 30←
Hjartdal 27←
Hjelle 26←
Hjelmelandsvagen 27↙
Hjerkinn 84↘
Hjerting 33↙
Hjo 28↘
Hjøllund 32↘
Hjørring 32←
Hjorted 31↖
Hjortkvarn 29↙
Hjörundfjord, *Norway* 26↙
Hjorundfjord, *Sweden* 84↘
Hjuksebo 27↖
Hjulsjö 28→
Hjulsta 29←
Hlohovec 90←
Hobro 32→
Höchberg 40→
Hochdorf 47←
Hochfelden 39↘
Hochfilzen 48→
Hochheim 40←
Höchst 40←
Höchstadt 40→
Hochstätten 39→
Hockenheim 40↙
Hockley Heath 22↗
Hockliffe 23←
Hoddesdon 23→
Hodenhagen 36→
Hódmezövásárhely 90↗
Hodnaberg 26↘
Hodnet 22↗
Hodošan 49↘
Hoek van Holland 35←
Hof, *Norway* 27↖
Hof, *Norway* 28↖
Hof, *W. Germany* 41←
Hofgeismar 36↘
Hofheim, *W. Germany* 40←
Hofheim, *W. Germany* 40→
Hofors 29↖
Höganäs 30↙
Högbo 29↖
Högboda 28→
Högbult 31↖
Högby 31←
Hogdal 28↙
Högerud 28←
Hogfors 28→
Hoghiz 92→
Högsby 31←
Högsäter, *Sweden* 28←
Högsäter, *Sweden* 28↙
Högsjö 84↗
Hogstorp 28↙
Hohenaschau 48→
Hohenau 49↗
Hohenbucka 37↘
Hohenburg 41↙
Hoheneggelsen 37←
Hohenems 47→
Hohenkammer 48↖
Hohenlimburg 40↖
Hohenmölsen 41↖
Hohenseeden 37→
Hohenwesdet 33↘
Hohenzethen 37↖
Hohultslatt 31←
Højby 33←
Højer 33↖
Højreby 33←
Hok 30↗
Hokhuvud 29↗
Hokksund 27↖
Hökön 30→
Hoe 26→
Hola 26↘
Holbæk 32↗
Holbeach 23↖
Holbøl 33↙
Hole, *Norway* 26↗
Hole, *Norway* 27↘
Hölen 27↖
Holič 90←
Hollabrunn 49↗
Höllen 27→
Hollfeld 41←
Hollingstedt 33↘
Höllviksnäs 30↙
Hollywood 25↖
Holm 84↗
Holme 21→
Holmedal 28←
Holmegil 28←
Holmenkollen 27↖
Holmes Chapel 21↙
Holmestrand 27↖
Holmfirth 21↙
Holmfors 85↘
Holmön 87↙
Holmöyane 26↙
Holmsbu 27↖
Holmsjö, *Sweden* 31←
Holmsjö, *Sweden* 84↖
Holmsund 87↙
Holmsveden 29↖
Hölö 29↘
Holöydal 84→
Holsbybrunn 31↖
Holsen 26↙
Holstebro 32↘
Holsted 33↙
Holsworthy 22↙
Holt, *Wales* 20↘
Holt, *England* 23↗
Holten 35→
Holtet 28↖
Holtzweissig 37↘
Holum 27→
Holwerd 35↗
Holycross 25←
Holyhead 20↘
Holywell, *Wales* 20↘
Holywell, *England* 22↘
Holywood 24↖
Holzgau 47→
Holzhausen 39→
Holzkirchen 48←
Holzminden 36↘
Homberg, *W. Germany* 39↗
Homberg, *W. Germany* 40↖
Homberg, *W. Germany* 40↗
Homborsund 27→
Homburg 39↘
Homelfjord 26↘
Hommelvik 84←
Hondón de las Nieves 71↙
Hondschoote 38↖
Hönefoss = Ringerike
Honfleur 43↗
Høng 33←
Honingham 23↗
Honiton 22↘
Honkajoki 87↙
Honley 21↙
Honnef 39↗
Hönningen 39→
Höno 30↖
Honningsvåg, *Norway* 26↙
Honningsvåg, *Norway* 86↖
Honrubia 70→
Hontalbilla 65↙
Hontoria del Pinar 65←
Hoofddorp 35←
Hoogeveen 35↗
Hoogezand 35↗
Hoogstraten 39↖
Hook 23←
Höör 30↘
Hoorn 35↖
Hope 18↗
Hope under Dinmore 22↗
Hopeman 19↖
Hôpital-du-Gros-Bois, l' 46→
Hopfgarten 48←
Hôpitaux, les 46↘
Hopland 26↙
Hoppestad 27↖
Hopseidet 86↖
Hopton 23↗
Hoptrup 33↙
Hor Stubña 90←
Horažďovice 41↘
Horb 47↖
Hörby 30↘
Horcajo de Santiago 70→
Horcajo de los Montes 69→
Horcajo Medianero 69↖
Horda, *Norway* 27↙
Horda, *Sweden* 30→
Horeb 22↖
Horezu 92↘
Horn, *Sweden* 31↖
Horn, *W. Germany* 36↘
Horn, *Austria* 49↗
Hornsjö, *Norway* 26↖
Hörnsjö, *Sweden* 87↙
Horsching, airport 49↖
Horst, *W. Germany* 33↘
Horst, *Netherlands* 35↘
Horgen 47←
Horgos 90↗
Hořice 88↘
Horken 28→
Horle 30→
Horley 23←
Hormakumpu 86←
Hornachos 69↙
Hornachuelos 73←
Hornbæk 32↗
Hornberg 47↖
Horncastle 21↘
Hornchurch 23→
Horndal 29↖
Horndean 23↙
Hörnefors 87↙
Horní Benešov 89↙
Hornindal 26↙
Horningsholm 29↘
Hornness 27→
Hornos 74↖
Hornsea 21→
Hornslet 32→
Hornstein 49↗
Hornu 38→
Hörnum 33↙
Hořovicé 41→
Horreby 33↖
Horred 30↖
Horsebridge 23↘
Horseleap Cross Roads 24→
Horsens 32→
Horsford 23↗
Horsforth 21←
Horsham 23←
Horsham St Faith 23↗
Horslunde 33←
Horsovski Tyn 41→
Horstmar 36←
Horta de San Juan 66↘
Horten 27↖
Hortézuela 65↘
Hortigüela 65←
Hortlax 86↘
Hörvik 30↘
Horwich 21↙
Hoslemo 27←
Hospital 25↖
Hospital de Orbigo 64→
Hospitalét 67↙
Hossa 86↗
Hossegor 65↗
Hostalrich 67→
Hosteland 26↘
Hostoun 41→
Hotagen 84←
Hotavlje 49↙
Hoting 84↖
Houches, les 46↘
Houdan 38↙
Houdelaincourt 46↖
Houeillès 50→
Houffalize 39←
Houghton-le-Spring 21←
Houlbjerg 32→
Houlgate 43↖
Hova 28↘
Hovås 30↖
Hovborg 33↙
Hovda 26↗
Hovdehytta 27←
Hove 23↙
Hövelhof 36↘
Hoven 32↘
Hovenaset 28↙
Hovet 26→
Hoviksnas 30↖
Hovin 27↖
Hovingham 21←
Hovmantorp 31←
Hövreslia 26↗
Hovringen 26↖
Hovsta 29←
Howden 21↘
Höxter 36↘
Hoya 36→
Hoyanger 26↙
Höydalen 26←
Hoyerswerda 41↗

Höyjord 27↖
Höylandet 84←
Hoym 37↙
Hoyo de Manzanares 69↗
Hoyos 68→
Hradec Králové 88↘
Hrádek 49↗
Hranice 89↙
Hrubieszów 89→
Hrvatska, prov. 90↘
Huaröd 30↘
Huben 48→
Hückeswagen 40↖
Hucknall Torkard 21↙
Huddersfield 21↙
Huddinge 29→
Huddunge 29←
Hude 36↗
Hudiksvall 84↗
Huedin 92↙
Huélamo 71↖
Huelgoat 42←
Huelma 73→
Huelva 72→
Huéneja 74←
Huercal de Almeria 74↙
Huércal-Overa 74→
Huerta de Rey 65←
Huerta de Valdecarábanos 70←
Huerto 66→
Huesa 74←
Huesca 66→
Huéscar 74←
Huete 70→
Huétor Tájar 73→
Hüfingen 47↖
Hugh Town 22↘
Hugnerud 28←
Hugulia 26↗
Huittinen 87↘
Huizen 35→
Hulabäck 30↖
Hull = Kingston upon Hull
Hullaryd 30↗
Hulltorp 30→
Hüls 39↗
Hulst 35↙
Hultanäs 31←
Hultsfred 31↖
Hultsjö 30↗
Hulsig 32←
Humanes 65↙
Humble 33←
Humenné 90↖
Hummelo 35→
Hummelvik 84↙
Humpolec 88↘
Humppila 87→
Hunge 84↗
Hunder 26↖
Hundested 32↗
Hundorp 26↖
Hundsjö 85↘
Hundvik 26↙
Hünfeld 40↗
Hungen 40←
Hungerford 23←
Hunmanby 21→
Hunnebostrand 28↙
Hunnedalen 27↘
Huntingdon 23↖
Huntley 22→
Huntly 19←
Hurbanovo 90←
Hurdal 28↖
Hurlers Cross 25↙
Hurlford 20↗
Hurst Green 23↘
Hurum 27↖
Hurup 32←
Husb-Långhundra 29→
Husbands Bosworth 23↖
Husby 29↖
Husby-Oppunda 29↙
Huşi 92↗
Husnes 27↙
Husum 33↘
Huttwil 47←
Huy 39←
Hvåle 26↗
Hvalstad 27↖
Hvarnes 27↖
Hvidbjerg 32↘
Hvide Sande 32↘
Hvitsen 27↖
Hvittingfoss 27↖
Hvož d'any 41→
Hycklinge 31↖
Hyde 21↙
Hyen 26↙
Hyères 52↘
Hylestad 27←
Hyllestad 26↙
Hyltebruk 30→
Hyndford Bridge 20↗
Hynnekliev 27→
Hynish 18↙
Hythe, *England* 23↙
Hythe, *England* 23→
Hyrynsalmi 86↗
Hyssna 30↖
Hyvinkää 87→

Ianca 92↗
Iaşi 92↖
Ibahernando 69←
Ibbenbüren 36←
Ibestad 85←
Ibi 71↙
Ibiza 74↘
Ibiza, island 74↘
Ibstock 23↖
Ichenhausen 47↗
Ichtershausen 41↖
Iciar 14↗
Idala 30↖
Idanha-a-Nova 68→
Idar Oberstein 39→
Idd 28↙
Idhra 96↗
Idkerberget 29↖
Idre 84→
Idrija 49↙
Idstein 40←
Ielč 88↘
Ieper 38↗
Ierápetra 95↙
Iesi 55↙
Ifach 15↘
Igea Marina 11↗
Igelbäcken 28↘
Igelfors 29↙
Iggesund 84↗
Ightham 23→
Igleb 28↘
Iglésias 61↙
Igls 13←
Igoumenítsa 94↖
Igrejinha 68↘
Igualada 67↙
Igüeña 64↖
Ihtiman 93↙
Iisalmi 87←
Iisvesi 87←
IJmuiden 35←
IJsselmeer, sea 35↗
IJzendijke 35↙
Ikaalinen 87↘
Ikast 32↘
Ilanz 47→
Iława 89↖
Ilchester 22↘
Île de France, reg. 38↙
Ileanda 92↙
Iles d'Hyères 52↘
Ilford 23→
Ilfracombe 22←
Ilhavo 68↖
Ilidza 91←
Iliókastron 96→
Ilirska Bistrica 55↖
Ilkeston 21↙
Ilkley 21←
Illana 70→
Illano 63↗
Illas 64↗
I'lle Rousse 62↖
Ille-sur-la-Têt 67↗
Illertissen 47↗
Illescas 69→
Illfeld 37↙
Illiers 45↖
Illingen 40↙
Illkirch-Graffenstaden 40↙
Illora 73→
Illueca 65↘
Ilmajoki 87↙
Ilmenau 41↖
Ilminster 22↘
Ilomantsi 87↖
Ilsenburg 37↙
Ilva Mică 92←
Ilz 49→
Iłża 89→
Imatra 87↗
Imdalen 26↖
Immenstadt 47→
Imola 54→
Imotski 91←
Impéria 53↙
Imroz 95↖
Imsa 26↖
Imsenden 26↖
Imst 47→
Inari 86←
Inca 75←
Inch 25↘
Inchigeelagh 25↘
Inchnadamph 18↗
Incio 63→
Incisa 54↘
Indal 84↗
Inderöy 84←
Indija 90↗
Indura 89↗
Inecik 93↗
Inegol 95↗
Ineu 92↘
Infantes 70↘
Infiesto 64↗
Inga 87→
Ingatestone 23→
Ingatorp 31↖
Ingedal 28←
Ingelheim 40←
Ingelmunster 38↗
Ingelstadt 30→
Ingierstrand 27↖
Ingleton 21←
Ingolstadt 41↙
Ingwiller 39↘
Iniesta 71←
Iniö 87↘
Inkberrow 22↗
Inkeroinen 87→
Innellan 18↘
Inner Sound 18←
Inner Villgraten 48↘
Inneralpbach 13→
Innerleithen 20↗
Innertkirchen 47↙
Innichen 48↘
Inning 48↖
Innsbruck 48←
Innset, *Norway* 84↘
Innset, *Norway* 85→
Innvik 26↙
Inönü 95↗
Inowrocław 89←
Insch 19←
Insh 18→
Insjön 28↗
Instefjord 26↘
Insterburg = Chernyakhovsk
Instow 22←
Insurăţei 92↗
Intepe 95↖
Interlaken 47↙
Intorsura Buzăului 92→
Intra-Verbania 47↙
Intróbio 47↘
Inver 24←
Inverallochy 19←
Inveraray 18↘
Inverbervie 19←
Inverey 19←
Invergarry 18→
Invergordon 18→
Inverie 18→
Inverinate 18→
Inverkeilor 19↙
Inverkeithing 19↙
Inverkip 18↘
Inverkirkaig 18↗
Inverlael 18↗
Inverlochlarig 18↘
Inverlussa 18↙
Invermoriston 18→
Inverness 18→
Inversanda 18↘
Invershin 18↗
Inversnaid 18↘
Inverurie 19←
Ioánnina 94↖
Ion Corvin 93↖
Ioneşti 92→
Ipáti 96←
Ipsala 93→
Ipsoús 96↘
Ipswich 23↗
Iráklion 95↙
Iria 96→
Irijo 63←
Iron Acton 22→
Ironbridge 22↗
Irsina 58→
Irun 65↗
Irúrzun 65↗
Irvine 20↗
Irvinestown 24→
Is-sur-Tille 46←
Isaba 66↖
Isari 96↘
Isbister 19↗
Iscar 64↘
Ischgl 47→
Ischia 57↙
Iselle 47↙
Iseo 54↖
Iserlohn 40↖
Isérnia 57←
Isigny 43↖
Isla Cristina 72→
Islam 55→
Isle-de-Noé 50↘
Isle-en-Dodon, l' 67↖
Isle-Jourdain, l' 51↙
Isle-sur-la-Sorgue, l' 52↙
Isleornsay 18←
Islip 23←
Ismaning 48↖
Isny 47→
Isojoki 87↙
Isokyrö 87↙
Isola, *France* 53←
Isola, *Italy* 53→
Isola del Gran Sasso 57↖
Isola della Scala 54↖
Isola di Capo Rizzoto 59←
Isperikh 93←
Ispica 60↗
Issambres, les 10→
Isselburg 36↙
Issigeac 50→
Issoire 51↗
Issoudun 45←
Issum 35→
Istán 73↙
Istanbul 93↗
Istha 40↗
Isthmía = Kirás Vrísi
Istiaía 96←
Istres 52↙
Itéa 96←
Ittiri 61↖
Itzehoe 33↘
Iungsdalen 26→
Ivalo 86←
Ivan 49→
Ivanec 49↘
Ivangrad 91↖
Ivanić Grad 90→
Ivaylovgrad 93→
Iveland 27→
Ivinghoe 23←
Iviza = Ibiza
Ivrea 53↖
Ivybridge 22↙
Ixworth 23↗
Izbica 89→
Izeda 63↘
Izmail 92↗
Izmir 95←
Iznalloz 73→
Iznatoraf 74↖
Iznik 95↗
Izola 55↖
Izsák 90↗
Izvorul Oltului 92→

Jaala 87→
Jaatila 86↙
Jabalquinto 73↗
Jablanac 55→
Jablonec 41↘
Jablonec-nad-Nisou 88↘
Jabłonowo 89↖
Jablunkov 89↙
Jabugo 72→
Jaca 66→
Jáchymov 41→
Jacobstad 87↙
Jadów 89→
Jädraås 29↖
Jadraque 65↘
Jaén 73→
Jagel 33↘
Jajce 91←
Ják 49↗
Jäkkvik 85↙
Jalance 71↙
Jalasjärvi 87↙
Jaligny 45↘
Jamestown 24→
Jämijärvi 87↘
Jämjö 31↙
Jämsä 87→
Jämsänkoski 87→
Jämshög 30↘
Jämtl Sikås 84↖
Janów Lub 89↘
Jánosháza 90←
Janville 45↖
Janzé 44↖
Jäppila 87←
Jaraba 65↘
Jarafuel 71↙
Jaraicejo 69←
Jaraiz 69←
Jarandilla 69←
Järbo, *Sweden* 28↙
Järbo, *Sweden* 29↖
Jaren 26↗
Jargeau 45↖
Järle 29←
Järna 29↘
Jarnac 50↗
Jarnages 45↙
Järnboås 28→
Järnforsen 31↖
Jarny 39↙
Jarocin 89←
Jaroměř 88↘
Jaroslavice 49↗
Jaroslaw 89↘
Järpås 28↙
Järpen 84←
Jarrow 21↖
Järsnäs 30↗
Järvenpää 87→
Järvsö 84↗
Jaša Tomić 90↗
Jasenak 55↗
Jasenica 55→
Jasło 89↘
Jasna 90↖
Jassy = Iaşi
Jastarnia 89↖
Jastrowie 88↗
Jászapáti 90↖
Jászberény 90↖
Jät 30→
Jativa 71↙
Jauge 50→
Jaun 46↘
Jávea 71↘
Jävenitz 37←
Javie, la 52→
Javier 66←
Jävre 86↘
Javron 44↗
Jawor 88→
Jaworzno 89↙
Jedburgh 21↖
Jędrzejów 89↙
Jeesio 86←
Jektevik 27↙
Jelenia Góra 88→
Jelling 32↘
Jelsa 16→
Jemnice 49↖
Jena 41↖
Jenbach 48←
Jennersdorf 49→
Jenny 31↖
Jeppo 87↙
Jeres del Marquesado 74←
Jerez de la Frontera 73↙
Jerez de los Caballeros 68↘
Jérica 71←
Jerichow 37→
Jerijärvi 85→
Jerzu 61→
Jesenice, *Czechoslovakia* 41→
Jesenice, *Yugoslavia* 49↙
Jessefors 28←
Jessen 37↘
Jessheim 28↖
Jessnitz 37↘
Jestetten 47←
Jeumont 38→
Jevenstedt 33↘
Jever 36↖
Jevnaker 26↗
Jezerane 55→
Jezero 91←
Jeznas 89↗
Jeżów 89←
Jhrhove 36↖
Jičín 88↘
Jihlava 88↘
Jijona 71↙
Jilava 93←
Jimbolia 90↗
Jimena 73→
Jimena de la Frontera 73↙
Jindřichův Hradec 88↘
Jirkov 41→
Jochberg 13↖
Jódar 74←
Joensuu 87↖
Joeström 85↙
Johanngeorgenstadt 41↗
Johannisholm 28↗
John o' Groat's 19↖
Johnston 22←
Johnstone 18↘
Johnstown 25←
Joigny 45↗
Joinville 46↖
Jokk 85↘
Jokkmokk 85↘
Jölster = Alhus
Jönåker 29↙
Jondal, *Norway* 27↖
Jondal, *Norway* 26↘
Jonköping 30↗
Jonsered 30↖
Jonstorp 30↙
Jonzac 50↗
Jordet 84→
Jorgastak 86←
Jörland 30↖
Jörn 86↘
Jörpeland 27↙
Josenfjorden 27↙
Josipdol 55↗
Jossa 40→
Josselin 42↘
Jössund, *Norway* 84←
Jössund, *Norway* 84↙
Jostedal 26←
Josvafö 90↖
Jotkajavrre 86↖
Joué-les-Tours 44→
Jougne 46↘
Joukokylä 86↗
Joutsa 87→
Joutsijärvi 86↙
Jouy 45↖
Jove 63↗
Joyeuse 52←
Juan-les-Pins 53↙
Juankoski 87↖
Juchhöh Töppen 41←
Judenberg 49←
Juelsminde 32→
Jugon 42→
Juillac 51↖
Jüken 86↘
Jukkasjärvi 85→
Jularbo 29↖
Jülich 39↗
Julita 29←
Jullouville 10↗
Jumeaux 51↗
Jumièges 43↗
Jumilhac-le-Grand 51↖
Jumilla 71↙
Jumkil 29→
Juncosa 67↙
Juneda 67↙
Jung 28↘
Jungfrau, mt. 47↙
Junglinster 39→
Juniville 38↘
Junkerdal 85↙
Junosuando 85→
Junquera de Ambia 63→
Junsele 84↖
Juntustranta 86↗
Juoksengi 86↙
Jura, mts. 46↘
Jurby 20→
Jurjevo 55→
Jurva 87↙
Jüterbog 37→
Jutis 85↙
Jutland = Jylland
Juuka 87↖
Juva 87↗
Juvisy 38↙
Jyderup 32→
Jylland, reg. 32↘
Jyllinge 32↗
Jyväskylä 87←

Kaamanen 86←
Kaarssen 37↖
Kaavi 87↖
Kaba 90↖
Kabelvåg 85←
Kačanik 91↗
Kåcksvik 30↗
Kadaň 41→
Kadarkút 90→
Kadhenoi 96↖
Kadrifakovo 93↘
Kåfalla 29←
Kåfjord 85→

La Estrada 63←
La Estrella 69→
La Felguera 64↗
La Fregeneda 63↘
La Fresneda 66↘
La Fuente de San Esteban 64↙
La Gallega 65←
La Garriga 67→
La Garrovilla 69↙
La Gineta 70↘
La Granja 65↙
La Guardia, *Spain* 63←
La Guardia, *Spain* 70←
La Gudiña 63→
La Haba 69↙
La Hermida 64↗
La Herradura 14↘
La Horcajada 69↖
La Horra 65↙
La Iglesuela 69↗
La Iglesuela del Cid 71↗
La Iruela 74←
La Jana 71↗
La Junquera 67→
La Lantejuela 73←
La Linea de la Concepción 73↙
La Luisiana 73←
La Maddalena 61↗
La Magdalena 64↗
La Merca 63→
La Molina 67←
La Muela 66↙
La Nava de Ricomalillo 69→
La Nava de Santiago 69↙
La Nucia 71↘
La Paca 74→
La Palma del Condado 72→
La Panadella 67←
La Peraleja 70↗
La Plaza 64↖
La Pobla de Lillet 67←
La Pola de Gordon 64↗
La Puebla 75←
La Puebla de Almoradier 70→
La Puebla de Cazalla 73←
La Puebla de Montalbán 69→
La Puerta de Segura 70↘
La Puebla de Valdavia 64→
La Puebla de Valverde 71←
La Puebla Nueva 69→
La Puebla de los Infantes 73←
La Puebla del Rio 73←
La Punt 47↘
La Rambla 73→
La Rinconada 73←
La Robla 64↗
La Roca de la Sierra 69↙
La Roche 12← (12 km S of Fribourg)
La Roda, *Spain* 70↘
La Roda, *Spain* 73→
La Romana 71↙
La Rua 63→
La Salceda 65↙
La Seca 64↘
La Selva 67↙
La Solana 70↘
La Spézia 54←
La Toja 63←
La Toledana 69→
La Tour-de-Peilz 12←
La Unión 75←
La Vecilla 64↗
La Vega 63→
La Vellés 64↘
La Ventosa 70→
La Victoria 73←
La Vid 65↙
La Villa de Don Fadrique 70→
La Yesa 71←
Laa 49↗
Laaben 49↗
Laage 33↗
Laanila 86←
Laapshe 40↖
Labajos 64↘
Labastide 67↗
Labastide-Murat 51←
Labbas 85↙
Labenne 50↙
Labin 55←
Labouheyre 50←
Labrède 50→
Labrit 50→
Labruguière 67↗
Läby 29→
Laç 91↗
Lac-Chambon 51↗
Lacanau-Océan 50←
Lacapelle-Marival 51←
Lacaune 51↘
Lacco Ameno 11↙
Laceby 21↘
Lachen 47←
Lachendorf 37←
Lack 24←
Läckö 28↘
Lacobeni 92←
Láconi 61←
Lacq 66↗
Ladelund 33↙
Ladhá 96→
Ladispoli 56→
Ladne Vode 91↖
Ladoeiro 68→
Ladoga, L. = Ladozhskoye Ozero
Ladozhskoye Ozero, lake 87↗
Lærdal 26→
Lærdalsöyri = Lærdal
Læsø, island 32←
Lævajoki 86↖
Lafnitz 49→
Lafrançaise 51←
Lagan 30→
Lagartera 69→
Lage, *Spain* 63↖
Lage, *W. Germany* 36→
Lågeros 26→
Laget 27↗
Lagg 18↙
Lagga 29→
Laggan 18→
Laghey 24←
Laghtgeorge 25↙
Lagny 38↘
Lago 54↗
Lagoa 72←
Lagoaça 63↘
Lagonegro 58↘
Lagos 72←
Lagovoúni 96↘
Laguardia 65→
Laguarres 66→
Laguarta 66→
Laguiole 51→
Lagunilla 69↖
Laharie 50↙
Lahinch 25↙
Laholm 30←
Lahr 47↖
Lahti 87→
Laichingen 47↗
Laignes 46↖
Laiguéglia 53↙
Laihia 87↙
Lainijaur 85↘
Lainio 85→
Lair 19←
Lairg 18↗
Laissac 51→
Laisvall 85↙
Laitila 87↘
Lajkovac 91↖
Lajeosa 68↗
Lajosmizse 90↖
Lakasjö 84↖
Lakaträsk 85↘
Lakhdenpokh'ya 87↗
Lákka 96↙
Lakkópetra 96↙
Lakselv 86↖
Laksfors 85↙
Lála 96↘
Lalin 63→
Lalinde 51←
Lalley 52→
Lalling 41↘
Lalm 26↖
Lálouka 96→
Lam 41↘
Lama Mocogno 54←
Lamalou 51↘
Lamarche 46↖
Lámari 96↖
Lamastre 52←
Lambach 48↗
Lámbaina 96↘
Lamballe 42→
Lamberhurst 23→
Lambéti 96↘
Lámbia 96↘
Lamborn 29↖
Lambourn 23←
Lamego 63↘
Lamía 96←
Lamlash 20↖
Lammhult 30→
Lammi 87→
Lamotte-Beuvron 45←
Lampertheim 40←
Lampeter 22↖
Lampinoú 94↗
Lämpsä 86↗
Lanaja 66→
Lanark 20↖
Lancaster 21←
Lanchester 21←
Lanchin 92←
Lanciano 57←
Lancieux 10↖
Lancin 52↗
Łańcut 89↘
Landau, *W. Germany* 40↙
Landau, *W. Germany* 41↘
Landeck 47→
Landedo 63→
Landerneau 42←
Landery 87↖
Landeryd 30→
Landeskogen 27→
Landete 71←
Landivisiau 42←
Landquart 47→
Landrecies 38→
Landså 85↘
Landsberg, *W. Germany* 48↖
Landsberg, *Poland* = Gorzów Wielkopolski
Landsbro 30↗
Landskrona 30↙
Landstuhl 39↘
Landvik 27→
Lane 28↙
Lanesborough 24→
Langa de Duero 65↙
Langadhás 93↘
Langádhia 96↘
Langangen 27↖
Långban 28→
Langeac 51↗
Langeland, island 33←
Längelmäki 87→
Langelsheim 37←
Langen 40←
Langenau 47↗
Langenberg, *E. Germany* 41↖
Langenberg, *W. Germany* 39↗
Langenbrücken 40↙
Langenburg 40↘
Langenes 85↖
Langenfeld, *Austria* 48←
Langenfeld, *W. Germany* 39↗
Langenlois 49↗
Langennagen 36→
Langensalza 40↗
Langensenbold 40←
Langensteinbach 40↙
Langenthal 47←
Langenzenn 40↘
Langenzersdorf 49↗
Langeoog 36↖
Långersud 28←
Langeskov 33←
Langesund 27↖
Langevåg 84↘
Långflon 28↖
Langfjord 85↗
Langholm 20↗
Langnau 47←
Langogne 52←
Langon 50→
Langquaid 41↙
Långradna 31↖
Langres 46↖
Langrune 43↖
Langsele 84↖
Langset 28↖
Langshyttan 29↖
Langtoft 21→
Långträsk 86↘
Languedoc, reg. 51↘
Langwathby 21←
Langworth 21↘
Lanjarón 73↘
Lanklaar 39↖
Länna, *Sweden* 29←
Länna, *Sweden* 29→
Lannavaara 85→
Lannemezan 66↗
Lannilis 42←
Lannion 42←
Lanouaille 51↖
Lanškroun 88↘
Lanslebourg 53↖
Lantadilla 65←
Lantzói 96↘
Lanusei 61→
Lanžhot 49↗
Lanzo d'Intelvi 13↘
Lanzo-Torinese 53↖
Laodicea, anc. site 95→
Laon 38↘
Lápa 96↙
Lapad 16↙
Lapalisse 45↘
Lapford 22↙
Lapinlahti 87←
Lappajärvi 87←
Lappe 29↙
Lappeenranta 87↗
Lappfjärd 87↙
Lappi 87↘
Lappträsk 86↙
Låpseki 95↖
Lapua 87↙
L'Aquila 57←
Laracha 63↖
Laragh 25↖
Laragne 52→
Larbert 18↘
Lärbro 31↗
Larceveau 66↖
Lårdal 27←
Lardero 65→
Laredo 65↖
Largentière 52←
Largo 19↙
Largs 18↘
Larimna 96←
Larino 58↙
Lárisa 94↗
Larkhall 18↘
Larkollen 27↖
Larne 24↖
Laroche 45↗
Laroquebrou 51←
Laroque d'Olmés 67↖
Lárraga 65→
Larrau 66↖
Larsjö 84↖
Larsmo 87↙
Larsnes 26↙
Laruns 66↗
Larv 30↗
Larvik 27↖
Las Cabezas 73↙
Las Caldas de Besaya 65↖
Las Mesas 70→
Las Navas de la Concepción 73←
Las Navas del Marqués 69↗
Las Negras 74↘
Las Pedroñeras 70→
Las Rotas 15↘
Las Rozas 69↗
Las Ventas con Peña Aquilera 69→
Las Vertientes 74←
Lasarte 65↗
Låsby 32→
Łasin 89↖
Łask 89↙
Lasmos 93→
Lassay 44↗
Lassbyn 85↘
Lassigny 38→
Lastéika 96↘
Lastra a Signa 54↙
Lästringe 29↘
Lastrup 36←
Lašva 91←
Latchingdon 23→
Laterza 58→
Lathen 36←
Latheron 19↖
Latikberg 84↖
Latina 56→
Latour-d'Auvergne 51↗
Latour-de-France 67↗
Latronico 58→
Latronquière 51←
Latterbach Spiez 47↙
Laubach 40↖
Lauchheim 40↘
Lauda 40→
Laudal 27→
Lauder 19↙
Lauenburg 37↖
Lauf 41↙
Laufach 40→
Laufen, *Switzerland* 46→
Laufen, *W. Germany* 48↗
Lauffen 40↙
Laugharne 22←
Lauingen 40↘
Laujar de Andarax 74↙
Lauka 87←
Laumes, les 46←
Launceston 22↙
Laupheim 47↗
Lauragh 25↘
Laureana di Borrello 59→
Laurencekirk 19←
Laurenzana 58→
Lauria 58↘
Laurière 45↙
Laurieston 20↗
Laurila 86→
Laurito 58↘
Lauritsala 87↗
Lausanne 46↘
Lauterbach 40↗
Lauterbrunnen 47↙
Lauterecken 39→
Lauvås 27→
Lauvsnes 84←
Lauvstad 26↙
Lauvvik 27↘
Lauzerte 51←
Lauzet, le 52→
Lauzun 50→
Laval 44↖
Lavamund 49↙
Lavandou, le 52↘
Lavangen 85→
Lavardac 50→
Lavaur 51↙
Lavelanet 67↖
Lavello 58↙
Lavelsloh 36→
Lavenham 23↗
Laveno 47↙
Lavia 87↘
Lavik 26↘
Lavington 22→
Lavinio Lido di Enea 56↗
Lavit 51↙
Lavoûte-Chilhac 51↗
Lavre 68↙
Lávrion 96↗
Lavsjö 84↖
Lawrencetown 24↗
Laxa 28↘
Laxey 20→
Laxford Bridge 18↗
Laxne 29←
Laxo 19↗
Laxvik 30←
Layna 65↘
Laza 63→
Lazarevac 91↖
Lazdijai 89↗
Lazonby 21←
Le Bourget, airport 38↙
Le Grand-Lucé 44↗
Le Mans 44↗
Le Poiré-sur-Vie 44←
Le Val-André 42→
Le Ville 54↘
Lea 22→
Leadburn 19↙
Leaden Roding 23→
Leadenham 21↘
Leadhills 20↗
Leamington Spa = Royal Leamington Spa
Leap 25↘
Leatherhead 23←
Łeba 89↖
Lebach 39↘
Lebenstedt 37←
Lebesby 86↖
Lębórk 89↖
Lebrija 73↙
Leça da Palmiera 63↙
Lecce 58↗
Lecco 47↘
Lécera 66↘
Lech 47→
Lechenich 39↗
Lechlade 23←
Lechovice 49↗
Leciñena 66→
Leck 33↙
Lectoure 50↘
Lecumberri 65↗
Łeczyca 89←
Ledaña 71←
Ledanca 65↘
Ledbury 22→
Ledenice 55↗
Ledesma 64↙
Lédignan 52↙
Ledmare 18↗
Lędyczek 89↖
Leeds 21↙
Leek 21↙
Leenane 24↘
Lee-on-the-Solent 23↙
Leer 36↖
Leeuwarden 35↗
Leganés 69↗
Legé 44←
Legnano, *Italy* 53↗
Legnano, *Italy* 54↗
Legnica 88→
Legrad 49↘
Léguevin 67↖
Lehnin 37→
Lehrte 36→
Lehtimäki 87←
Leibnitz 49↘
Leicester 23↖
Leiden 35←
Leigh 21↙
Leigh-on-Sea 23→
Leighlinbridge 25↖
Leighton Buzzard 23←
Leikanger 26↙
Leikong 26↙
Leimbach 37→
Leinefelde 37↙
Leintwardine 22↗
Leipojärvi 85→
Leipzig 37↘
Leira 26↗
Leiranger 85←
Leirbotn 85↗
Leirgulen 26↙
Leiria 68←
Leiro 63←
Leirpollen 86↖
Leirvassbu 26←
Leirvik, *Norway* 26↘
Leirvik, *Norway* 27↙
Leisnig 41↗
Leiston 23↗
Leith 19↙
Leitrim 24→
Leivonmäki 87→
Leiza 65↗
Lekeryd 30↗
Lekhainá 96↘
Lékhovon 94↖
Lekhyttan 28→
Leknes 26↙
Leksand 28↗
Leksvik 84←
Lekvattnet 28↖
Lelystad 35→
Léman, Lac, lake 46↘
Lemberg = L'vov
Lembèye 66↗
Lembruch 36→
Lemgo 36→
Lemmer 35↗
Lempäälä 87→
Lempdes 51↗
Lemreway 18↖
Lemvig 32↘
Lena, *Norway* 26↗
Lena, *Sweden* 30↖
Lencroître 44→
Lend 48→
Lendava 49↘
Lengau 48→
Lengenfeld 41↖
Lengerich 36←
Lenggries 48←
Lenham 23→
Lenhovda 31←
Leningrad 87↗
Lenk 46↘
Lennartsfort 28←
Lenningden 26↗
Lenno 13↘
Lennoxtown 18↘
Leno 54↖
Lens 38→
Lensahn 33→
Lensvik 84↙
Lentini 60↗
Lentura 86↗
Lenvik 85↗
Lenzburg 47←
Lenzen 37↖
Lenzerheide-Lai 47↘
Leoben 49←
Leominster 22↗
Léon, *France* 50↙
León *Spain* 64→
Leonberg 40↙
Leondári 94↗
Leondárion 96→
Leonessa 56↗
Leonfelden 49↖
Leonforte 60→
Leonídhion 96→
Leopoldsburg 39↖
Leovo 92↗
Lepe 72→

Lepenoú 96↙
Leppävirta 87↖
Lépreon 96↘
Lepsény 90←
Lequeitio 65↗
Lerbäck 28↘
Lercara Friddi 60←
Lérici 54←
Lérida 66→
Lerin 65→
Lerissós 94↗
Lerma 65←
Lermoos 48←
Lerum 30↖
Lerwick 19↗
Lés 67↖
Les Fins 46→
Les Haudères 47↙
Lesa 13↙
Lesaca 65↗
Lescar 66↗
Lesce 55→
Lesconil 42↙
Lesično 49↘
Lesja 26←
Lesjaskog 84↘
Lesjöfors 28→
Lesko 89↘
Leskovac 93↙
Leskovik 94↖
Leslie 19↙
Lesmahagow 20↗
Lesna 49↗
Lesnica 90↗
Lesparre 50↖
Lespérau 51→
Lessay 43↖
Lessebo 31←
Lestelle-Bétharram 66↗
Lestijärvi 87←
Leswalt 20↖
Leszno 88→
Letafors 28↖
Letchworth 23←
Letino 57←
Letmathe 36↙
Letnitsa 93←
Letterkenny 24←
Letterston 22←
Leu 93↙
Léuca 59↖
Leuchars 19↙
Leuglay 46↖
Leuk 47↙
Leukerbad 47↙
Leupoldstein 41←
Leutkirch 47↗
Leuven 39↖
Leuze 38↗
Levádhia 96←
Levan 91↗
Levanger 84←
Lévanto 53→
Lévanzo 60↙
Levede 31↗
Leveld 26→
Leven, *England* 21→
Leven, *Scotland* 19↙
Levene 28↙
Levens Bridge 21←
Levens l'Escarène 53↙
Leverburgh 18↖
Leverkusen 39↗
Levice 90←
Levico 48↙
Levídhi 96→
Lévie 62←
Levier 46→
Levkás 96↙
Lévki 96↘
Levkímmi 96↘
Lévktra 96←
Levoča 89↘
Levroux 45←
Levski 93←
Lewes 23↘
Leyburn 21←
Leyland 21↙
Leysdown 23→
Leysin 46↘
Leżajsk 89↘
Lezat 67↖
Lezay 44↘
Lezhë 91↗
Lézignan 67↗
Lezoux 45↘
Lezuza 70↘
Li 86→
Lianokládhion 96←
Liapádhes 96↘
Liatorp 30→
Liberec 88↘
Libochovice 41→
Libohovë 94↖
Libourne 50→
Librilla 74→
Libros 71←
Licata 60→
Lich 40←
Lichfield 23↖
Lichtenau 40↙
Lichtenfels 41←
Lichtenstein 41↖
Licki Osik 55→
Licko Jesenica 55→
Liddel 21↖
Liden, *Sweden* 84↗
Liden, *Sweden* 86↘
Lidgate 23↙
Lidhult 30→
Lidingo 29→
Lido 54↗
Lido degli Estensi 54→
Lido degli Scachi 11↖
Lido del Savio 11↖
Lido delle Nazioni 11↖
Lido di Camaiore 54↙
Lido di Classe 11↖
Lido di Iésolo 54↗
Lido di Mortelle 60↖
Lido di Óstia 56→
Lidzbark 89↖
Liebenau, *Austria* 49↖
Liebenau, *W. Germany* 36→
Liebenstein 40↗
Liebenwalde 37↗
Liedakka 85→
Liège 39↖
Liegnitz = Legnica
Lieksa 87↖
Lienz 48↘
Lieokhoríkion 96←
Lier 27↖
Lierfoss 28←
Liérganes 65↖
Liestal 47←
Lieşti 92↗
Liétor 70↘
Liezen 49←
Liffol-le-Grand 46↖
Lifton 22↙
Ligga 85↘
Lignano Sabbiadoro 55↖
Lignières 45←
Ligny-en-Barrois 39↙
Ligny-le-Châtel 45↗
Ligoúrion 96→
Ligueil 44→
Liguria, prov. 53→
Likenäs 28↗
Liknes = Kvinesdal 27→
Likókhia 96→
Likoúri 96→
Lilaía 96←
L'Ile-Bouchard 44→
Lilla Edet 30↖
Lille 38↗
Lillebonne 43↗
Lillehammer 26↗
Lillerod 32↗
Lillers 38←
Lillesand 27→
Lillestrom 28←
Lillhärdal 84→
Lillholmsjön 84←
Lillholmträsk 86↘
Lilliholm 86↘
Lillo 70←
Lillsved 29→
Lima 28↗
Limanowa 89↘
Limavady 24↖
Limbach 41↗
Limburg 40←
Limedsforsen 28↗
Limén Vathéos 95←
Limenária 94↗
Limerick 25←
Limín = Thásos
Liminka 86→
Limmared 30↗
Limmen 35↖
Límnes 96→
Límni 96←
Limogárdhi 96←
Limoges 51↖
Limogne 51←
Limone Piemonte 53←
Limousin, reg. 51↖
Limoux 67↖
Limpsfield 23→
Linaälv 85→
Linares 73↗
Linares de Mora 71↖
Linares de Riofrío 69↖
Linariá 96↖
Linate, airport 53↗
Lincoln 21↘
Lindale 20→
Lindås, *Norway* 26↘
Lindås, *Sweden* 31←
Lindau 47→
Linderas 30↗
Linderhof 13←
Linderöd 30↘
Lindesnas 28↗
Lindfors 28→
Lindlar 40↖
Lindone 30↖
Líndos 95↘
Lindoso 63←
Lindsberg 29←
Lindved 32↘
Lingbo 84↗
Lingen 36←
Lingfield 23←
Linghed 29↖
Linguaglossa 60↖
Linhamn 30↙
Linköping 29↙
Linlithgow 19↙
Linneryd 30→
Linsell 84→
Linthal 47←
Linton 23↗
Lintzel 37↖
Linyola 67←
Linz, *W. Germany* 39→
Linz, *Austria* 49↖
Liomseter 26↖
Lion 43↖
Lion d'Angers, le 44←
Lioran, le 51↗
Lipa 90↘
Liperi 87↖
Liphook 23←
Lipiany 88↗
Lipkany 92←
Lipljan 91↖
Lipník 89↙
Lipniţa 93↖
Lipno 89←
Lippborg 36↙
Lippstadt 36↘
Lipsk 89↗
Liptovský Mikuláš 89↙
Lipusz 89↖
Liria 71←
Lirkía 96→
Lisboa 68↙
Lisbon = Lisboa
Lisburn 24↖
Liscarney 24↘
Liscarrol 25↘
Lisdoonvarna 25↙
Liseux 43↗
Liskeard 22↙
Lisle 51↙
L'Isle-sur-Serein 45→
Lismore 25→
Lisnaskea 24→
Liss 23↙
Lissaréa 96↘
Lisskogsåsen 28↗
List 33↙
Lista 29←
Listerby 31↙
Listowel 25↙
Lit 84↖
Litcham 23↗
Litchfield 23←
Lithakiá 94←
Litija 49↙
Litlos 26→
Litókhoron 94↗
Litoměřice 41↗
Litomyšl 88↘
Litschau 49↖
Litslena 29←
Little Fakenham 23↗
Little Ouseburn 21←
Little Walsingham 23↗
Littleborough 21↙
Littlehampton 23↙
Littleport 23↗
Littlestone-on-Sea 23↘
Littleton 25←
Litvínov 41↗
Livada 92↙
Livádhi 96↗
Livárji 96↘
Livanátais 96←
Livarot 43→
Liverpool 20↘
Livigno 47↘
Livingston 19↙
Livno 91←
Livorno 54↙
Livorno-Ferraris 53↖
Livron 52←
Lixnaw 25↙
Lixoúrion 94←
Lizard 22↘
Lizumer Hütte 48←
Lizy 38↘
Ljosland 27→
Ljubinje 91←
Ljubljana 49↙
Ljubuški 91←
Ljugarn 31↗
Ljung 30↖
Ljunga 84↗
Ljungaverk 84↗
Ljungby 30→
Ljungbyhed 30↘
Ljungbyholm 31←
Ljungdalen 84→
Ljunghusen 30↙
Ljungsbro 29↙
Ljungskile 28↙
Ljüsdal 84↗
Ljusne 84↗
Ljusterö 29→
Ljutomer 49↘
Llafranch 15↖
Llagostera 67→
Llanaelhaiarn 20↘
Llanarth 22↖
Llanaves 64↗
Llanbadarnfynydd 22↗
Llanbedr 22↖
Llanbedrog 20↘
Llanberis 20↘
Llanbister 22↗
Llanboidy 22←
Llanbryde 19←
Llandaff 22→
Llanddarog 22←
Llandegla 20↘
Llandeilo 22←
Llandissilio 22←
Llandovery 22←
Llandrillo 20↘
Llandrindod Wells 22↗
Llandudno 20↘
Llandyssul 22↖
Llanelli 22←
Llanelltyd 22↖
Llanerchymedd 20↘
Llanes 64↗
Llanfair Caereinion 22↗
Llanfair Talhaiarn 20↘
Llanfairfechan 20↘
Llanfarian 22↖
Llanfyllin 22↗
Llangadfan 22↗
Llangadog 22←
Llangefni 20↘
Llangeler 22←
Llangelynin 22↖
Llangollen 20↘
Llangurig 22↗
Llangynog 22↗
Llanidloes 22↗
Llanilar 22↖
Llanrhidian 22←
Llanrhystyd 22↖
Llanrwst 20↘
Llansá 67→
Llansantffraid-ym-Mechain 22↗
Llanstephan 22←
Llanthony 22→
Llantrisant 22→
Llantwit Major 22→
Llanwrda 22←
Llanwrtyd Wells 22↖
Llanybydder 22↖
Llanymynech 22↗
Llavorsi 67←
Llera 69↙
Llés 67←
Llívia 67←
Llodio 65↖
Llombay 71←
Lloret de Mar 67→
Llosa de Ranes 71↙
Lluchmayor 75↙
Llwyngwril 22↖
Llyswen 22→
Loanhead 19↙
Loano 53→
Loans 20↗
Löbau 88→
Löbejün 37↘
Lobenstein 41←
Löberöd 30↘
Łobez 88↗
Löbnitz 33↗
Lobón 69↙
Lobonäs 84↗
Loburg 37→
Locarno 47↙
Lochailort 18→
Lochaline 18↙
Lochboisdale 18←
Lochbuie 18↙
Lochcarron 18→
Lochearnhead 18↘
Lochem 35→
Loches 44→
Lochgair 18↘
Lochgelly 19↙
Lochgilphead 18↘
Lochgoilhead 18↘
Lochinver 18↗
Lochlee 19←
Lochmaben 20↗
Lochmaddy 18←
Lochranza 20↖
Lochwinnoch 18↘
Lochy Bridge 18→
Lockerbie 20↗
Locknevi 31↖
Locle, le 46→
Lochmariaquer 42↘
Locminé 42↘
Locri 59→
Löddeköpinge 30↙
Loddon 23↗
Lodé 61↗
Löderup 30↘
Lodève 51↘
Lodi 53↗
Lödingen 85←
Lodosa 65→
Lödöse 30↖
Łódź 89←
Loeches 70↖
Loen 26←
Löfallstrand 27↙
Lofer 48→
Lofsdalen 84→
Loftahammar 31↖
Lofthus 26↘
Loftus 21→
Loga 36↖
Loggerheads 21↙
Logis du Pin, le 53↙
Logroño 65→
Logrosán 69←
Logstør 32↙
Løgumkloster 33↙
Lohals 33←
Lohausen, airport 39↗
Lohijärvi 86↙
Lohja 87→
Lohmen 33↗
Lohne 36→
Lohnsfeld 39→
Lohr 40→
Lohtaja 87←
Loimaa 87↘
Loire, river 45←
Loja 73→
Loka 28→
Lokalahti 87↘
Lokeren 38↗
Loket 41→
Lokka 86→
Løkken, *Denmark* 32←
Lökken, *Norway* 84↘
Lolland, island 33←
Lollar 40↖
Lom, *Bulgaria* 93↙
Lom, *Norway* 26←
Łomazy 89→
Lombardia, prov. 53↗
Lombez 67↖
Lomello 53↗
Lomma 30↙
Lommatzsch 41↗
Lommel 39↖
Lomond, Loch, lake 18↘
Lomsjö 84↖
Łomża 89↗
Londinières 38←
London 23←
Londonderry 24←
Lone 27↙
Long Ashton 22→
Long Bennington 21↘
Long Compton 23←
Long Eaton 21↙
Long Marston 21←
Long Melford 23↗
Long Preston 21←
Long Stratton 23↗
Long Sutton 23↗
Longá 94←
Longaníkos 96→
Longarone 48↘
Longástra 96→
Longbridge Deverill 22→
Longeau 46←
Longega 48↙
Longemer 46↗
Longford 24→
Longhorsley 21↖
Longhoughton 21↖
Longjumeau 38↙
Longny 44↗
Longobucco 59←
Longridge 21↙
Longside 19←
Longton 21↙
Longtown 20↗
Longué 44→
Longuyon 39↙
Longwy 39↙
Lonigo 54↗
Löningen 36←
Lons-le-Saunier 46↙
Lonsboda 30→
Lönsdal 85↙
Lönset 84↘
Loo Bridge 25↘
Looe 22↙
Loon op Zand 35→
Lopar 55→
Lopcombe Corner 23←
Lopera 73→
Lopovo 91↖
Loppa 85↗
Lopud 16↙
Lora 26←
Lora del Río 73←
Lorca 74→
Lorch, *W. Germany* 39→
Lorch, *W. Germany* 40↘
Lörenskog 28←
Lorentzen 39↘
Lorenzana 63↗
Loreo 54→
Loreto 55↙
Lorette 52↖
Loriga 68↗
Loriol 52←
Lormes 45→
Lörrach 47←
Lorraine, reg. 39↙
Lorrha 25←
Lorris 45↖
Lorsch 40←
Los 84↗
Los Alcázares 75←
Los Arcos 65→
Los Arenales del Sol 15↙
Los Balbases 65←
Los Barrios 73↙
Los Battios de Luna 64↗
Los Berengueles 14↘
Los Boliches 14↙
Los Corrales 73←
Los Corrales de Buelna 65↖
Los Cortijos 69→
Los Dolores 74→
Los Hinojosos 70→
Los Monteros 14↙
Los Navalmorales 69→
Los Navalucillos 69→
Los Palacios y Villafranca 73←
Los Rabanos 65↘
Los Santos 69↖
Los Santos de Maimona 69↙
Los Villares 73→
Los Yébenes 69→
Losacino 64↙
Losar de la Vera 69←
Losheim 39→
Loshult 30→
Lorient 42↙
Losna 84↘
Losser 36←
Lossiemouth 19↖
Lostwithiel 22↙
Löt 31←
Lote 26↙
Loten 28↖
Lotorp 29↙
Lötzen = Gizycko

Loudéac 42→
Loudon 44→
Loué 44↗
Loughborough 23↖
Loughor 22←
Loughrea 25←
Louhans 46↙
Louisburgh 24↘
Louká 96→
Loulé 72←
Louny 41→
Loupe, la 45↖
Lourdes 66↗
Loures 68↙
Loures Barousse 66↗
Louriçal 68←
Lourinha 68←
Lourosa 68↗
Lousã, *Portugal* 68←
Lousã, *Portugal* 68↙
Lousada 63↙
Lousiká 96↙
Loutoúfi 96←
Louth, *England* 21↘
Louth, *Rep. of Ireland* 24↗
Loutrá Aidhipsoú 96←
Loutrá Ipátis 96←
Loutrá Killínis 96↘
Loutráki 96→
Loutraki Spailiá 96→
Loutrón 96↙
Loútsai 96↘
Louvesc, la 52↖
Louvière, la 38→
Louviers 38↙
Louvigné-du-Désert 44↖
Lövanger 86↘
Lovasberény 90←
Lövberg 86↘
Lövberga 84↖
Lovech 93←
Lóvere 47↘
Lövestad 30↘
Lövhögen 84→
Loviisa = Lovisa
Lovikka 85→
Lovinac 55→
Lovios 63←
Lovisa 87→
Lövlund 26↖
Lövnäs, *Sweden* 84→
Lövnäs, *Sweden* 85↘
Lövo 49→
Lovosice 41→
Lovran 55↖
Lovrin 90↗
Lövsjon, *Sweden* 28↗
Lövsjön, *Sweden* 84↖
Lövstabruk 29↗
Löwenberg 37↗
Löwenstein 40↘
Lower Austria = Niederösterreich
Lower Saxony = Niedersachsen
Lowestoft 23↗
Lowick, *England* 20→
Lowick, *England* 21↖
Łowicz 89←
Lozarevo 93↖
Loznica 91↖
Lozorno 49↗
Lozoyuela 65↙
Luarca 64↖
Lubań 88→
Lubartów 89→
Lubawa 89↖
Lübbecke 36→
Lübben 88→
Lübeck 33→
Lubenec 41→
Lubian 63→
Lubień 89↖
Lubin 88→
Lublin 89→
Lubliniec 89↙
Lubrin 74→
Lubsko 88→
Lübtheen 37↖
Lübz 37↗
Luc 43↖
Luc-en-Diois 52→
Luc, le 52↘
Lucan 25↖
Lucainena de las Torres 74→
Lucca 54↙
Lucena 73→
Lucena del Cid 71→
Lučenec 90↖
Lucera 58↙
Lucerne, Lake of = Vierwaldstätter-See

Lüchow 37↖
Luciana 69↘
Lucido 59←
Lucie-de-Tallan 62←
Lucija 16↖
Lucillo 64←
Luckau 88→
Luckenwalde 37→
Luco de Jiloca 66↙
Luçon 44↙
Ludag 18←
Ludborough 21↘
Ludbreg 49↘
Lüdenscheid 40↖
Ludford 21↘
Ludgershall 23←
Ludgo 29↙
Lüdinghausen 36↙
Ludlow 22↗
Luduș 92→
Ludvika 29↖
Ludwigsburg 40↘
Ludwigshafen 40↙
Ludwigslust 37↖
Ludwigsstadt 41←
Luesia 66←
Lugagnano Val d'Arda 53→
Lugano 47↙
Lugau 41↗
Lugnås 28↘
Lugo, *Italy* 54→
Lugo, *Spain* 63↗
Lugoj 92↘
Lugones 64↗
Lugton 18↘
Luhanka 87→
Luib 18←
Lukova 91↗
Lukovit 93←
Lukovo Sugarje 55→
Łuków 89→
Luksfjell 27↖
Lukve 48↘
Lüleburgaz 93↗
Luleå 88↘
Lulsgate 22→
Lulworth Cove 22↘
Lumbier 66↖
Lumbrales 63↘
Lumbreras 65→
Lumbres 38↖
Lumby 33←
Lumijoki 86→
Lummelunda 31↗
Lumparland 87↘
Lumphanan 19←
Lumpiaque 66←
Lumscheden 29↖
Lumsden 19←
Lun 55→
Luna 66←
Lunano 54↘
Lund 30↘
Lunda 29→
Lunde, *Norway* 26↙
Lunde, *Norway* 27↖
Lunde, *Norway* 27→
Lunden 33↘
Lunderseter 28↖
Lunderskov 33↙
Lundin Links 19↙
Lundsbrunn 28↘
Lüneburg 37↖
Lunel 52↙
Lünen 36↙
Lunéville 39↘
Lungern 47←
Lungro 59←
Lungsjön 84↖
Lungsund 28↙
Lunna 19↗
Lunner 26↗
Lunz 49←
Luogosanto 61↗
Luque 73→
Lurcy-Lévy 45→
Lure 46→
Lurgan 24↗
Lurisia 53←
Luröy 85↙
Lushnjë 91↗
Lusignan 44↘
Lusk 24↗
Luso 68↖
Luspebryggan 85→
Luss 18↘
Lussac-les-Châteaux 44↘
Lussan 52←
Lustenau 47→
Luster 26←

Lutjenburg 33→
Lutnes 28↖
Luton 23←
Lutterworth 23↖
Lützen 41↖
Lutzerath 39→
Lützow 33→
Luumaki 87↗
Luusua 86↙
Luvia 87↘
Luvos 85↘
Luvozero 86↗
Luxembourg 39→
Luxeuil-les-Bains 46↗
Luz 66↗
Luzarches 38↙
Luzech 51←
Luzern 47←
Lužice 49↗
Luzy 45→
L'vov 89↘
Lwówek 88→
Lwówek Slaski 88→
Lyaskelya 87↗
Lyaskovets 93←
Lybster 19↖
Lychen 37↗
Lyck = Elk
Lycke 30↖
Lyckeby 31↙
Lycksele 84↖
Lydd 23↘
Lydford 22↙
Lydham 22↗
Lydney 22→
Lye 31↗
Lyfjord 85↗
Lygna 26↗
Lykkia 26→
Lyme Regis 22↘
Lymington 23↙
Lympne 23→
Lyndhurst 23↙
Lyneham 22→
Lyness 19↖
Lyngdal, *Norway* 27↖
Lyngdal, *Norway* 27→
Lyngen 85↗
Lyngseidet 85↗
Lynmouth 22←
Lynton 22←
Lyon 52↖
Lyons-la-Forêt 38↙
Lyonshall 22↗
Lyrestad 28↘
Lyseba 27↙
Lysebotn 27↙
Lysekil 28↙
Lyss 46→
Lysvik 28→
Lytchett Minster 22↘
Lytham 20↘
Lytham St Annes 20↘
Lythe 21→
Lyubimets 93→

Maam Bridge 24↘
Maam Cross 24↘
Maaninka 87←
Maarianhamma = Mariehamn
Maas 24←
Maaseik 39↖
Maassluis 35←
Maastricht 39↖
Mablethorpe 21↘
Macael 74→
Mação 68→
Macclesfield 21↙
Macduff 19←
Maceda 63→
Macedo de Cavaleiros 63↘
Macedonia = Makedonija
Maceira 68←
Macelj 49↘
Macerata 55↙
Machault 39↙
Machecoul 44←
Machine, la 45→
Machrihanish 20↖
Machynlleth 22↖
Mâcin 92↗
Macinaggio 62↖
Mackmyra 29↖
Macomér 61←
Mâcon 46↙
Macosquin 24↖
Macotera 64↘
Macroom 25↘
Macugnaga 47↙
Macure 55↘

Madangsholm 30↗
Maddaloni 57↙
Maderuelo 65↙
Madésimo 47↘
Madley 22→
Madonna di Campiglio 48↙
Madrid 70↖
Madridejos 70←
Madrigal de la Vera 69←
Madrigal de las Atlas Torres 64↘
Madrigalejo 69↙
Madrigueras 71←
Madroñera 69←
Maella 66↘
Maentwrog 20↘
Măeruș 92→
Maesteg 22←
Maestu 65→
Mafra 68↙
Magallón 66←
Magaña 65→
Magaz 65←
Magdeburg 37←
Magellarë 91↗
Magenta 53↗
Maggiore, Lago, lake 47↙
Maghera 24↖
Magherafelt 24↖
Maghull 20↘
Magione 56↗
Maglaj 91←
Maglehem 30↘
Máglie 58↗
Magnac-Laval 45↙
Magnor 28←
Magny-en-Vexin 38↙
Magouládhes 96↘
Magoúliana 96→
Maguilla 69↙
Maguire's Bridge 24→
Mahalás 96↙
Mahón 75→
Mahora 71↙
Mähring 41←
Maia 63↙
Maiche 46→
Máida 59←
Maiden Newton 22↘
Maidenhead 23←
Maidstone 23→
Maillezais 44↙
Mailly-le-Camp 38↘
Mainburg 41↙
Maine, reg. 44↖
Mainsriddle 20↗
Maintenon 38↙
Mainua 87←
Mainz 40←
Maiorca 68←
Maiori 11↘
Mairena del Alcor 73←
Maison-Neuve 46↖
Maisons Blanches, les 44↘
Maisons-Laffite 38↙
Maissau 49↗
Maissin 39←
Maizières-les-Vic 39↘
Majdan 89↘
Majdanpek 93↙
Majorca = Mallorca
Majšperk 49↘
Makarska 91←
Makedonija, prov. 91↗
Makhaira 96↙
Makkaur 86↗
Makó 90↗
Makov 89↙
Maków Mazowiecki 89↗
Makrakómi, *Greece* 96↖
Makrakómi, *Greece* 96←
Mákri, *Greece* 93→
Mákri, *Greece* 96→
Makrikáppa 96↖
Makrísia 96↘
Makrokhóri 96↖
Malá 73→
Mala Krsna 91↖
Malacky 49↗
Maladeta, Pico de la, mt. 66↗
Málaga 73↘
Malagón 69↘
Malahide 24↗
Malalbergo 54→
Malandríni 96→
Malandrínon 96←
Malangen, *Norway* 85↗
Malangen, *Sweden* 28↖
Mălani 92←
Mälaren, lake 29←
Målaskog 30→

Malatianoi 96↖
Malåtrask 85↘
Malaucène 52→
Malaunay 38↙
Malax 87↙
Malbork 89↖
Malcésine 54↖
Malchin 33↗
Malchow 37↗
Maldegem 38↗
Maldon 23→
Malé 48↙
Máleme 94↘
Malène, la 51→
Malente 33→
Mälerås 31←
Máles 95↙
Malesco 47↙
Malesherbes 45↖
Malesiádha 96↙
Malesína 96←
Malexader 31↖
Malgovik 84↖
Malgrat 67→
Mali Losinj 55→
Malicorne-sur-Sarthe 44↗
Målilla 31↖
Malin 24←
Malin More 24↙
Malingsbo 29←
Malinska 55↗
Maljevac 55↗
Malkara 93↗
Malkenes 26↘
Malko Tŭrnovo 93↗
Mallaig 18←
Mallaranny 24↘
Mallén 66←
Mallersdorf 41↙
Malles 47↘
Mallorca, island 75↘
Mallow 25→
Mallwyd 22↖
Malmbäck 30↗
Malmberget 85→
Malmby 29←
Malmedy 39↙
Malmei 27↘
Malmesbury 22→
Malmköping 29←
Malmö 30↙
Malmön 28↙
Malmslätt 29↙
Malnes 85←
Malo 54↗
Malo-les-Bains 38↖
Målav 32↗
Måloy 26↙
Malparitda de Cáceres 69←
Malpartida de la Serena 69↙
Malpartida de Plasencia 69←
Malpas 21↙
Malpensa, airport 53↗
Malpica, *Portugal* 68→
Malpica, *Spain* 63↖
Malpica, *Spain* 69→
Mals 47↘
Målsåker 29→
Målselv 85↗
Målset 26↘
Malsjö 28↗
Målsryd 30↗
Maltby 21↙
Maltby le Marsh 21↘
Malton 21→
Malung 28↗
Malungsfors 28↗
Malvaglia 47↙
Malveira 68↙
Malvern Wells 22↗
Malzieu, le 51→
Mambaloú 96↗
Mamers 44↗
Mammern 47←
Mamonovo 89↖
Mána 96→
Manacor 75↙
Mánari 96→
Mancha Real 73→
Manchester 21↙
Manciano 56↖
Manciet 50↘
Mandal 27→
Mándas 61→
Mandayona 65↘
Mandello 47↘
Mandoúdhion 96↖
Mándra, *Greece* 96↖
Mándra, *Greece* 96↘
Manduria 58↗

Manérbio 54↖
Manfredonia 58↙
Mangalia 93↖
Manged 28←
Manger 26↘
Mångsbodarna 28↗
Mangualde 68↗
Maniago 48↘
Manilva 73↙
Manisa 95←
Manises 71←
Månkarbo 29↗
Manlleu 67→
Mannersdorf 49↗
Mannheim 40↙
Manningtree 23→
Manolás 96↙
Manoppello 57←
Manorbier 22←
Manorhamilton 24→
Manosque 52↘
Manresa 67←
Mans, le 44↗
Mansfield 21↙
Mansilla 65→
Mansilla de las Mulas 64→
Mansle 44↘
Manteigas 68↗
Mantes 38↙
Manthiréa 96→
Mantinea, anc. site 96→
Mantorp 29↙
Mantova 54↖
Mäntsälä 87→
Mantta 87→
Mäntyharju 87→
Mäntyluoto 87↘
Manzanares 70↙
Manzanares el Real 70↖
Manzanera 71←
Manzat 45↘
Maqueda 69→
Mår 26→
Maracena 73→
Maranchón 65↘
Marano 54←
Marans 44↙
Mărășești 92↗
Maratea 58↘
Marateca 68↙
Marathón 96↖
Marathos 96↙
Marathópolis 96↘
Marazion 22↘
Marba 26→
Marbach, *Austria* 49↖
Marbach, *W. Germany* 40↘
Marbäck 30↗
Marbella 73↙
Marbesa 14↙
Marburg 40↖
Marby 84→
Marčana 55←
Marcaria 54↖
March 23↗
Marchamalo 70↗
Marche 39←
Marche, reg., *France* 45↙
Marche, prov., *Italy* 55↙
Marchegg 49↗
Marchena 73←
Marchstein 49↖
Marchtrenk 49↖
Marchwiel 20↘
Marciac 66↗
Marciana Marina 56↖
Marcigny 45↘
Marcilla 65→
Marcillac-Vallon 51→
Marcilly 45↗
Marck, airport 38↖
Marckolsheim 47↖
Marco de Canavezes 63↙
Marden 23→
Mårdsel 85↘
Mårdsele 86↘
Mårdsjö 84↖
Mårdsjön 84↖
Marebello 11↗
Marennes 44↙
Maresfield 23↘
Marèttimo 60↙
Mareuil, *France* 50↗
Mareuil, *France* 44↙
Margam 22←
Margarítion 94↖
Margate 23→
Margharita di Savoia 58←
Mărghita 92↙
Margina 92↘

Margone 53↖
Marí 96→
Maria 74→
Maria Laach 39→
Mariager 32→
Mariannelund 31↖
Mariánské-Lázně 41→
Mariazell 49←
Maribo 33↖
Maribor 49↘
Maridalen 27↖
Mariedam 29↙
Mariefried 29←
Mariehamn 87↘
Marienberg 41↗
Marienbourg 39←
Marienburg = Malbork
Marienwerder = Kwidzyn
Mariestad 28↘
Marifjöra 26←
Marignane 52↘
Marigny le Châtel 45↗
Marija Bistrica 49↘
Marin 63←
Marina di Badolato 59→
Marina di Campo 56↖
Marina di Carrara 54↙
Marina di Gioiosa 59→
Marina di Grosseto 56↖
Marina di Massa 54↙
Marina di Pisa 54↙
Marina di Ravenna 54→
Marina Montegiordano 58→
Marina Romea 11↖
Marinaleda 73←
Marine de Sisca 62↖
Marines 38↙
Marinetta 54↙
Maringues 45↘
Marinha Grande 68←
Marino 56→
Marisholm 30↘
Maristuen 26→
Maritsa 93→
Markabygd 84←
Markaryd 30→
Markátes 96↖
Markdorf 47→
Market Bosworth 23↖
Market Deeping 23↖
Market Drayton 21↙
Market Harborough 23↖
Market Rasen 21↘
Market Warsop 21↙
Market Weighton 21→
Markethill 24↗
Markham Moor 21↘
Markhus 27↙
Markinch 19↙
Markitta 85→
Markkleeberg 41↖
Markneukirchen 41←
Markópoulon 96↗
Markranstädt 37↘
Marks Tey 23→
Markšuhl 40↗
Markt Bibart 40→
Markt Indersdorf 48↖
Markt Rettenbach 47↗
Markt Schwaben 48↖
Marktbreit 40→
Marktheidenfeld 40→
Marktoberdorf 47↗
Marktredwitz 41←
Markyate 23←
Marl 36↙
Marlborough 23←
Marle 38→
Marlow, *England* 23←
Marlow, *E. Germany* 33↗
Marma, *Sweden* 29↗
Marma, *Sweden* 84↗
Marmande 50→
Mármara 96←
Marmaraereğlisi 93↗
Marmári 96↖
Marmaris 95→
Marmelete 72←
Marmolada, Gruppo della, mt. 48↙
Marmolejo 73↗
Marmoutier 39↘
Marnach 39←
Marnås 29↖
Marnay 46←
Marne 33↘
Marnheim 40←
Marnitz 37↖
Maroldsweisach 40↘
Marolles les-Braults 44↗
Maromme 38↙
Maróstica 54↗
Marquina 65↗
Marquise 38↖
Marraskoski 86↙
Marságlia 53→
Marsala 60↙
Marsciano 56↗
Marseillan 51↘
Marseille 52↘
Marseille-en-Beauvaisis 38←
Marshbrook 22↗
Marshfield 22→
Marske 21←
Mårsta 29→
Marstal 33←
Marstrand 30↖
Martel 51←
Martelange 39←
Martello 47↘
Martfeld 36→
Martigny 46↘
Martigny-les-Bains 46↖
Martigues 52↙
Martim Longo 72←
Martin 89↙
Martin de la Jara 73←
Martin de Yeltes 64↙
Martin del Río 66↙
Martin Muñoz 64↘
Martina, *Italy* 58↗
Martina, *Switzerland* 47→
Martinengo 53↗
Martinez 69↖
Martinganca 68←
Martínon 96←
Martock 22↘
Martoft 33←
Marton 21↘
Martorell 67↙
Martos 73→
Martres-Tolosane 67↖
Martrou 44↙
Martti 86←
Marttila 87↘
Marum 35↗
Maruševec 49↘
Marvão 68→
Mårvatn 26→
Marvejols 51→
Marybank 18→
Maryculter 19←
Marykirk 19←
Maryport 20→
Marzahne 37→
Mas-d'Azil, le 67↖
Mas de la Matas 66↘
Mas Neuf 51↘
Masamagrell 71→
Maschen 36↗
Maseqoso de Tajuña 65↘
Masevaux 46→
Masfjorden 26↘
Masham 21←
Masi 85↗
Másino 47↘
Masku 87↘
Masnou 67↘
Massa 54↙
Massa Lombarda 54→
Massa Lubrense 57↙
Massa Marittima 56↖
Massa Martana 56↗
Massafra 58→
Massanet de Cabrenys 67→
Massat 67↖
Masseube 66↗
Massiac 51↗
Massif Central, mts. 51↗
Mästerby 31↗
Mašun 55↖
Matadepera 67←
Matamorosa 65↖
Mataporquera 65↖
Matapozuelos 64↘
Mataránga 96↙
Mataró 67↘
Matélica 55↙
Matera 58→
Mátészalka 92↙
Matfors 84↗
Matha 44↘
Mathry 22←
Máti 96→
Matilla de los Caños 64↘
Matlock 21↙
Matlock Bath 21↙
Matozinhos 63↙
Matre, *Norway* 26↘
Matre, *Norway* 27↙
Matrei 48←
Matrei in Osttirol 48→
Matterhorn, mt. 47↙
Mattersburg 49→
Mattmar 84←
Mátzani 96→
Matzen 49↗
Maubeuge 38→
Maubourguet 66↗
Mauchline 20↗
Maud 19←
Maulbronn 40↙
Mauld 86↙
Mauléon 44←
Mauléon-Soule 66↖
Mauriac 51↖
Mauron 42↘
Maurs 51←
Mautern 49↗
Mauterndorf 48→
Mauvezin 51↙
Mauvoisin 46↘
Mauze 44↙
Mavríkion 96↙
Mavrommáti 96←
Mavronéri 96←
Mawnan 22↙
Maxmo 87↙
Maxwellheugh 21↖
Maxwelltown 20↗
Mayals 66↘
Maybole 20↗
Mayen 39→
Mayenne 44↗
Mayeraille, la 43↗
Mayet 44↗
Mayet-de-Montagne, le 45↘
Mayfield, *England* 21↙
Mayfield, *England* 23↘
Mayorga 64→
Mayrhofen 48←
Mazamet 67↗
Mazara del Vallo 60↘
Mazaráki 96↙
Mazaricos 63↖
Mazarrón 74→
Mazéres 67↖
Mazzarino 60→
Mealhada 68↖
Méan 39←
Measham 23↖
Meaux 38↘
Mechelen 39↖
Mechernich 39→
Mecidiye 95↖
Mecina Bombaron 74↙
Meda 63↘
Medak 55→
Medebach 40↖
Medellin 69↙
Medemblik 35↖
Medevibrunn 28↘
Medgidia 93↖
Mediaş 92→
Medicina 54→
Medina de las Torres 69↙
Medina de Pomar 65↖
Medina de Rioseco 64→
Medina del Campo 64↘
Medina Sidonia 73↙
Medinaceli 65↘
Médoc, reg. 50↖
Medulin 55←
Meerane 41↖
Meersburg 47→
Meeth 22↙
Méga Khorió 96↙
Méga-Pondiás 96↙
Megalópoliso 96↘
Mégara 96→
Megaspílaion, anc. site 96←
Megève 46↘
Mehadia 92↘
Mehamn 86↖
Mehede 29↗
Meheia 27↖
Mehun-sur-Yèvre 45←
Meigle 19↙
Meijel 35↘
Meikleour 19↙
Meilen 12↖
Meimoa 68→
Méina 13↙
Meine 37←
Meinerzhagen 40↖
Meiningen 40↗
Meira 63↗
Meiringen 47↙
Meisenheim 39→
Meissen 41↗
Meitingen 40↘
Mélambes 94↘
Melba 85←
Melbourn 23↗
Melbourne 23↖
Melchtal 12↖
Meldal 84↘
Melderstein 85↘
Meldola 54→
Meldorf 33↘
Melegnano 53↗
Melendugno 58↗
Melfi 58↙
Melgaço 63←
Melgar de Fernamental 65←
Melhus 84←
Melida 65→
Melides 72↖
Meligalá 96↘
Melisey 46→
Melíssi 96←
Melito di Porto Salvo 59↘
Melksham 22→
Mellansjö 84↗
Mellbystrand 30←
Melle 36→
Mellerud 28↙
Mellid 63↗
Mellor Brook 21↙
Mellrichstadt 40→
Melmerby 21←
Melnitsa Podolskaya 92←
Melöy 85←
Melrose 21↖
Mels 12↗
Melsbroek = National, airport
Melsungen 40↗
Meltaus 86↙
Melton Mowbray 23↖
Melun 45↖
Melvaig 18↖
Melvich 18↗
Mélykút 90↗
Membrilla 70↙
Membrio 68→
Membrolle, la 44→
Memmingen 47↗
Memuruba 26←
Menággio 47↘
Menai Bridge 20↘
Menasalbas 69→
Mendavia 65→
Mende 51→
Menden 36↙
Mendrisio 47↙
Menemen 95←
Menen 38↗
Menetai 95↙
Menfi 60↘
Mengen 47↗
Mengeringhausen 40↖
Mengibar 73→
Mengshol 26↗
Mennetou 45←
Mennock 20↗
Menonen 87→
Menorca, island 75→
Mentana 56→
Menthon 46↘
Menton 53↙
Meopham 23→
Méounes 52↘
Meppen 36→
Meppel 35↗
Mequinenza 66↘
Mer 45←
Meråker 84←
Meran 48↙
Merano 48↙
Merate 53↗
Mérbakas 96→
Mercadal 75→
Mercatale 54↘
Mercato San Severino 58↘
Mercato Saraceno 54↘
Mercurea 92↘
Merdrignac 42→
Mere 22→
Mere Corner 21↙
Méribel-les-Allues 12↙
Merida 69↙
Merikarvia 87↘
Mering 48↖
Merkebekk 27↖
Merkendorf 40↘
Merkenes 85↙
Merkine 89↗
Merlebach 39↘
Merópi 96↘
Merrow 23←
Mers 38←
Mersch 39←
Merseburg 37↘
Merthyr Tydfil 22→
Mértola 72←
Merton 22↙
Méru 38↙
Merville 38↖
Méry 45↗
Merzenstein 49↖
Merzig 39↘
Mesagne 58↗
Mesão Frio 63↘
Meschede 40↖
Meselefors 84↖
Meshaw 22↙
Meslay-du-Maine 44↗
Mesle, le 44↗
Mesnalien 26↗
Mesocco 47↘
Mésola 54→
Mesolóngion 96↙
Mesón del Viento 63↖
Mesorrákh 96→
Messina 60↖
Messíni 96↘
Messkirch 47↖
Mestanza 69↘
Mestlin 37↖
Město Teplá 41→
Mestre 54↗
Métabief 46↘
Metaxádha 96↘
Metelen 36←
Méthana 96↗
Metheringham 21↘
Methil 19↙
Methlick 19←
Methóni, anc. site 94↙
Methven 19↙
Metković 91←
Metlika 55↗
Metohija = Kosovo i Metohija
Metovnica 93↙
Métsovon 94↖
Metten 41↘
Mettingen 36←
Mettlach 39↘
Mettmann 39↗
Metz 39↘
Meulan 38↙
Meung 45↖
Meursault 46←
Meuselwitz 41↖
Mevagissey 22↙
Mexborough 21↙
Mexiátais 96←
Meximieux 46↙
Mey 19↖
Meyenburg 37↗
Meymac 51↖
Meyrm 46↘
Meyrueis 51→
Mezdra 93↙
Mèze 51↘
Mezhdurech'ye 89↗
Mézières-en-Brenne 45←
Mezimostí 88↘
Mézin 50↘
Mezőcsáto 90↖
Mezőhegyes 90↗
Mezőkovacsháza 90↗
Mezőkövesd 90↖
Mezőtúr 90↖
Mezquitilla 14↘
Mezzolombardo 48↙
Miajadas 69↙
Miastko 89↖
Michalovce 90↖
Micheldorf 49←
Michelstadt 40←
Michendorf 37→
Michurin 93↗
Mickleton 23↖
Mid Calder 19↙
Mid Yell 19←
Middelburg 35↙
Middelfart 33←
Middelharnis 35←
Middelkerke Bad 38↗
Middleham 21←
Middlesbrough 21←
Middleton 21↙
Middleton in Teesdale 21←
Middleton Stoney 23←
Middletown, *N. Ireland* 24↗
Middletown, *Wales* 22↗
Middlewich 21↙
Midhéa 96→
Midhurst 23↙
Midleton 25→
Midsomer Norton 22→
Midtgulen 26↙
Midtskog 28↖
Midyé 93↗
Miechów 89↙
Miedes 66↙
Międzylesie 88↘
Międzyrzec 89→
Międzyrzecz 88→
Miejska Gorka 89←
Mielan 66↗
Mielec 89↘
Miercurea Ciuc 92→
Mieres 64↗
Mieron 85↗
Miersterhorst 37←
Miesbach 48←
Mieszkowice 88→
Migliarino 54→
Miguel Esteban 70→
Miguelturra 70↙
Mihăileni 92←
Mikhaylovgrad 93↙
Mikhói 96↘
Mikkeli 87→
Mikołajki 89↗
Mikołów 89↗
Mikre 93←
Mikrománi 96↘
Mikrón Pondiás 96↙
Mikulov 49↗
Milagro 65→
Miland 27←
Milano 53↗
Milano Marittima 54→
Milazzo 60↖
Mildenhall 23↗
Miléai, *Greece* 94↗
Miléai, *Greece* 96←
Milesi 96↖
Mileto 59→
Miletus, anc. site 95←
Milevsko 88↘
Milford, *England* 23←
Milford, *Rep. of Ireland* 24←
Milford Haven 22←
Milford on Sea 23↙
Milhão 63↘
Miliá 96←
Miliaís 96↘
Milicz 89←
Milín 41→
Milis 61←
Mill 35→
Millas 67↗
Millau 51→
Millom 20→
Millport 18↘
Millstreet 25↘
Milltown Malbay 25↙
Milly 45↖
Milmarcos 65↘
Milnathort 19↙
Milngavie 18↘
Milnthorpe 21←
Miloi 96→
Milos 94↘
Miłosław 89←
Milovaig 18←
Milówka 89↙
Miltach 41↘
Miltenberg 40→
Milton Abbot 22↙
Milton Ernest 23↖
Milverton 22↘
Mimizan 50←
Mimizan-Plage 50←
Mina de São Domingos 72→
Minas de Riotinto 72→
Minaya 70→
Mindelheim 47↗
Minden 36→
Mindilógli 96↙
Mindin 44←
Mindrummill 21↖
Mîneci Ungureni 92→
Minehead 22→
Minervino Murge 58←
Minffordd 22↖
Mingary 18←
Minglanilla 71←
Mingorria 69↗
Mini 91←
Minićevo 93↙
Minnesund 28↖
Minorca = Menorca
Minori 11↘
Minsen 36↖
Mińsk Mazowieckie 89→

Minsterley 22↗
Minsterworth 22→
Mintlaw 19←
Mira, *Italy* 54↗
Mira, *Portugal* 68↖
Mira, *Portugal* 68←
Mira, *Spain* 71←
Miraflores de la Sierra 65↙
Miramare 11↗
Mirambeau 50↗
Miramont 50→
Miranda de Arga 65→
Miranda de Ebro 65→
Miranda do Corvo 68←
Miranda do Douro 64↙
Mirande 66↗
Mirandela 63↘
Mirandilla 69↙
Mirándola 54←
Mirano 54↗
Miravci 93↘
Mirebeau, *France* 44→
Mirebeau, *France* 46←
Mirecourt 46↗
Mirepoix 67↖
Miribel 46↙
Mirna 49↙
Mirotice 41↘
Miroşi 93←
Mirosławiec 88↗
Mirow 37↗
Mirsíni 96↘
Mirtiá, *Greece* 96↙
Mirtiá, *Greece* 96↘
Mirto Crosia 59←
Misano Adriatico 11↗
Misilmeri 60←
Miskolc 90↖
Mistelbach 49↗
Misterhult 31↖
Misterton 22↘
Mistras, anc. site 96→
Mistretta 60←
Mistros 96↖
Mitandersfors 28←
Mitchell 22↙
Mitchelstown 25→
Mítikas 96↖
Mitilíni 95↖
Mittelberg, *Austria* 47→
Mittelberg, *Austria* 48←
Mittenwald 48←
Mitterbach 49←
Mitterding 48↗
Mitterndorf 49↗
Mittersheim 39↘
Mittersill 48→
Mitterteich 41←
Mittewald 48↘
Mittweide 41↗
Mizil 92→
Mjåvatn 27→
Mjell 26↙
Mjöbäck 30↖
Mjölby 29↙
Mjölfjell 26→
Mjöndalen 27↖
Mladá Boleslav 88↘
Mladenovac 91↖
Mława 89↖
Mlini 16↙
Mnichovo Hradiště 88↘
Mníšek, *Czechoslovakia* 41→
Mníšek, *Czechoslovakia* 90↖
Mo, *Norway* 26↘
Mo, *Norway* 27←
Mo, *Norway* 28↖
Mo, *Sweden* 28↙
Mo, *Sweden* 84↖
Moaña 63←
Mocejón 69→
Mochales 65↘
Mochtin 41↘
Mockfjärd 28↗
Mockrehna 37↘
Möckern 37→
Möckmühl 40↘
Modane 53↖
Modbury 22↙
Modena 54←
Módhion 96←
Modica 60↗
Modigliana 54→
Mödling 49↗
Modnejar 70↗
Modřany 41→
Modriča 90→
Modrý Kameň 90↖
Moeiu 92↙
Moelv 26↗
Moena 48↙
Moers 39↗
Moffat 20↗
Mogadouro 63↘
Mogen 27←
Mogente 71↙
Mogielnica 89→
Mogilev Podolskiy 92↖
Móglia 54←
Mogliano Véneto 54↗
Moglicë 91↗
Mogofores 68↖
Moguer 72→
Mohács 90→
Moheda 30→
Mohedas 69↖
Mohill 24→
Moholm 28↘
Moi 27↘
Moimenta da Beira 63↘
Moira 24↗
Moirans-en-Montagne 46↙
Moisdon 44←
Moissac 51←
Moisund 27→
Moita, *Portugal* 68↙
Moita, *Portugal* 68↗
Mojácar 74→
Mojados 64↘
Moklinta 29←
Mõksy 87←
Mol 39↖
Mola di Bari 58←
Moláoi 94↘
Mold 20↘
Moldava nad Bodvou 90↖
Molde, *Norway* 27↙
Molde, *Norway* 84↘
Moldova Nouă 92↘
Møldrup 32↘
Moledo 63←
Molfetta 58←
Molières 51←
Molina de Aragón 65↘
Molina de Segura 74↗
Molinella 54→
Molinicos 70↘
Molinos 66↘
Molins de Rey 67↙
Moliterno 58↘
Molkom 28→
Mölle 30←
Möllebrucke 48↘
Molledo 65↖
Möllenbeck 37↗
Mollerusa 67←
Mollesjok 86↖
Mollet 67↘
Mollina 73→
Mollington 23↖
Mölln 33→
Mollösund 30↖
Mölltorp 28↘
Mölnbo 29↘
Mölndal 30↖
Mölnlycke 30↖
Moloja 47↘
Mólos 96←
Moloy 46←
Molsheim 40↙
Moltrasio 13↘
Molveno 48↙
Molygrove 22↖
Mombeltrán 69↗
Mombuey 64←
Monchilgrad 93→
Mömlingen 40←
Mommark 33←
Momo 53↗
Mon 28↙
Møn, island 33↖
Mon Idée 38→
Monasterace Marina 59→
Monasterevin 25↖
Monastier, le 52←
Monastir = Bitola
Monastiráki, *Greece* 96↙
Monastiráki, *Greece* 96↘
Monástra 96↘
Moncada, *Spain* 67↙
Moncada, *Spain* 71←
Moncalieri 53←
Moncalvo 53↗
Monção 63←
Moncarapocho 72←
Mönchdorf 49↖
Mönchen-Gladbach 39↗
Mönchhof 49↗
Monchique 72←
Moncófar 71→
Moncontour, *France* 42→
Moncontour, *France* 44→
Moncoutant 44↙
Monda 73↘
Mondariz 63←
Mondelange 39↘
Mondello 60←
Mondium de Basto 63↘
Mondoñedo 63↗
Mondoubleau 44↗
Mondragon 52←
Mondragone 57↙
Mondsee 48→
Monéglia 53→
Monegrillo 66→
Monein 66↗
Monemvasía 94↘
Monesterio 72↗
Monestier-de-Clermont 52→
Moneymore 24↖
Moneyneany 24↖
Monfero 63↗
Monflanquin 51←
Monforte, *Portugal* 68→
Monforte, *Portugal* 68↘
Monforte de Lemos 63→
Monforte del Cid 71↙
Monfrotinho 68→
Monkebüll 33↙
Monheim 40↘
Moniaive 20↗
Monifieth 19↙
Monistrol, *France* 52↖
Monistrol, *France* 52←
Monistrol, *Spain* 67←
Monk Fryston 21↙
Monkton 20↗
Monleras 64↙
Monmouth 22→
Monnikendam 35←
Monódhri 96↖
Monólithos 95↘
Monopoli 58↗
Monor 90↖
Monóvar 71↙
Monpazier 51←
Monreal 65→
Monreal del Campo 66↙
Monreale 60←
Monroy 69←
Monroyo 66↘
Mons 38→
Monschau 39→
Monségur 50→
Monsélice 54↗
Monsheim 40←
Mönsterås 31←
Monsummano 54↙
Mont-de-Marsan 50↘
Mont-Dore, le 51↗
Mont-Louis 67←
Mont-St-Michel, le 44↖
Mont-sous-Vaudrey 46←
Montabaur 39→
Montacute 22↘
Montagnac 67↗
Montagnana 54↗
Montaigu 44←
Montaigut 45↘
Montalbán 66↙
Montalbano Ionico 58→
Montalbo 70→
Montalcino 56↖
Montalcone 55↖
Montaldo di Cósola 53→
Montalegre 63↙
Montallegro 60→
Montalto di Castro 56←
Montalvão 68→
Montamarta 64↘
Montana-Vermala 46↘
Montañana 66→
Montánchez 69←
Montargil 68↙
Montargis 45↖
Montastruc-la-Conseillère 51↙
Montauban 51↙
Montauban-de-Bretagne 42→
Montbard 46←
Montbazens 51←
Montbazon 44→
Montbéliard 46→
Montbenoit 46→
Montblanch 67↙
Montbrison 52↖
Montbron 50↗
Montbrun-les-Bains 52→
Montceau-les-Mines 46↙
Montcenis 46→
Montchanin 46↙
Montcornet 38→
Montcuq 51←
Montdidier 38←
Monte Carlo 53↙
Monte Estoril 14←
Monte Rael 68←
Monte Redondo 68←
Monte San Savino 54↘
Monte Sant' Ángelo 58←
Monteagudo de las Vicarias 65↘
Montealagre del Castillo 71↙
Montebello 54↗
Montebelluna 54↗
Montebourg 43↖
Montebruno 53→
Montecatini Terme 54↙
Montécchio 54←
Montécchio Maggiore 54↗
Montecerignone 54↘
Montech 51↙
Montechiaro d'Asti 53←
Montecorvino Rovella 58↘
Montefalco 56↗
Montefiascone 56↗
Montefrio 73→
Montegiórgio 57↖
Montegrotto Terme 54↗
Montehermoso 69←
Montejicar 73→
Montélimar 52←
Montella 58↘
Montellano 73←
Montelpulciano 56↗
Montemar 14↙
Montemayor 73→
Montemolin 73↖
Montemor-o-Novo 68↙
Montemor-o-Velho 68←
Montendre 50↗
Montenegro = Crna Gora
Montepiano 54←
Montereau 45↗
Monterotondo 56→
Monterrey 63→
Monterroso 63→
Monterrubio de la Serena 69↙
Montesano Santa Marcellana 58↘
Montesarchio 57↙
Montesquieu-Volvestre 67↖
Montevarchi 54↘
Montevécchio Marina 61←
Montfaucon, *France* 39↙
Montfaucon, *France* 44←
Montfaucon, *France* 52↖
Montfleur 46↙
Montfort 50↘
Montfort-l'Amaury 38↙
Montfort-sur-Meu 42↘
Montgenèvre 53←
Montgiscard 67↖
Montgomery 22↗
Montguyon 50↗
Montherme 39←
Monthey 46↘
Monthureux-sur-Saône 46↗
Monti 61↗
Montichiari 54↖
Montier-en-Der 46↖
Montignac 51←
Montigny-le-Roi 46↖
Montigny-sur-Aube 46↖
Montijo, *Portugal* 68↙
Montijo, *Spain* 69↙
Montilla 73→
Montillana 73→
Montivilliers 43↗
Montjean 44←
Montluçon 45↙
Montluel 46↙
Montmarault 45↘
Montmartin-sur-Mer 43←
Montmélian 52↗
Montmirail 38↘
Montmoreau 50↗
Montmorillon 44↘
Montoire 44↗
Montorio al Vomano 57↖
Montoro 73↗
Montpaon 51↘
Montpellier 52↙
Montpezat-de-Quercy 51←
Montpezat-sous-Bauzon 52←
Montpon 50→
Montréal, *France* 45→
Montréal, *France* 67↖
Montredon-Labessonié 51↘
Montrejeau 66↗
Montrésor 45←
Montreuil 38←
Montreuil-Bellay 44→
Montreux 46↘
Montrevault 44←
Montrevel-en-Bresse 46↙
Montrichard 45←
Montroig 67↙
Montrond-les-Bains 52↖
Montrose 19←
Montroy 71←
Montsalvy 51→
Montseny 67→
Montserrat, Monasterio de 67↙
Montsûrs 44↗
Montuenga 64↘
Montuiri 75↙
Monturque 73→
Monza 53↗
Monzón 66→
Monzón de Campos 65←
Moosburg 48↖
Mór 90←
Mor Budějovice 88↘
Mora, *Sweden* 28↗
Mora, *Portugal* 68↙
Mora, *Spain* 70←
Mora de Ebro 66↘
Mora de Rubielos 71↖
Mora la Nueva 66↘
Moraga 89↖
Moraira 15↘
Moral de Calatrava 70↙
Moraleda de Zafayona 73→
Moraleja 69←
Moraleja del Vino 64↘
Morales de Toro 64↘
Moralina 64↙
Morar 18←
Morata de Jalon 65↘
Morata de Jiloca 65↘
Morata de Tajuña 70↖
Moratalla 74↗
Moravská Třebová 88↘
Morbach 39→
Morbegno 47↘
Mörbisch 49→
Mörbylanga 31←
Morcenx 50↙
Morciano di Romagna 54↘
Morcone 57↙
Morcote 13↙ (9 km SW of Lugano)
Morcott 23↖
Mordiford 22→
Morebattle 21↖
Morecambe 20→
Morella 66↘
Morena, Sierra, mts. 73↖
Mores 61↖
Moret 45↗
Moreton in Marsh 23←
Moretonhampstead 22↙
Moretta 53←
Moreuil 38←
Morez 46↘
Morgat 42←
Morgedal 27←
Morges 46↘
Morgins 46↘
Morgongåva 29←
Morhange 39↘
Moriani Plage 62←
Morillo de Monclús 66→
Moringen 36↘
Morjärv 85↘
Mork 26←
Mörkö 29↘
Mörkri 26←
Morlaas 66↗
Morlaix 42←
Morley 21↙
Mörlunda 31↖
Mormanno 58→
Mormant 38↘
Morón de Almazán 65↘
Morón de la Frontera 73←
Morović 90↗
Morpeth 21↖
Morriston 22←
Mörrum 30↘
Morsbach 39↗
Mörsch 40↙
Morskogen 28↖
Morsvik 85←
Mortagne 44↗
Mortagne-sur-Sèvre 44←
Mortágua 68↖
Mortain 43←
Mortara 53↗
Morteau 46→
Mortehoe 22←
Mörtfors 31↖
Mortimer's Cross 22↗
Morton 23↖
Mortsel 39↖
Moruo 30←
Morville 22↗
Morwenstow 22↙
Morzine 46↘
Mos 63←
Mosbach 40↙
Mosby 27→
Moščenička Draga 16↗
Mosjöen 85↙
Moskene 85←
Moskijärvi 85→
Moskhokariá 96←
Moskhokhórion 96←
Moskog 26↙
Moskosel 85↘
Mosonszolnok 49→
Mosqueruela 71↖
Moss 27↖
Mossbank 19↗
Mossley 21↙
Moss-side 24↖
Mosstodloch 19←
Most 41→
Mostar 91←
Mosterhamn 27↙
Mosterton 22↘
Mostiska 89↘
Móstoles 69↗
Mostrim 24→
Mostu 28↖
Mostyn 20↘
Mosvik 84←
Mota del Cuervo 70→
Mota del Marqués 64↘
Motala 28↘
Mothe-Achard, la 44↙
Mothe-St-Héraye, la 44↘
Motherwell 18↘
Motilla del Palancar 70→
Motjärnshyttan 28→
Motrico 65↗
Motril 73↘
Motta di Livenza 54↗
Motte-du-Caire, la 52→
Mottištone 23↙
Mottola 58→
Mottram 21↙
Mouchard 46←
Moúdhros 95↖
Moudon 46↘
Moulins 45↘
Moulins-Engilbert 45→
Moulins-la-Marche 43→
Mount Bellew 24→
Mountfield 24←
Mountmellic 25←
Mountrath 25←
Mount's Bay 22↘
Mountshannon 25←
Mountsorrel 23↖
Moura 72↗
Mourão 68↘
Mouríkion 96←
Mourmelon-le-Grand 38↘
Mourne Abbey 25→
Mouscron 38↗
Moustiers-Ste-Marie 52↘
Mouthe 46↘
Moutier 46→
Moutiers 52↗
Moûtiers les Mauxfaits 44↙
Mouzáki 96↘
Moville 24←
Moy, *Scotland* 18→
Moy, *N. Ireland* 24↗
Möyå 67←
Moycullen 25↙
Moyeuvre-la-Grande 39↙
Moyuela 66↙
Mozárbez 64↘
Mrkonjič Grad 91←
Mrkopalj 55↗
Mrocza 89↖
Mrzeżyno 88↗
Mrzle Vodice 55↗
Mszezonów 89←
Mur-de-Barrez 51→
Murat 51↗
Múccia 56↗
Much 40↖
Much Wenlock 22↗
Muchalls 19←
Mücheln 41↖
Mückenberg 41↗
Muckross 25↘
Mudanya 95↗
Müden 37←

Muel 66↙
Muff 24←
Mugaire 65↗
Mugardos 63↖
Mügeln 41↗
Mugla, *Spain* 63↖
Muğla, *Turkey* 95→
Mugron 50↘
Mühlacker 40↙
Mühlbach, *Austria* 49↗
Mühlbach, *Austria* 13↖
Mühlberg 37↘
Mühldorf 48↗
Mühlen Eichsen 33→
Mühlenbeck 37→
Mühlhausen 40↗
Muhos 86→
Muine Bheag 25↖
Muir of Ord 18→
Muirdrum 19↙
Muirkirk 20↗
Mukachevo 92↙
Muker 21←
Mula 74↗
Mulhacen, Cerro de, mt. 74←
Mülheim 39↗
Mulhouse 46→
Mulhyttan 28→
Mullagh 24↗
Mullaghmore 24→
Müllheim 47↖
Mullinavat 25←
Mullingar 24→
Mullion 22↘
Mullsjö 30↗
Mulseryd 30↗
Multia 87←
Multrå 84↖
Multyfarnham 24→
Mumbles 22←
Mumby 21↘
Muñana 69↗
Münchberg 41←
Müncheberg 88→
München 48↖
Münchhausen 40↖
Munderkingen 47↗
Mundesley 23↗
Mundford 23↗
Mundheim 26↘
Munera 70↘
Munguia 65↗
Muniesa 66↙
Munich = München
Munka Ljungby 30↙
Munkedal 28↙
Munkflohögen 84↖
Munkfors 28→
Munksund 86↘
Münnden 36↘
Münnerstadt 40→
Muñogalindo 69↗
Münsingen, *Switzerland* 47←
Münsingen, *W. Germany* 47↗
Munsö 29→
Münster, *W. Germany* 36←
Münster, *W. Germany* 36↗
Munster, *France* 46↗
Münster, *Switzerland* 47↙
Münstereifel 39→
Münzenberg 40←
Münzkirchen 48↗
Muonio 86←
Muonionahusta 85→
Mur-de-Bretagne 42→
Muradiye 95←
Murano 54↗
Murat-sur-Vebre 51↘
Muratli 93↗
Murau 49←
Murca 63↘
Murchante 65→
Murcia 74→
Mure, la 52→
Mürefte 93↗
Muret 67↖
Murgeni 92↗
Murguia 65↗
Muri 47←
Murias de Paredes 64↖
Murillo de Rio Leza 65→
Murillo el Fruto 65→
Murjek 85↘
Murnau 48←
Muro, *Italy* 62←
Muro, *Spain* 75←
Muro de Alcoy 71↙
Muro Lucano 58↘
Murol 51↗
Muros 63↖
Muros de Nalón 64↖
Murovano Kurilovtsy 92↖
Murowana Goslina 89←
Mürren 47↙
Murrhardt 40↘
Murska Sobota 49↘
Mursko Središće 49↘
Murten 46→
Murtosa 68↖
Murtovaara 86↗
Murum 30↗
Murvica 55→
Mürzzuschlag 49→
Musel 64↗
Musken 85←
Musselburgh 19↙
Mussidan 50→
Mussomeli 60→
Mussy 46↖
Mustadfors 28↙
Mustafa Kemalpaşa 95↗
Mustèr = Disentis/Mustèr
Mustola 86←
Muszyna 89↘
Muthill 18↘
Mutters 13←
Mutterstadt 40↙
Mutzig 46↗
Muurame 87←
Muurola 86↙
Muy, le 52↘
Muyezeno 87↖
Muzillac 42↘
Mycenae, anc. site 96→
Mydland 27↘
Mykland 27→
Myklemyr 26←
Myllykoski 87→
Mynämäki 87↘
Myra, anc. site 95↘
Myrdal 26→
Myre 85↖
Myrheden 86↘
Mysen 28←
Myślenice 89↙
Myśliborz 88↗
Myssjo 84→
Mysubyttseter 26←
Mysuseter 26↖
Myszyniec 89↗
Mýto 41→

Nå 26↘
Na Logu 48↘
Naantali 87↘
Naarden 35→
Naas, *Rep. of Ireland* 25↖
Nääs, *Sweden* 30↖
Nabburg 41↙
Náchod 88↘
Nacka 29→
Naddvik 26→
Nådendal = Naantali
Nădlac 90↗
Nadvornaya 92←
Nærbö 27↘
Naeröy, *Norway* 26↘
Naeröy, *Norway* 84←
Næstved 33↖
Näfels 47←
Naggen 84↗
Nago 54↖
Nagold 47↖
Nagu 87↘
Nagybajom 90→
Nagycenk 49→
Nagykálló 92↙
Nagykanízsa 90→
Nagykáta 90↖
Nagykörös 90↖
Nagyvárad = Oradea
Nahe 33→
Naila 41←
Nailbridge 22→
Nailsea 22→
Nailsworth 22→
Nairn 18→
Najac 51←
Nájera 65→
Näkkälä 86←
Nakkila 87↘
Nakksjo 27↖
Nakskov 33←
Nakło 89↖
Näläntö 87←
Nalliers 44↙
Namdalseid 84←
Namna 28↖
Nampa 86↙
Namsos 84←
Namur 39←
Namysłów 89←
Nanclares de Oca 65↗
Nancy 39↘
Nangis 45↗
Nans 46→
Nant 51↘
Nantes 44←
Nanteuil-le-Haudouin 38↘
Nantgarw 22→
Nantiat 45↙
Nantua 46↙
Nantwich 21↙
Náousa 94↗
Nápoli 57↙
Napoule, la 53↙
När 31→
Narberth 22←
Narbonne 67↗
Narborough, *England* 23↖
Narborough, *England* 23↗
Nardo 58↗
Narkaus 86↙
Narken 85→
Narni 56↗
Naron 63↖
Närpes 87↙
Narvik 85←
Näs, *Sweden* 28↗
Näs, *Sweden* 29↖
Năsăud 92←
Näsberg, *Sweden* 28↗
Näsberg, *Sweden* 85↘
Nasbinals 51→
Naseby 23↖
Näshulta 29←
Našice 90→
Naso 60↖
Nassau 39→
Nassenfels 41↙
Nasserreith 47→
Nassjo 30↗
Nassundet 28→
Näsviken 84↗
National, airport 39↖
Nätra 84↖
Nattavaara 85↘
Nättraby 31↙
Naturno 48↙
Naturns 48↙
Naucelle 51→
Nauders 47→
Nauen 37→
Naul 24↗
Naumburg, *E. Germany* 41↖
Naumburg, *W. Germany* 40↗
Naunhof 41↖
Naupaktos, anc. site 96↙
Naustdal 26↙
Nauste 84↘
Nava 64↗
Nava de Roa 65↙
Nava de la Asunción 64↘
Nava del Rey 64↘
Navacepeda de Tormes 69↖
Navacerrada 69↗
Navaconcejo 69←
Navafria 65↙
Navahermosa 69→
Naval 66→
Navalcán 69→
Navalcarnero 69↗
Navalmanzano 65↙
Navalmoral 69↗
Navalmoral de la Mata 69←
Navamorcuende 69→
Navalvillar del Pela 69↙
Navan = An Uaimh
Navarcles 67←
Navarino = Pílos
Navaríti 96→
Navarredonda de la Sierra 69↗
Navarrenx 66↗
Navarrés 71↙
Navas de Oro 65↙
Navas de San Juan 74↖
Navas del Madroño 69←
Navas del Rey 69↗
Navascués 66↖
Navasfrias 68↗
Navatalgordo 69↗
Nävekvarn 29↙
Navenby 21↘
Naverstad 28↙
Navés 67←
Navia 63↗
Navia de Suarna 63↗
Navis 48←
Návpaktos 96↙
Návplion 96→
Nävragöl 31←
Náxos 95←
Nay 66↗
Nayland 23→
Nazaré 68←
Naze, The, *England* 23→
Naze, The, *Norway* = Lindesnes
Nazilli 95→
Néa Artáki 96↖
Néa-Epídhavros 96→
Néa Filippiás 94↖
Néa Kíos 96→
Néa Moudhaniá 94↗
Néa Palátia 96↖
Néa Psará 96↖
Néa Víssi 93→
Néa Zoryiáni 96↖
Neagh, Lough, lake 24↖
Neap 19↗
Neápolis, *Greece* 94↖
Neápolis, *Greece* 94↘
Neápolis, *Greece* 95→
Neath 22←
Nečemice 41→
Neckarelz 40↙
Neckargemünd 40↙
Neckarsulm 40↘
Neda 63↗
Nedalen 84→
Nedelišće 49↘
Nédha 96↘
Nédhousa 96→
Nedstrand 27↙
Needham Market 23↗
Neerdar 40↖
Neermoor 36↖
Nefyn 20↘
Negotin 93↙
Negotino 93↘
Negreira 63↖
Negreni 92→
Nègrepelisse 51←
Negru Vadă 93↖
Neheim-Hüsten 36↙
Nehoiasu 92→
Neiden 86↖
Neira de Jusa 63↗
Neisse = Nysa
Nejdek 41→
Neksø 31↙
Nelas 68↗
Nelson 21↙
Neméa 96→
Nemours 45↖
Nenagh 25←
Neokhór 46↘
Neokhóri 96↙
Néon Karlóvasi 95↖
Néon Petrítsi 93↘
Nepomuk 41↘
Nérac 50→
Neresheim 40↘
Nereto 57↖
Néris 45↙
Nerja 73↘
Nérondes 45→
Nerpio 74↗
Nerva 72→
Nervi 53→
Nes, *Netherlands* 35↗
Nes, *Norway* 26←
Nes, *Norway* 26↙
Nes, *Norway* 26↗
Nes, *Norway* 28↖
Nes, *Norway* 84↙
Nesbyen 26↗
Nesflaten 27↙
Neslandsvath 27↖
Nesle 38→
Nesna 85↙
Nesodden 27↖
Ness, Loch, lake 18→
Nesscliff 22↗
Nesseby 86↖
Nesselwang 47→
Nesset 84↘
Nestáni 96→
Nesterov, *Ukraine* 89↘
Nesterov, *U.S.S.R.* 89↗
Néston 20↘
Nestórion 94↖
Nesttun 26↘
Nether Stowey 22→
Netheravon 22→
Nethybridge 19←
Netlandsnes 27→
Nettlebed 23←
Nettuno 56→
Netzschkau 41↖
Neu Isenburg 40←
Neu Pölla 49↖
Neu-Ulm 47↗
Neubrandenburg 37↗
Neubukow 33↗
Neuburg 41↙
Neuchâtel 46→
Neuchâtel, Lac de, lake 46→
Neudorf 40↙
Neudörfl 49→
Neuenbürg, *W. Germany* 36↖
Neuenbürg, *W. Germany* 40↙
Neuenkirchen, *W. Germany* 36←
Neuenkirchen, *W. Germany* 36↗
Neuenkirchen, *W. Germany* 36→
Neuf-Brisach 46↗
Neufchâteau, *Belgium* 39←
Neufchâteau, *France* 46↖
Neufchâtel 38←
Neuffen 40↘
Neuhaus, *E. Germany* 37↖
Neuhaus, *E. Germany* 41←
Neuhaus, *W. Germany* 36↗
Neuhaus, *W. Germany* 36↘
Neuhausen 47←
Neuhausen ob Eck 47↖
Neuillé-Pont-Pierre 44→
Neuilly-l'Évêque 46↖
Neuilly-St-Front 38↘
Neukalen 33↗
Neukirch 41↗
Neukirchen, *Austria* 48↗
Neukirchen, *W. Germany* 40↗
Neukirchen, *W. Germany* 41↘
Neukloster 33↗
Neuland 36↗
Neulengbach 49↗
Neulussheim 40↙
Neumarkt-St Veit 48↗
Neumarkt, *Austria* 49←
Neumarkt, *W. Germany* 41↙
Neumünster 33→
Neung-sur-Beuvron 45←
Neunburg 41↙
Neunkirchen, *Austria* 49→
Neunkirchen, *W. Germany* 39↘
Neuruppin 37↗
Neuses 40→
Neusiedl 49↗
Neuss 39↗
Neustadt, *E. Germany* 37→
Neustadt, *E. Germany* 41↖
Neustadt, *E. Germany* 41↗
Neustadt, *W. Germany* 33→
Neustadt, *W. Germany* 40↗
Neustadt, *W. Germany* 41←
Neustadt, *W. Germany* 47↖
Neustadt am Rübenberge 36→
Neustadt an dem Waldnaab 41←
Neustadt an der Aisch 40→
Neustadt an der Donau 41↙
Neustadt an der Weinstrasse 40↙
Neustadt-Glewe 37↖
Neustettin = Szczecinek
Neustrelitz 37↗
Neuves-Maisons 39↙
Neuveville, la 46→
Neuvic-d'Ussel 51↖
Neuville 46↙
Neuville-aux-Bois 45↖
Neuvy-St-Sépulcre 45↙
Neuvy-sur-Barangeon 45←
Neuwied 39→
Nevada, Sierra, mts. 74←
Nevernes 84←
Nevers 45→
Nevesinje 91←
Neviges 39↗
Neville's Cross 21←
Nevlunghavn 27↖
New Abbey 20↗
New Aberdour 19←
New Alresford 23←
New Brighton 20↘
New Buckenham 23↗
New Buildings 24←
New Chapel 23←
New Cumnock 20↗
New Deer 19←
New Galloway 20↗
New Holland 21↘
New Hunstanton 21↘
New Luce 20↖
New Machar 19←
New Mill 21↙
New Mills 21↙
New Pitsligo 19←
New Quay 23←
New Romney 23↘
New Ross 25↖
New Scone 19↙
New Selma 18↘
New Radnor 22↗
Newark-on-Trent 21↘
Newbiggin-by-the-Sea 21↖
Newbliss 24→
Newborough 20↘
Newbridge, *Scotland* 20↗
Newbridge, *Wales* 22→
Newbridge on Wye 22↗
Newburgh, *Scotland* 19←
Newburgh, *Scotland* 19↙
Newbury 23←
Newby Bridge 20→
Newcastle 24↗
Newcastle Emlyn 22↖
Newcastle under Lyme 21↙
Newcastle upon Tyne 21↖
Newcastle West 25↙
Newcastleton 21↖
Newent 22→
Newgale 22←
Newhaven 23↘
Newhouse 18↘
Newlyn 22↘
Newmains 18↘
Newmarket, *England* 23↗
Newmarket, *Rep. of Ireland* 25↘
Newmilns 20↗
Newnham 22→
Newport, *England* 22↗
Newport, *England* 23↙
Newport, *England* 23→
Newport, *Rep. of Ireland* 24↘
Newport, *Rep. of Ireland* 25←
Newport, *Wales* 22→
Newport on Tay 19↙
Newport Pagnell 23↖
Newquay 22↙
Newry 24↗
Newton 21←
Newton Abbot 22↙
Newton Aycliffe 21←
Newton Ferrers 22↙
Newton-le-Willows 21↙
Newton Mearns 18↘
Newton St Boswells 21↖
Newton Stewart 20↗
Newton-upon-Trent 21↘
Newtongrange 19↙
Newtonmore 18→
Newtown, *Rep. of Ireland* 25←
Newtown, *England* 22↗
Newtown, *Wales* 22↗
Newtown Butler 24→
Newtown Hamilton 24↗
Newtown Stewart 24←
Newtownards 24↖
Newtownbarry = Bunclody
Newtowncunningham 24←
Newtyle 19↙
Nexon 51↖
Neyland 22←
Niata 94→
Nibbiano 53→
Nibe 32←
Nicastro 59←
Nice 53↙
Nickelsdorf 49↗
Nicolosi 60↗
Nicosia 60←
Nicotera 59↘
Nidda 40←
Niebla 72→
Niebüll 33↙
Nieder Wöllstadt 40←
Niederanven 39→
Niederau 13→
Niederaula 40↗
Niederbronn 39↘
Niedereisenhausen 40↖
Niederkaufungen 40↗
Niederkrüchten 39↗
Niedermarsberg 36↘
Niederolm 40←
Niederorschel 40↗
Niederösterreich, prov. 49↖
Niederrickenbach 12↖
Niedersachsen, prov. 36←
Niederweidbach 40↖
Niederwölz 49←
Niederzerf 39→
Nieheim 36↘
Niemcża 88↘
Niemegk 37→
Niemis 86↙
Niemodlin 89↙
Nienburg 36→
Niendorf 33→
Niepołomice 89↙
Nierstein 40←

Pendálofon 94↖
Pendeória 96←
Pendine 22←
Pendlebury 21↙
Penedono 63↘
Penela 68←
Penhas da Saúde 68↗
Peniche 68←
Penicuik 19↙
Peñíscola 71↗
Penistone 21↙
Penkridge 22↗
Penmaen-mawr 20↘
Penmon 20↘
Pennal 22↖
Pennan 19←
Penne 57↖
Penningby 29→
Penrhyndeudraeth 20↘
Penrith 21←
Penryn 22↙
Pensford 22→
Pentre-Foelas 20↘
Penybont 22↗
Penzance 22↘
Penzberg 48←
Péra Mélana 96→
Perafita 67←
Perakhóra 96→
Peraleda de Zaucejo 69↙
Peraleda de la Mata 69←
Peralejos de las Truchas 70↗
Perales del Alfambra 66↙
Perales del Puerto 69←
Peralta 65→
Peralta de Alcofea 66→
Pérama 94↘
Peraseinäjoki 87↙
Peratiá 96↙
Perchtoldsdorf 49↗
Percy 43←
Pérdhika 96↗
Perdhikóvrisi 96↙
Perechin 92↙
Peredo 63↘
Pereiro 72→
Perelló 67↙
Pereruela 64↘
Pérfugas 61↖
Perg 49↖
Pergamum, anc. site 95↖
Pérgine 48↙
Pérgola 55↙
Perho 87←
Periana 73↘
Périers 43↖
Périgueux 51↖
Perivóli, *Greece* 96←
Perivóli, *Greece* 96↘
Perivólia 96→
Periyiáli 96→
Perleberg 37↖
Perlez 90↗
Përmet 94↖
Pernik 93↙
Perniö 87↘
Pernitz 49→
Pero 53↗
Péronne 38→
Perosa-Argent 53←
Pérouges 46↙
Perpignan 67↗
Pedreguer 71↘
Perranporth 22↙
Perros-Guirec 42←
Persberg 28→
Pershagen 29→
Pershore 22↗
Perstorp 30↘
Perth 19↙
Perthus, le 67→
Pertisau 48←
Pertuis 52↘
Pertunma 87→
Perúgia 56↗
Perusic 55→
Péruwels 38→
Pesaguero 64↗
Pésaro 55↙
Pescara 57↖
Pèschici 58←
Peschiera del Garda 54↖
Péscia 54↙
Pescina 57←
Pescolanciano 57←
Pescopagano 58↘
Peshkopi 91↗
Pesmes 46←
Pesquera de Duero 65↙
Pessac 50→
Pessin 37→
Pessoux 39←
Peşteana Jiu 92↘
Péta 94↖
Petäjävesi 87←
Petalax 87↙
Pétange 39↙
Peteranec 49↘
Peterborough 23↖
Peterchurch 22→
Peterhead 19←
Peterlee 21←
Petersfield 23↙
Petilia 59←
Petkula 86←
Petkus 37→
Pétra, *Greece* 96←
Petra, *Spain* 75←
Petralia-Sotto-Soprana 60←
Petrel 71↙
Petrína 96→
Petrinja 90→
Petromagoúla 96←
Petroseni 92↘
Petrovac, *Yugoslavia* 91↖
Petrovac, *Yugoslavia* 91→
Petrovaradin 90↗
Petrovo Selo 55→
Petsákoi 96↙
Pettigo 24←
Petworth 23↙
Peurasuvant 86←
Peuraure 85↘
Pevensey 23↘
Pévki 96↘
Pewsey 23←
Peyrehorade 66↖
Peyriac-Minervois 67↗
Peyrolles 52↘
Pézenas 67↗
Pfaffenhofen 48↖
Pfäffikon 47←
Pfarrkirchen 48↗
Pfeffenhausen 41↙
Pforzheim 40↙
Pfronten 47→
Pfullendorf 47↗
Pfullingen 47↗
Pfunds 47→
Pfungstadt 40←
Phalsbourg 39↘
Philadelphía, anc. site 95→
Philippeville 39←
Philippopolis = Plovdiv
Piacenza 53↗
Piana, *France* 62←
Piána, *Greece* 96→
Piana degli Albanesi 60←
Piancastagnáio 56↗
Piani 48↘
Piani Resinelli 13↘
Piano cel Vóglio 54←
Pianottoli-Caldarello 62↙
Pias 72↗
Piaseczno 89→
Piaski 89→
Piatra Neamț 92←
Piazza Armerina 60→
Piazza al Sérchio 54←
Piazza Brembo 47↘
Piazzatore 13↘ (3 km SE of Mezzoldo)
Picardie, reg. 38←
Picasent 71←
Pickering 21→
Picquigny 38←
Pídhima 96↘
Piedicroce 62←
Piedimonte d'Alife 57↙
Piedmont = Piemonte
Piedrabuena 69↘
Piedrafita 63→
Piedrahita 69↖
Piedralaves 69↗
Piedras Albas 68→
Piégut 50↗
Pieksämäki 87←
Piemonte, prov. 53←
Pieniężno 89↖
Pienza 56↗
Piera 67↙
Piercebridge 21←
Pierowall 19↖
Pierre-Buffière 51↖
Pierre-de-Bresse 46←
Pierrefeu 52↘
Pierrefitte-Nestalas 66↗
Pierrefonds 38↘
Pierrefort 51→
Pierrelatte 52←
Piešt'any 90←
Piesting 49→
Pietarsaari = Jacobstad
Pietra Ligure 53→
Pietrasanta 54↙
Pietroşiţa 92→
Pieve di Bono 47↘
Pieve di Cadore 48↘
Pieve di Teco 53←
Pievepelago 54←
Pieux, les 42↗
Pigádhia 95↙
Pignataro 57↙
Pihlajavesi 87←
Piła 88↗
Pilalístra 96↘
Pilar de la Horadada 75←
Pilas 72→
Pilatipudas 87←
Pilatus, mt. 47←
Píli 96↖
Pílion 94↗
Pilis 90↖
Pilisvörösvar 90←
Pílos 94←
Pilton 22→
Pilzno 89↘
Pina 66↘
Piñar 73→
Pinarella 11↗
Pinarello 62←
Pinarhisar 93↗
Pinchbeck 23↖
Pinchbeck West 23↖
Pincota 92↘
Pindea de la Sierra 65←
Pindhos, mts. 94↖
Pindus = Pindhos
Pinerolo 53←
Pineto 57↖
Piney 46↖
Pinhão 63↘
Pinheiro 63↙
Pinhel, *Portugal* 64↙
Pinhel, *Portugal* 68←
Pinhoe 22↘
Pinkafeld 49→
Pinneberg 33→
Pino 62↖
Pinofranqueado 69↖
Pinos Puente 73→
Pinoso 71↙
Pinto 70↖
Pinwherry 20↖
Piombino 56↖
Piotrków 89←
Piove di Sacco 54↗
Piovene Rocchette 54↗
Piraeus = Piraiévs
Piraiévs 96↗
Piran 55↖
Pirdop 93←
Pírgos, *Greece* 94↘
Pírgos, *Greece* 95↙
Pírgos, *Greece* 96↘
Pirí 96←
Pirineos, mts. 66↖
Pirkkala 87→
Pirmasens 39↘
Pirna 41↗
Pirnmill 20↖
Pirot 93↙
Pirovac 55↘
Pirttitörmä 86→
Piryí 95←
Pisa 54↙
Pıscopi = Tílos
Písogne 47↘
Pissos 50→
Pisticci 58→
Pistoia 54↙
Pisz 89↗
Piteå 86↘
Piteşti 92→
Pithiviers 45↖
Pitigliano 56↗
Pitlochry 18→
Pitmedden 19←
Pitsá 96←
Pitscottie 19↙
Pittenweem 19↙
Piyaí 94↖
Pizarra 73↘
Pizzighettone 54↖
Pizzo 59←
Pjätteryd 30→
Plá de Santa María 67↙
Plaffeien 46↘
Plan-de-Baix 52→
Plan-d'Orgon 52↙
Plan-du-Var 53↙
Planá 41→
Plancoët 42→
Planiteron 96→
Plansee 13←
Plasencia 69←
Plassen 84→
Plat 16↙
Plataiaí 96←
Platamon 93↘
Platanákion 96→
Platanistós 96↗
Plátanos, *Greece* 96←
Plátanos, *Greece* 96↘
Platí, *Greece* 94↗
Platí, *Greece* 96↘
Platiána 96↘
Platíkambos 94↗
Plattling 41↘
Plau 37↗
Plaue 37→
Plauen 41←
Plavca 55→
Plavnica 91↗
Playa de Aro 67→
Pleasley 21↙
Pléaux 51↖
Pléheuf 10↖
Plélan-le-Grand 42↘
Plémet-la-Pierre 42→
Plencia 65↗
Plešivec 90↖
Plestin 42←
Pleszew 89←
Plettenberg 40↖
Pleumartin 44→
Pleven 93←
Pliego 74→
Plitvice 55→
Pljevlja 91↖
Ploaghe 61↖
Ploče 91←
Plochingen 40↘
Płock 89←
Plockton 18→
Ploërmel 42↘
Ploieşti 92→
Plombières-les-Bains 46↗
Plön 33→
Płońsk 89←
Płoty 88↗
Plouaret 42←
Plouay 42↘
Ploubalay 42→
Ploudalmézeau 42←
Plouescat 42←
Plouha 42→
Ploutokhóri 96↘
Plovdiv 93→
Pluckley 23→
Plumb Bridge 24←
Pluvigner 42↘
Plymouth 22↙
Plymstock 22↙
Plzeň 41→
Pniewy 88→
Po, river 53↗
Pobla de Segur 67←
Poblet, Monasterio de 67↙
Pocinho 63↘
Pocklington 21→
Podbořany 41→
Poddębice 89←
Poděbrady 88↘
Podensac 50↘
Podgora 16→
Podgorica = Titograd
Podkova Charnichevo 93→
Podlapača
Podnanos 55↖
Podplat 49↘
Podravska Slatina 90→
Podsused 49↘
Podturen 49↘
Podu Iloaiei 92↖
Podujevo 91↖
Poetto 61↘
Poggibonsi 54↙
Póggio Mirteto 56→
Póggio Rusco 54←
Pogradec 91↗
Pohorelá 90↖
Pohořelice 88↘
Poiana Brasov 92→
Poiana Teiului 92←
Poiares 68←
Poikela 86←
Poinsat 45↙
Poirino 53←
Poissy 38↙
Poitiers 44↘
Poitou, reg. 44↙
Poix 38←
Poix-Terron 39←
Pojan 91↗
Pojo 87→
Pokka 86←
Pokój 89↙
Pola de Allande 64↖
Pola de Laviana 64↗
Pola de Lene 64↗
Pola de Siero 64↗
Pola de Somiedo 64↖
Polán 69→
Polanow 88↗
Połczyn Zdrój 88↗
Polegate 23↘
Polesella 54→
Polgár 90↖
Polgárdi 90←
Poliána 96→
Polička 88↘
Polidhroson 96←
Polientes 65↖
Poligny 46←
Políkastron 93↘
Políkhnitos 95↖
Politiká 96↖
Políyiros 94↗
Poljak 55→
Polla 58↘
Põllau 49→
Pollensa 75←
Pollfoss 26↙
Polmak 86↖
Polperro 22↙
Polvijärvi 87↖
Polzeath 22↙
Pomar 66→
Pomarance 54↙
Pomarez 50↘
Pomarkku 87↘
Pombal 68←
Pomeroy 24↖
Pomézia 56→
Pommersfelden 40→
Pomorski 88↗
Pomoy 46→
Pomposa 54→
Ponferrada 64←
Pons, *France* 50↗
Pons, *Spain* 67←
Ponsacco 54↙
Pont, le 46↘
Pont-à-Mousson 39↙
Pont-Audemer 43↗
Pont Canavese 53↖
Pont-d'Ain 46↙
Pont-Faverger 38↘
Pont-de-Chéruy 52↗
Pont-de-Claix 52↗
Pont-de-Dore 51↗
Pont-de-l'Arche 38↙
Pont-de-Montvert, le 52←
Pont-de-Roide 46→
Pont-de-Salars 51→
Pont de Suert 67←
Pont-de-Vaux 46↙
Pont-de-Veyle 46↙
Pont-du-Château 51↗
Pont du Gard 52↙
Pont du Travo, le 62←
Pont-l'Abbé 42↙
Pont l'Évêque 43↗
Pont-St-Esprit 52←
Pont St Martin 53↖
Pont-sur-Yonne 45↗
Pontacq 66↖
Pontailler 46←
Pontão 68←
Pontardaw 22←
Pontardulais 22←
Pontarion 45↙
Pontarlier 46→
Pontassieve 54↘
Pontaubault 43←
Pontaumur 45↙
Pontcharra 52↗
Pontchâteau 42↘
Ponte da Barca 63←
Ponte de Lima 63↙
Ponte de Sôr 68→
Ponte delle Arche 48↙
Ponte di Legno 47↘
Ponte di Piave 54↗
Ponte-Leccia 62←
Ponte nelle Alpi 48↘
Ponte Nova 48↙
Ponte Nuovo 62←
Ponte Tresa 47↙
Pontebba 48↘
Pontecorvo 57←
Pontedécimo 53→
Pontedera 54↙
Pontefract 21↙
Ponteland 21↖
Pontelongo 54↗
Pontepetri 54↙
Ponterwyd 22↖
Pontet, le 52↙
Pontevedra 63←
Pontevico 54↖
Pontgibaud 51↗
Pontigny 45↗
Pontivy 42↘
Pontoise 38↙
Pontones 74↖
Pontoon 24↘
Pontorson 44↖
Pontrémoli 54←
Pontresina 47↘
Pontrhydfendigaid 22↖
Pontrieux 42→
Pontrilas 22→
Ponts-de-Cé 44→
Pontypool 22→
Pontypridd 22→
Ponza 56↘
Pool 21←
Poole 22↘
Poolewe 18↗
Pooley Bridge 20→
Poperinge 38↗
Pópoli 57←
Popovo 93←
Poppi 54↘
Poprad 89←
Porcuna 73→
Pordenone 54↗
Poreč 55↖
Pori 87↘
Porjus 85→
Porkkala 87→
Porlezza 13↘
Porlock 22←
Pörnbach 41↙
Pornic 44←
Pornichet 42↘
Póros 96↗
Porquerolles 10←
Porrentruy 46→
Porreras 75↙
Porretta Terme 54←
Porriño 63←
Porsa 85↗
Porsabygget 30→
Port Appin 18↘
Port Askaig 18↙
Port Bannatyne 18↘
Port Bou 67→
Port Carlisle 20↗
Port Charlotte 18↙
Port-Cros 10←
Port-de-Bouc 52↙
Port de Chiavari 62←
Port Dinorwic 20↘
Port Ellen 18↙
Port-en-Bessin 43↖
Port Erin 20←
Port Eynon 22←
Port Glasgow 18↘
Port Isaac 22↙
Port-la-Nouvelle 67↗
Port Lligat 15↗
Port Logan 20←
Port Louis 42↘
Port-Navalo 42↘
Port of Ness 18↖
Port-St-Louis 52↙
Port St Mary 20→
Port-sur-Saône 46→
Port Talbot 22←
Port-Vendres 67→
Port William 20→
Portacloy 24↘
Portadown 24↗
Portaferry 24↗
Portalegre 68→
Portarlington 25←
Porte-Ste-Marie 50→
Portel 68↘
Portel, le 38↖
Portelandolfo 57↗
Portezuelo 69←
Portglenone 24↖
Portgordon 19←
Porthcawl 22←
Porthleven 22↘
Portillo 64↘
Portillo de Toledo 69→

Rava Russkaya 89↘
Raved 33↙
Ravello 57↙
Ravenglass 20→
Ravenna 54→
Ravensburg 47↗
Ravenscar 21→
Ravna Gora 55↗
Rawa Mazowiecka 89←
Rawcliffe 21↙
Rawicz 88→
Rawtenstall 21↙
Rayleigh 23→
Rayol, le 52↘
Razgrad 93←
Razlog 93↘
Rea 65←
Reading 23←
Reale, la 61↖
Réalmont 51↙
Rear Cross 25←
Rearsby 23↖
Reawick 19↗
Reay 18↗
Reay Forest 18↗
Rebais 38↘
Rebolla de Jadraque 65↘
Reboly 87↖
Rebordelo 63↘
Recanati 55↙
Recaş 92↘
Recco 53→
Recea 93↙
Recess 24↘
Recey 46←
Rechnitz 49→
Recke 36←
Recklinghausen 36↙
Recoaro Terme 54↗
Recogne 39←
Recologne 46←
Recz 88↗
Redalen 26↗
Red Bull 21↙
Red Dial 20→
Redbourn 23←
Redcar 21←
Redditch 22↗
Redhill 23←
Redlin 37↗
Redon 42↘
Redondo 68↘
Redondela 63←
Redruth 22↙
Redslared 30↖
Rees 36↙
Reeth 21←
Reftele 30→
Regen 41↘
Regensburg 41↙
Réggio di Calábria 59↘
Reggio nell'Emilia 54←
Reggiolo 54←
Reghin 92→
Regna 29↙
Régua 63↘
Reguengos de Monsaiaz 68↘
Rehau 41←
Rehburg 36→
Rehden 36→
Rehna 33→
Reichenau, *Austria* 49→
Reichenau, *Switzerland* 47→
Reichenau, *W. Germany* 47←
Reichenbach 41↖
Reichshoffen 39↘
Reiersdal 27→
Reigate 23←
Reillo 70→
Reims 38↘
Reinach 47←
Reine 85←
Reinfeld 33→
Reinheim 40←
Reinosa 65↖
Reinsvoll 26↗
Reisjärvi 87←
Reiss 19↖
Reistad 27↖
Rejmyra 29↙
Rejowiec 89→
Rekarne 29←
Reliquias 72←
Relleu 71↘
Remagen 39→
Rémalard 44↗
Remels 36↖
Remich 39↘
Remiremont 46↗
Remollen 52→
Remoulins 52↙
Remschied 39↗
Rémuzat 52→
Rena 84→
Renazé 44↖
Rendal 84→
Rendina 96↙
Rendsburg 33↖
Renfrew 18↘
Rengínion 96←
Reni 92↗
Renko 87→
Renkum 35→
Renneba 84↘
Rennerod 40↖
Rennes 44↖
Rennes-les-Bains 67↗
Renteria 65↗
Renvyle 24↘
Réole, la 50→
Repcelak 90←
Replot 87↙
Reposaari 87↘
Repvag 86↖
Requéna 71←
Réquista 51↘
Resana 54↗
Resele 84↖
Resen 91↗
Resende 63↘
Reşiţa 92↘
Resolven 22←
Ressons-sur-Matz 38↘
Reszel 89↗
Retamal 69↙
Reteag 92↙
Rethel 38↘
Rethem 36→
Réthi 96→
Réthimnon 94↘
Retournac 52↖
Retuerta de Bullaque 69→
Retz 49↗
Reus 67↙
Reusel 35↘
Reutlingen 47↗
Reutte 47→
Revel 67↖
Revholmen 28←
Revigny 39↙
Revilla del Campo 65←
Revin 39←
Revnice 41→
Revničov 41→
Revsbotn 86↖
Revsudden 31←
Revsund 84↗
Rexbo 29↖
Rhaunen 39→
Rhayader 22↗
Rhein, river 40←
Rhein/Main, airport 40←
Rheinbach 39↗
Rheinberg 36↙
Rheinböllen 39→
Rheine 36←
Rheinfelden 47←
Rheinfelden Baden 47←
Rheinhausen 39↗
Rheinland-Pfalz, prov. 39→
Rheinsberg 37↗
Rheinzabern 40↙
Rheydt 39↗
Rhiconich 18↗
Rhine = Rhein, etc.
Rhineland-Palatinate = Rheinland-Pfalz
Rho 53↗
Rhodes = Ródhos
Rhodope Mts. = Rodopi Planina
Rhondda 22→
Rhône, river 52↙
Rhoose 22→
Rhosili 22←
Rhosneigr 20↘
Rhuddlan 20↘
Rhydspence 22↗
Rhyl 20↘
Rhynie 19←
Riala 29→
Riaño 64↗
Rians 52↘
Riaza 65↙
Ribadavia 63←
Ribadeo 63↗
Ribadesella 64↗
Ribaflecha 65→
Ribarroja 71←
Ribas de Fresser 67←
Ribe 33↙
Ribeauville 46↗
Ribécourt 38↘
Ribeira 63←
Ribeira de Pena 63↘
Ribemont 38→
Ribera 60→
Ribera del Fresno 69↙
Ribérac 50↗
Ribnitz-Damgarten 33↗
Říčany Jílové 88↘
Riccia 58↙
Riccione 54↘
Riceys, les 46↖
Richelieu 44→
Richhill 24↗
Richmond, *England* 23←
Richmond, *England* 21←
Richtenberg 33↗
Rickmansworth 23←
Ricla 66↙
Ricobayo 64↙
Riddarhyttan 29←
Riddes 46↘
Riding Mill 21↖
Ridley 21↙
Riec-sur-Belon 42↙
Ried, *Austria* 47→
Ried, *Austria* 48↗
Riedau 48↗
Riedenburg 41↙
Riedlingen 47↗
Riem, airport 48↖
Riesa 37↘
Riesi 60→
Rietberg 36↘
Rieti 56↗
Rieumes 67↖
Rieupeyroux 51←
Rieux, *France* 67↖
Rieux, *France* 67↗
Riez 52↘
Rigánion 96↙
Rigi, mt. 47←
Rignano Flamínio 56→
Riipi 86←
Rijeka 55↗
Rijssen 35→
Rillington 21→
Rimavská Sobota 90↖
Rimbo 29→
Rimforsa 31↖
Rímini 54↘
Rîmnicu Sărat 92↗
Rincón de Soto 65→
Rincón de la Victoria 73↘
Rindal 84↘
Ringarum 29↙
Ringe 33←
Ringebu 26↖
Ringenpen 26↗
Ringerike 26↗
Ringford 20↗
Ringkøbing 32↘
Ringládhes 96↘
Ringnes 26↗
Ringöy 26↘
Ringsend 24↖
Ringsted 33↖
Ringstorp 29↙
Ringwood 22↘
Rinkaby 30↘
Rinteln 36→
Rio Maior 68←
Riobarba 63↗
Riodeva 71←
Riofrio de Aliste 64←
Riom 45↘
Riom-ès-Montagnes 51↗
Rionero, *Italy* 57←
Rionero, *Italy* 58↘
Rios 63→
Rioseco de Tapia 64→
Riotorto 63↗
Riovéggio 54←
Rioz 46→
Ripatransone 57↖
Ripats 85→
Ripley, *England* 21←
Ripley, *England* 21↙
Ripley, *England* 23←
Ripoll 67←
Ripollet 67↙
Ripon 21←
Ripponden 21↙
Riquewihr 46↗
Risan 91←
Risbäck 84↖
Risberg 28↗
Risberget 28↖
Riscle 50↘
Risede 84↖
Riseley 23←
Rish 93↖
Risnes, *Norway* 26↘
Risnes, *Norway* 27→
Rîsnov 92→
Risör 27↗
Rissa 84↙
Ristiina 87↗
Ristijärvi 86↗
Risträsk 85↘
Ritsóna 96↖
Riudoms 67↙
Riva 54↖
Riva Azzurra 11↗
Rivabella 11↗
Rivanazzano 53→
Rivarolo Ligure 53→
Rive-de-Gier 52↖
Rivergaro 53→
Riverhead 23→
Riverstown 24→
Rives 52↗
Rivesaltes 67↗
Rivière Thibouville, la 43↗
Rivisóndoli 57←
Rivoli 53↖
Rixheim 46→
Rízais 96→
Rízoma 94↖
Rjånes 26↙
Rjukan 27←
Roa 26↗
Roade 23↖
Roadford 25↙
Roadhead 21↖
Roanne 45↘
Róbbio 53↗
Röbel 37↗
Roberton 20↗
Robertsbridge 23↘
Robertsfors 86↘
Robleda 69↖
Robledillo de Trujillo 69←
Robledo de Chavela 69↗
Robres 66→
Roc-Amadour 51←
Rocca di Mezzo 57←
Rocca di Papa 56→
Rocca San Casciano 54↘
Rocca Sinibalda 56→
Roccadáspide 58↘
Roccamonfina 57↙
Roccaraso 57←
Roccastrada 56↖
Roccella Ionica 59→
Rochdale 21↙
Roche-Bernard, la 42↘
Roche-Derrien, la 42→
Roche-en-Ardenne, la 39←
Roche-Posay, la 44→
Roche-sur-Yon, la 44↙
Rochechouart 44↘
Rochefort, *Belgium* 39←
Rochefort, *France* 44↙
Rochefort-en-Terre 42↘
Rochefort-Montagne 51↗
Rochefoucauld, la 44↘
Rochelle, la 44↙
Rocheservière 44←
Rochester, *England* 21↖
Rochester, *England* 23→
Rochford 23→
Rochfortbridge 24→
Rochlitz 41↗
Rociana 72→
Rock 22↙
Rockcliffe 20↗
Rockenhausen 39→
Rockesholm 28→
Rockingham 23↖
Rockneby 31←
Rocroi 39←
Rodach 40→
Rodaljice 55↘
Rödberg 26↗
Rødby 33←
Rødbyhavn 33←
Rødding 33↙
Rødekro 33↙
Rodel 18↖
Rodellar 66→
Roden 35↗
Rodenkirchen 36↗
Rodewald 36→
Rodewisch 41↖
Rodez 51→
Rodholívos 94↗
Ródhos 95↘
Ródhos, island 95↘
Rodi Gargánico 58↙
Roding 41↙
Rödjdafors 28↖
Rödje 29←
Rodna 92←
Rodopi Planina, mts. 93↘
Rodosto = Tekirdağ
Rödöy 85↙
Roermond 35↘
Roeselare 38↗
Rofors 28↘
Rogatec 49↘
Rogätz 37←
Rogliano, *France* 62↖
Rogliano, *Italy* 59←
Rognan 85←
Rogne 26↗
Rogoźno 89←
Rogsta 84↗
Rohrbach, *Austria* 41↘
Rohrbach, *Austria* 49→
Rohrberg 37←
Rohrbrunn 40→
Roisel 38→
Rojales 75↖
Röjan 84→
Rok 28↘
Roke 30↘
Roknäs 86↘
Rokycany 41→
Rolampont 46↖
Röldal 27↙
Rolfstorf 30←
Rollag 27↖
Rolle 46↘
Rollesby 23↗
Rolsberga 30↘
Rolsöy 28←
Roma, *Italy* 56→
Roma, *Sweden* 31↗
Romagna, reg. 54→
Romagnano 53↗
Roman 92←
Romano, Carpineto = Carpineto Romano
Romans 52↗
Romanshorn 47→
Romarheim 26↘
Rome = Roma
Romford 23→
Römhild 40→
Romilly 45↗
Rommenas 28←
Romont 46↘
Romorantin 45←
Romsen 29↖
Romsey 23↙
Römskog 28←
Ronaldsway, airport 20→
Roncegno 48↙
Roncesvalles 66↖
Ronchamp 46→
Ronchi 55↖
Ronciglione 56→
Ronco, *Italy* 53→
Ronco, *Switzerland* 13↙ (5 km W of Ascona)
Roncobilaccio 54←
Ronda 73↙
Rondablikk 26↖
Rønde 32→
Rondissone 53↖
Rondoassba 26↖
Ronehamn 31→
Rönnäng 30↖
Rønne 31↙
Ronneburg 41↖
Ronneby 31↙
Rönninge 29→
Rönnöfors 84←
Rönö 29↙
Ronse 38↗
Roordahuizum 35↗
Roosendaal 35↙
Ropczyce 89↘
Ropeid 27↙
Roquebillière 53↙
Roquefort 50↘
Roquemaure 52←
Roquestéron 53↙
Roquetas 66↘
Roquetas de Mar 74↙
Roquevaire 52↘
Rörbäcksnäs 28↗
Rörö 30↖
Röros 84→
Rorschach 47→
Rörstad 85←
Rörvik, *Norway* 26↙
Rörvik, *Norway* 84←
Rosa, Monte, mt. 47↙
Rosal de la Frontera 72→
Rosapenna 24←
Rosarno 59↘
Rosas 67→
Rosbeg 24←
Roscoff 42←
Roscommon 24→
Roscrea 25←
Rosegreen 25←
Rosehall 18↗
Rosehearty 19←
Roseldorf 49↗
Rosell 66↘
Rosemarkie 18→
Rosendal 27↙
Rosenfeld 47↖
Rosenheim 48←
Roseto degli Abruzzi 57↖
Rosice 88↘
Rosières 38→
Rosiers, les 44→
Roşiorii de Vede 93←
Roskilde 32↗
Roslags-Kulla 29→
Rosmaninhal 68→
Rosmuc 25↙
Rosporden 42↙
Rösrath 39↗
Ross 24↗
Ross-on-Wye 22→
Rossa 47↙
Rossano 59←
Rossas 63↙
Rosscarbery 25↘
Rossett 20↘
Rossfjord 85↗
Rossiglione 53→
Rosslare 25↖
Rosslare Harbour 25↗
Rosslau 37→
Rosslea 24→
Rossleben 41↖
Rossnes 26↘
Rössvassbukt 85↙
Rosswein 41↗
Rosta 85↗
Rostånga 30↘
Rostock 33↗
Rostrenen 42→
Rostrevor 24↗
Rostuša 91↗
Röstvangen 84→
Rosyth 19↙
Rot 84→
Rota 72→
Rotenburg 36↗
Roth, *W. Germany* 39↗
Roth, *W. Germany* 41↙
Rötha 41↖
Rothbury 21↖
Rothenburg ob der Tauber 40↘
Rothéneuf 42→
Rotherham 21↙
Rothes 19←
Rothesay 18↘
Rothienorman 19←
Rothwell 23↖
Rötgen 39↗
Rotonde, la 51→
Rotondella 58→
Rott 48↖
Rottach 48←
Rottenbuch 48←
Rottenburg, *W. Germany* 41↙
Rottenburg, *W. Germany* 47↖
Rottenmann 49←
Rotterdam 35←
Rottna 28→
Rottne 30→
Rottneros 28→
Rottweil 47↖
Roubaix 38↗
Roudnice 41→
Rouen 38↙
Rouffach 46↗
Roughton 23↗
Rouillac 44↘
Roujan 67↗
Roundstone 25↙
Roundwood 25↖
Roupakia 96↘
Roussillon, reg. 67→
Rovaniemi 86↙
Rovato 54↖
Rovde 84→
Rovereto 54↖
Rövershagen 33↗
Roverud 28↖
Roviais 96←
Rovigo 54→

Rovinj 55←
Rowardennan 18↘
Rowsley 21↙
Royal Leamington Spa 23↖
Royal Tunbridge Wells 23→
Royan 50↖
Royat 51↗
Roybridge 18→
Roye 38→
Royère 51↖
Royken 27↖
Royknes 27→
Röyrvik 84←
Roysheim 26←
Royston 23←
Röyttä 86→
Różan 89↗
Rozay-en-Brie 38↘
Rozier, le 51→
Rožmitál 41→
Rožňava 90↖
Roznov 92←
Rozoy-sur-Serre 38→
Rozprza 89←
Roztoky 41→
Rozwadów 89↘
Ruabon 20↘
Rubbestadneset 27↙
Rubena 65←
Rubery 22↗
Rubí 67↙
Rubiães 63←
Rubielos de Mora 71←
Rucăr 92→
Ruda 31←
Rude 33↖
Rüdesheim 39→
Rudki 89↘
Rudkøbing 33←
Rudna 88→
Rudnik 89↘
Rudolstadt 41↖
Rue 38←
Rueda 64↘
Ruelle 50↗
Ruffec 44↘
Ruffieux 46↙
Rufford 20↘
Rugby 23↖
Rugeley 22↗
Rugles 43→
Rühimäki 87→
Ruhla 40↗
Ruhpolding 48→
Ruidera 70↘
Rukkedalen 26↗
Rulbo 84↗
Rülzheim 40↙
Rum 49↗
Ruma 90↗
Rumblingbridge 19↙
Rumeln 39↗
Rumilly 46↘
Rummelsburg = Miastko
Rumont 39↙
Runcorn 21↙
Runderoth 40↖
Rungsted 32↗
Runhallen 29←
Rünkhofen 47←
Runnymede 23←
Runsten 31←
Runswick 21→
Ruokolahti 87↗
Ruovesi 87←
Rupe 55↘
Rus 73↗
Ruschuk = Ruse
Ruse, *Bulgaria* 93←
Ruše, *Yugoslavia* 49↘
Rush 24↗
Rushall 22→
Rushden 23↖
Rusksele 86↘
Ruskträsk 84↖
Rüsselsheim 40←
Russi 54→
Rust 49→
Rute, *Spain* 73→
Rute, *Sweden* 31↗
Rutherglen 18↘
Ruthin 20↘
Ruthven 18→
Rüti 47←
Rutino 58↘
Rutledal 26↘
Ruurlo 35→
Ruvo di Puglia 58←
Ružomberok 89↙
Rybnik 89↙
Rybnitsa 92↖
Rychnov 88↘
Ryd 30→
Rydaholm 30→
Ryde 23↙
Rydöbruk 30→
Rydsnäs 31↖
Rye 23↘
Ryfoss 26→
Rygg 26↙
Rygge 27↖
Rygnestad 27←
Ryki 89→
Rymanów 89↘
Rýmařov 89↙
Ryn 89↗
Ryngestad 27←
Rypin 89↖
Ryr 28↙
Ryshkany 92↖
Ryssby 30→
Ryton 21↖
Ryttern 29←
Rzepin 88→
Rzeszów 89↘

Saal 41↙
Saalbach 48→
Saalburg 41←
Saales 46↗
Saalfeld 41↖
Saalfelden 48→
Saanen 46↘
Saanenmöser 12←
Saarbrücken 39↘
Saarburg 39→
Saarenkylä 86↙
Saari 87↗
Saarijärvi 87←
Saarland, prov. 39↘
Saarlouis 39↘
Saas-Fee 47↙
Šabac 90↗
Sabadell 67↙
Sabáudia 56↘
Sabbioneta 54←
Sabero 64↗
Sabiñánigo 66→
Sabinov 89↘
Sabiote 74↖
Sablé 44↗
Sables d'Olonne, les 44↙
Sables-d'Or 42→
Saboia 72←
Sabres 50→
Sabrosa 63↘
Sabugal 68↗
Sabuncu 95↗
Säby 30↗
Sacavém 68↙
Sacecorbo 65↘
Sacedón 70↗
Săcel 92←
Săcele 92→
Sachseln 47←
Sachsenhausen 40↖
Sacile 54↗
Säckingen 47←
Sacra di San Michele 53↖
Sacramenia 65↙
Săcueni 92↙
Sada 63↖
Sádaba 66←
Saddell 20↖
Sadgora 92←
Sæbö, *Norway* 26↙
Sæbö, *Norway* 26→
Sæby 32←
Sægrov 26↙
Saelices 70→
Saerbeck 36←
Sævrasvåg 26↘
Săfara 72↗
Šafárikovo 90↖
Säffle 28→
Saffron Walden 23→
S'Agaró 67→
Sågen 28↗
Sagfjord 85←
Sagmyra 29↖
Sagone 62←
Sagres 72←
Sagunto 71→
Sagvåg 27↙
Sahagún 64→
Šahy 90←
Saignelégier 46→
Saija 86↙
Saillagouse 67←
Saimaa, lake 87↗
St Abbs 19↙
St Aegidi 48↗
St-Affrique 51↘
St Agnes 22↙
St-Agrève 52←
St-Aignan 45←
St-Alban-sur-Limagnole 51→
St Albans 23←
St-Amand-de-Vendôme 45←
St-Amand-en-Puisaye 45→
St-Amand-les-Eaux 38→
St-Amand-Montrond 45←
St Amans 51→
St-Amans-Soult 67↗
St Amant 51↗
St Ambroix 52←
St-André-de-Cubzac 50→
St-André-de-l'Eure 38↙
St-André-les-Alpes 52↘
St. Andrews 19↙
St. Ann's 20↗
St Anthème 52↖
St Anthony 22↙
St Anton 47→
St-Antonin 51←
St Asaph 20↘
St Athan 22→
St-Auban 53↙
St-Auban-sur-l'Ouvèze 52→
St-Aubin, *France* 43↖
St-Aubin, *Switzerland* 46→
St-Aubin-du-Cormier 44↖
St-Aulaye 50↗
St Austell 22↙
St-Avold 39↘
St Aygulf 53↙
St Bartholomä 13↖
St-Béat 67↖
St Bees 20→
St-Benoît-du-Sault 45↙
St-Benoît-en-Woëvre 39↙
St-Benoît-sur-Loire 45↖
St-Bertrand-de-Comminges 66↗
St Blaise 46↗
St Blasien 47←
St Blazey 22↙
St Bonnet-le-Château 52↖
St Brévin 44←
St-Briac 42→
St Briac-sur-Mer 10↗
St Briavels 22→
St Brides 22←
St Brides Major 22→
St-Brieuc 42→
St Buryan 22↘
St Calais 44↗
St Cannat 52↘
St-Cast 42→
St-Cergue 46↘
St Chamond 52↖
St-Chély-d'Apcher 51→
St-Chinian 67↗
St Ciers 50↗
St-Clar 51↙
St-Claude-sur-Bienne 46↙
St Clears 22←
St Columb Major 22↙
St Cyprien 51←
St-Cyr-sur-Mer 52↘
St Cyrus 19←
St David's 22←
St Denis 38↙
St-Didier-en-Velay 52↖
St-Dié 46↗
St Dier 51↗
St Dizier 46↖
St Donat 52↖
St-Éloy-les-Mines 45↘
St-Étienne 52↖
St-Étienne-de-Montluc 44←
St-Étienne-de-St-Geoirs 52↗
St-Étienne-de-Tinée 53←
St-Étienne-les-Orgues 52→
St Fargeau 45→
St-Félicien 52↖
St Fergus 19←
St Fillans 18↘
St-Florent 45←
St-Florent-le-Vieil 44←
St Florentin 45↗
St Flour 51→
St Fons 52↖
St François-Longchamp 12↙
St-Fulgent 44←
St Gallen 47→
St Gallenkirch 47→
St-Galmier 52↖
St.-Gaudens 67↖
St-Gaultier 45↙
St-Gély-du-Fesc 52↙
St Gengoux 46↙
St-Geniez-d'Olt 51→
St-Genis-Laval 52↖
St-Genis-Pouilly 46↘
St Georgen, *Austria* 49←
St Georgen, *W. Germany* 47↖
St-Georges-sur-Loire 44←
St Geours-de-Maremne 50↙
St-Germain 38↙
St-Germain-de-Joux 46↙
St Germain-des-Fossés 45↘
St-Germain-Laval 52↖
St-Germain-Lembron 51↗
St-Germain-les-Belles 51↖
St-Germain-l'Herm 51↗
St Germans 22↙
St Gertraud 48↙
St Gervais 46↘
St Gervais-d'Auvergne 45↘
St-Gildas-des-Bois 42↘
St Giles 23←
St Gilgen 13↗
St-Gilles 52↙
St-Gilles-sur-Vie 44↙
St Gingolph 46↘
St-Girons 67↖
St Goar 39→
St Goarshausen 39→
St-Gobain 38→
St-Gorgon 46→
St-Guénole 42↙
St-Guilhem-le-Desert 51↘
St Heddinge 33↖
St Helens, *England* 21↙
St Helens, *England* 23↙
St Helier 42↗
St-Hilaire-du-Harcouet 43←
St-Hippolyte-du-Fort 52←
St-Honoré 45→
St-Hubert 39←
St-Imier 46→
St Ingebert 39↘
St Ives, *England* 22↘
St Ives, *England* 23↖
St Jacut-de-la-Mer 10↖
St Jakob 49↙
St Jean-Cap-Ferrat 10↙
St-Jean-d'Angély 44↘
St-Jean-de-Bournay 52↗
St-Jean-de-Losne 46←
St-Jean-de-Luz 65↗
St-Jean-de-Maurienne 52↗
St-Jean-de-Monts 44←
St-Jean-de-Sixt 56↘
St-Jean-du-Bruel 51↘
St-Jean-du-Gard 52←
St-Jean-en-Royans 52→
St Jean-le-Thomas 10↗
St-Jean-Pied-de-Port 66↖
St Johann, *Austria* 48→
St Johann, *Austria* 49↖
St Johann, *Austria* 49←
St Johann im Tirol 48→
St John's 20→
St John's Chapel 21←
St Johnston 24←
St-Juéry 51↙
St-Julien 46↘
St-Julien-Chapteuil 52↖
St-Junien 44↘
St Just 22↘
St-Just-en-Chevalet 45↘
St-Juste-en-Chaussée 38↙
St Justin 50↘
St Keverne 22↙
St-Lary 66↗
St-Laurent-de-la-Salanque 67↗
St-Laurent-de-Médoc 50↗
St-Laurent-du-Jura 46↘
St-Laurent-du-Pont 52↗
St Lawrence 23↙
St Léonard 51↖
St Leonards 23↘
St Leonhard 48←
St-Lô 43←
St Louis, airport 47←
St-Loup 46↗
St-Lunaire 42→
St-Lys 67↖
St-Maixent 44↘
St-Malo 42→
St-Marcellin 52↗
St Margaret's at Cliffe 23→
St Margaret's Hope 19↖
St Margrethen 47→
St-Mars-la-Jaille 44←
St-Martin 44↙
St-Martin-de-Londres 52↙
St-Martin-de Valamas 52←
St-Martin-Vésubie 53←
St-Martory 67↖
St Mary's 19↖
St Mathieu 51↖
St-Maurice, *France* 51↘
St-Maurice, *Switzerland* 46↘
St Mawes 22↙
St-Maximin 52↘
St-Médard-en-Jalles 50→
St-Méen 42→
St Mellons 22→
St Méloir-des-Ondes 10↗
St Michael 49→
St-Michel-de-Maurienne 52↗
St-Mihiel 39↙
St Monance 19↙
St Moritz 47↘
St-Nazaire 44←
St Nectaire 51↗
St Neots 23↖
St-Nicolas 51↙
St-Nicolas-de-Port 39↘
St-Nicolas-du-Pélem 42→
St Niklaas 38↗
St Niklaus 47↙
St Nikolai 49←
St Oedenrode 35→
St Omer 38↖
St Pair-sur-Mer 10↗
St-Palais, *France* 50↖
St-Palais, *France* 66↖
St Pardoux 51↖
St-Paul, *France* 50↙
St Paul, *France* 53↙
St-Paul-Cap-de-Joux 51↙
St-Paul-de-Fenouillet 67↗
St-Paulien 52↖
St-Pé 66↗
St-Péray 52←
St-Père-en-Retz 44←
St Peter 33↘
St Peter Port 42↗
St-Philbert 44←
St-Pierre, *France* 44↙
St. Pierre, *France* 51↘
St-Pierre-de-la-Fage 51↘
St-Pierre Église 43↖
St Pierre-en-port 43↗
St-Pierre-le-Moutier 45→
St-Pierre-sur-Dives 43←
St-Point 46→
St Pol 38←
St-Pol-de-León 42←
St Pölten 49↗
St-Pons 67↗
St Pourçain 45↘
St-Privat 51↖
St-Quay-Portrieux 42→
St Quentin 38→
St-Rambert, *France* 46↙
St Rambert, *France* 52↖
St-Rambert-d'Albon 52↖
St-Raphaël 53↙
St-Rémy 52↙
St Romain 43↗
St-Rome-de-Cernon 51↘
St-Rome-de-Tarn 51→
St-Saulge 45→
St Sauveur, *France* 45→
St-Sauveur, *France* 53←
St Sauveur, *France* 66↗
St-Sauveur-le-Vicomte 43↖
St Savin, *France* 44↘
St-Savin, *France* 50↗
St-Savinien 44↙
St-Seine-l'Abbaye 46←
St-Sernin-sur-Rance 51↘
St Servan 42→
St-Sever 50↘
St Stefan 48↘
St-Sulpice-la-Pointe 51↙
St-Sulpice-les-Feuilles 45↙
St-Symphorien 50→
St-Symphorien-d'Ozon 52↖
St-Symphorien-sur-Coise 52↖
St Teath 22↙
St Tönis 39↗
St-Trivier-de-Courtes 46↙
St-Trojan 44↙
St Tropez 53↙
St-Truiden 39↖
St Ulrich 48↙
St-Vaast-la-Hougue 43↖
St Valentin 49↖
St Valéry-en-Caux 43↗
St Vallier, *France* 52↖
St-Vallier, *France* 53↙
St-Varent 44→
St Vaury 45↙
St Veit 49↙
St Vincent 53↖
St-Vith 39←
St Wendel 39↘
St Weonards 22→
St Wolfgang 13↗
St Yorre 45↘
St-Yrieix 51↖
St Zacharie 52↘
Ste-Adresse 43↗
Ste Baume, la 52↘
Ste-Croix 46→
Ste Engrâce 66↗
Ste-Enimie 51→
Ste-Foy-la-Grande 50→
Ste Gauburge-Ste Colombe 43→
Ste-Geneviève 51→
Ste-Hermine 44↙
Ste Livrade 50→
Ste-Marie-aux-Mines 46↗
Ste-Maure 44→
Ste-Maxime 52↘
Ste Ménehould 39↙
Ste-Mère-Église 43↖
Ste Odile 46↗
Ste-Sévère 45↙
Saintes 50↗
Saintes-Maries, les 52↙
Saintfield 24↗
Saintonge, reg. 50↗
Saivomuotka 85→
Sajószentpéter 90↖
Sakskøbing 33↖
Saksum 26↗
Säkylä 87↘
Šala, *Czechoslovakia* 90←
Sala, *Sweden* 29←
Sala Consilina 58↘
Salamanca 64↘
Salamis 96↗
Salån 28↗
Salangen 85→
Salar 73→
Sălard 92↙
Salardu 67↖
Salas, *Spain* 64↖
Salas, *Yugoslavia* 93↙
Salas de los Infantes 65←
Sälboda 28←
Salbohed 29←
Salbris 45←
Sălciua 92↘
Salcombe 22↙
Saldaña 64→
Sale 21↙
Saleby 28↘
Salemi 60↙
Salen, *Scotland* 18←
Salen, *Scotland* 18↙
Sälen, *Sweden* 84→
Salernes 52↘
Salerno 58↘
Salers 51↗
Salford 21↙
Salgótarján 90↖
Salhus, *Norway* 26↘
Salhus, *Norway* 27↙
Salies-de-Béarn 66↖
Salies-du-Salat 67↖
Salignac 51←
Salihli 95→
Salinas 64↗
Salinas de Añana 65↖
Salinas de Oro 65↗
Saline 54↙
Salins-les-Bains 46→
Salir 72←
Salisbury 23←
Săliste 92↘
Salla 86↙
Sallanches 46↘
Sallent, *Spain* 66↗
Sallent, *Spain* 67←
Salles-Curan 51→
Sally Gap 25↖
Salmela 86↙
Salmeron 70↗
Salmivaara 86↙
Salmünster 40→
Salo, *Finland* 87↘
Salò, *Italy* 54↖
Salobreña 73↘
Saloinen 86→
Salon 52↘
Salonica = Thessaloníki
Salonta 92↙
Salorino 68→
Salou 67↙
Salsadella 71↗
Salsåker 84↗
Salses 67↗
Salsomaggiore 54←

Skogstorp 29←
Skokloster 29→
Sköldinge 29↙
Skøldstrup 32→
Skole 89↘
Sköllesta 29↙
Skopí 96→
Skopje 91↗
Skorovatn 84←
Skorped 84↖
Skotfoss 27↖
Skotiní 96→
Skotterud 28←
Skottorp 30←
Skoura 96→
Skourokhóri 96↘
Skövde 28↘
Skradin 55↘
Skredsvik 28↙
Skreia 28↖
Skrein 26↗
Skripoú 96←
Skromberga 30↙
Skröven 85↘
Skudeneshavn 27↙
Skui 27↖
Skulerud 28←
Skultorp 28↘
Skultuna 29←
Skulyany 92↖
Skurup 30↘
Skute 26↗
Skutskär 29↖
Skutvik 85←
Skwierzyna 88→
Skylberg 28↘
Skyttorp 29→
Slagelse 33←
Slaggyford 21↖
Slagnäs 85↘
Slane 24↗
Slangerup 32↗
Slano 91←
Slany 41→
Sląska 88→
Slåtafly 31←
Slatina 93←
Slättberg 29↖
Slåtten 86↖
Slåttevik 27↙
Slavkov 88↘
Slavonia, reg. 90→
Slavonice 49↖
Slavonska Požega 90→
Sławno 88↗
Sławoborze 88↗
Sleaford 21↘
Sleights 21→
Slesiń 89←
Slidre 84↘
Slife 31↗
Sligachan 18←
Sligo 24→
Sliven 93←
Slivnitsa 93↙
Slobozia, *Rumania* 92↗
Slobozia, *Rumania* 93←
Słońsk 88→
Slottsbron 28→
Slough 23←
Slovenia = Slovenija
Slovenija, prov. 90↘
Slovenjgradec 49↙
Slovenska Bistrica 49↘
Słubice 88→
Sluknow 88→
Slunchev Bryag = Sunny Beach
Slunj 55↗
Słupca 89←
Słupsk 89↖
Smådalseter 26←
Smål Taberg 30↗
Smålandsstenar 30→
Smedby 31←
Smederevska Palanka 91↖
Smedjebacken 29←
Smedland 27←
Smedsbo 29↖
Smedstorp 30↘
Smigiel 88→
Smilcic 55→
Smilde 35↗
Smîrdiosa 93←
Smiřice 88↘
Smögen 28↙
Smolyan 93→
Smörhamn 26↙
Smyadovo 93↖
Smygehamn 30↘
Smyrna = İzmir
Snäckgärdsbeden 31↗
Snaefell, mt. 20→
Snaith 21↙
Snarum 27↖
Snåsa 84←
Sneek 35↗
Sneem 25↘
Snejbjerg 32↘
Sniadowo 89↗
Snina 89↘
Snössvallen 84→
Snöstorp 30←
Snowdon, mt. 20↘
Snyatyn 92←
Soanlakhti 87↖
Soave 54↖
Sobernheim 39→
Soběslav 88↘
Sobótka 88→
Sobrado 63↗
Sobrado de Paiva 63↙
Sobral de Monte Agraço 68↙
Sobrance 90↖
Sobreira Formosa 68→
Søby 33←
Sočerga 55↖
Socuellamos 70→
Sodankylä 86←
Söderåkra 31←
Söderala 84↗
Söderbärke 29←
Söderby-Karl 29→
Söderfjärden 84↗
Söderfors 29↖
Söderhamn 84↗
Söderköping 29↙
Södertälje 29→
Sodra Finnskoga 28↖
Sodupe 65↖
Soest 36↙
Soestdijk 35→
Sofádhes 94↗
Sofia = Sofíya
Sofikón 96→
Sofíya 93↙
Sögel 36←
Soglio 47↘
Sogndal, *Norway* 27↘
Sogndal, *Norway* 84↘
Sogndalsfjöra 26←
Söğüt 95↗
Soham 23↗
Soignies 38→
Soini 87←
Soissons 38↘
Söjtör 49↘
Söke 95←
Sokhós 94↗
Sokna 26↗
Sokółka 89↗
Sokolniki 89←
Sokolov 41←
Sokołów Podlaski 89→
Sola 27↘
Solana de los Barros 69↙
Solana del Pino 69↘
Solares 65↖
Solbad Hall 48←
Solberg 84↖
Solberga 30↗
Solbergelva 27↖
Solca 92↖
Solda 47↘
Sölden 48←
Solec Kujawski 89↖
Solenzara 62←
Solesmes 38→
Solf 87↙
Solheim 26↘
Solheimstuhl 26→
Solheimsvik 27↙
Solihull 22↖
Solingen 39↗
Sollana 71→
Sollentuna 29→
Sóller 75←
Sollerön 28↗
Solli 28←
Sollia 84→
Solliès-Pont 52↘
Sollihogda 27↖
Solna 29→
Solórzano 65↖
Solosancho 69↗
Solothurn 47←
Solrød 33↖
Solsona 67←
Solstad 84←
Solsvik 26↘
Soltau 36↗
Solumsmo 27↖
Solund 26↘
Solvalla 29→
Sölvesborg 30↘
Solvorn 26←
Soma 95↖
Sombernon 46←
Sombor 90↗
Somcuța Mare 92↙
Somero 87→
Somersham 23↖
Somerton 22→
Sommariva del Bosco 53←
Sommen 30↗
Sömmerda 41↖
Sommesous 38↘
Sommières 52↙
Somogyszob 90→
Somogyvár 90→
Somosierra 65↙
Sompuis 39↙
Son, *Norway* 27↖
Son, *Spain* 63←
Son Servera 75→
Soncillo 65↖
Soncino 53↗
Söndeled 27↗
Sønder Brody 33←
Sønder Dalby 33↖
Sønder Felding 32↘
Sønder Omme 32↘
Sønderborg 33←
Sondershausen 40↗
Sønderso 33←
Søndervig 32↘
Sondre Holand 28←
Sóndrio 47↘
Songe 27↗
Songeons 38↙
Sonkajärvi 87↖
Sonneberg 41←
Sonogno 47↙
Sonseca con Casalgordo 69→
Sönsterud 28↖
Sonthofen 47→
Sontra 40↗
Sopeira 67←
Sopelana 14↖
Sopot 89↖
Soppero 85→
Sopron 49→
Sör Aurdal = Bagn
Sör Flatanger 84←
Sör-Odal 28↖
Sora 57←
Sorbas 74→
Sörberget 84→
Sörby 28→
Sörbygden 84↗
Sore 50→
Sörenberg 47←
Sörfjord 85↙
Sörfold 85←
Sörfors 84↖
Sorgono 61←
Sorgues 52↙
Sörheim 26←
Soria 65→
Soriano nel Cimino 56↗
Sorihuela 69↖
Sorihuela del Guadalimar 70↘
Sorisdale 18←
Sörkedalen 27↖
Sörli 84←
Sörmark 28↖
Sörmjöle 87↙
Sorn 20↗
Sorø 33↖
Soroki 92↖
Soroksár 90←
Sörreisa 85↗
Sorrento 57↙
Sorsele 85↘
Sörskog 29↖
Sorso 61↖
Sort 67←
Sortavala 87↗
Sortino 60↗
Sortland 85→
Sorum 26↗
Sörumsand 28←
Sorunda 29↘
Sörvær 85↗
Sörvattnan 84→
Sos del Rey Católico 66←
Sösdala 30↘
Sosnovo 87↗
Sosnowiec 89↙
Soso 86→
Sospel 53↙
Šoštanj 49↙
Sotillo 69↗
Sotkamo 87↖
Sotoserrano 69↖
Sotra 26↘
Sotresgudo 65←
Sotta 62↙
Sottunga 87↘
Sotuélamos 70↘
Souillac 51←
Souilly 39↙
Soulac 50↖
Soultz-sous-Forêts 40↙
Sound, The 22↙
Souppes 45↖
Sourdeval 43←
Soure 68←
Soúrpi 94↗
Sousel 68↘
Soustons 50↙
Souterraine, la 45↙
South Brent 22↙
South-haa 19↗
South Harting 23↙
South Hayling 23↙
South Kessock 18→
South Mimms 23←
South Molton 22↙
South Queensferry 19↙
South Shields 21↖
South Wootton 23↗
Southam 23↖
Southampton 23↙
Southborough 23→
Southend 20↖
Southend on Sea 23→
Southery 23↗
Southport 20↘
Southsea 23↙
Southwell 21↘
Southwold 23↗
Souvigny 45↘
Sover 48↙
Soverato 59←
Soveria Mannelli 59←
Sovestad 30↘
Sowerby Bridge 21↙
Spa 39←
Spaichingen 47↖
Spalato = Split
Spalding 23↖
Spångenäs 31↖
Spangenberg 40↗
Spangereid 27→
Sparbu 84←
Sparkford 22↘
Sparreholm 29↙
Sparta, *Greece*, anc. site 96→
Sparta, *Italy* 60↖
Spárti 96→
Spartílla 96↘
Spártos 96↙
Spáta 96↗
Spean Bridge 18→
Spello 56↗
Spennymoor 21←
Sperkhiás 96←
Sperlonga 57↙
Spetisbury 22↘
Spétsai 96→
Speyer 40↙
Spezzano Albanese 59←
Spezzano della Sila 59←
Spiddal 25↙
Spiekeroog 36↖
Spíli 94↘
Spilimbergo 48↘
Spilling 27→
Spilsby 21↘
Spinazzola 58←
Spincourt 39↙
Spind 27↘
Spindleruv Myln 88↘
Spišská Belá 89↘
Spišská Nová Ves 90↖
Spišské Podhradie 89↘
Spiterstulen 26←
Spittal 48↘
Spittal of Glenshee 19←
Spitz 49↖
Split 91←
Splügen 47↘
Spoláita 96↙
Spoleto 56↗
Spondigna 47↘
Spondinig 47↘
Spotorno 53→
Sprakensehl 37←
Spremberg 88→
Sprendlingen 40←
Spresiano 54↗
Springe 36→
Sproxton 21←
Spurn Head 21↘
Spydeberg 28←
Squinzano 58↗
Squire's Gate, airport 20↘
Srbica 91↖
Srbija, prov. 91↖
Srbobran 90↗
Srem 89←
Srem Raca 90↗
Sremska Mitrovica 90↗
Sremski Karlovci 90↗
Sroda 89←
Stabbestad 27↗
Stabio 13↙ (4 km SW of Mendrisio)
Stade 36↗
Stadhampton 23←
Stádhion 96→
Stadra 28→
Stadthagen 36→
Stadtilm 41↖
Stadtkyll 39→
Stadtlauringen 40→
Stadtlengsfeld 40↗
Stadtlohn 36←
Stadtoldendorf 36↘
Stadtroda 41↖
Stäfa 47←
Staffanstorp 30↘
Staffelstein 41←
Staffin 18←
Stafford 22↗
Stafsberg 28↖
Stai 26↖
Stainach 49←
Staines 23←
Stainz 49↙
Stair 20↗
Stalac 91↖
Stalbridge 22↘
Stalcerji 55↗
Stalham 23↗
Stallarholmen 29←
Ställberg 28→
Ställdallen 28→
Stâlpeni 92→
Stalybridge 21↙
Stamford 23↖
Stamford Bridge 21→
Stammham, *W. Germany* 41↙
Stammham, *W. Germany* 48↗
Stamna 96↙
Stamnes 26↘
Stamsund 85←
Standal 26↙
Standish 21↙
Standon 23←
Stanford-le-Hope 23→
Stånga 31↗
Stange 28↖
Stangvik 84↘
Stanhope 21←
Stanjel 55↖
Stanke Dimitrov 93↙
Staňkov 41→
Stanley, *England* 21↖
Stanley, *Scotland* 19↙
Stánόs 96↙
Stans 47←
Stansted Mountfitchet 23→
Staphorst 35↗
Stapleford, *England* 21↙
Stapleford, *England* 22→
Stará L'ubovňa 89↘
Stara Pazova 90↗
Stara Planina, mts. 93↙
Stara Zagora 93←
Stargard, *E. Germany* 37↗
Stargard, *Poland* 88↗
Stårheim 26↙
Stari Grad 16→
Starigrad-Paklenica 55→
Starina 89↘
Staringrad 55→
Starnberg 48↖
Staro Oryakhovo 93↖
Starogard 89↖
Starokazache 92↗
Starup 32↘
Starý Plzenec 41→
Stary Sambor 89↘
Stary Smokovec 90↖
Staszów 89↘
Stathelle 27↖
Statsbuöyen 26↖
Staunton 22→
Stavali 26→
Stavanger 27↙
Stavby 29→
Stavely 21↙
Stavelot 39←
Stavenisse 35←
Staveren 35↗
Stavern 27↖
Staverton 23↖
Stavrodhrómi 96↘
Stavrós 94↗
Stavroúpolis 93→
Stavsjö 28↖
Stavsjöbruk 29↙
Stavsnäs 29→
Stawiski 89↗
Staxton 21→
Steane 27←
Stechelberg 47↙
Steenbergen 35←
Steenwijk 35↗
Stefaneşti 92↖
Stefáni 96↖
Stegaros 26→
Stege 33↖
Stegeborg 29↙
Steierdorf Anina 92↘
Steiermark, prov. 49←
Steigen 85←
Stein, *Austria* 49←
Stein, *Scotland* 18←
Stein, *Switzerland* 47←
Stein, *W. Germany* 41↙
Stein-am-Rhein 47←
Steinach, *Austria* 48←
Steinach, *E. Germany* 41←
Steinau 40→
Steinfeld 36←
Steingaden 48←
Steinhausen 47↗
Steinheim, *W. Germany* 40←
Steinheim, *W. Germany* 47↗
Steinkjer 84←
Steinsbole 26→
Stemshang 84↙
Stenåsa 31←
Stenay 39↙
Stend 26↘
Stendal 37←
Steneby 28↙
Steni 96↖
Steninge 30←
Stenón 96→
Stensele 84↖
Stensjön 30↗
Stenstorp 28↘
Stenstråsk 85↘
Stenungsund 30↖
Stérna 96→
Sternberg 33↗
Sternberk 89↙
Sterringi 26←
Sterup 33←
Sterzing 48↙
S'tęszew 88→
Stet 91↖
Stettin = Szczecin
Stevenage 23←
Steveníkoi 96←
Stevenston 20↗
Steventon 23←
Stewarton 20↗
Stewartstown 24↖
Steyning 23↙
Steyr 49↖
Stiarnhov 29↙
Stibb Cross 22↙
Stickney 21↘
Stiens 35↗
Stigamo 30↗
Stigen 28↙
Stigliano 58→
Stigsjö 84↗
Stigtomta 29↙
Stilís 96←
Stillington 21←
Stilo 59→
Stilton 23↖
Stimånga 96→
Stintino 61↖
Štip 93↘
Stíra 96↖
Stirling 18↘
Stirovaca 55→
Stjärnfors 28→
Stjärnsund 29↖
Stjördalshalsen 84←
Stock 23→
Stockach 47↖
Stöckalp 12↖
Stockaryd 30→

Tazones 64↗
Tczew 89↖
Teaca 92←
Teano 57↙
Teba 73↙
Tébar 70→
Tebay 21←
Teckomatorp 30↙
Tecuci 92↗
Tedburn St Mary 22↙
Teesside 21←
Tefenni 95→
Teg 87↙
Tegea, anc. site 96→
Tegel, airport 37→
Tegelen 35↘
Tegelsmora 29↗
Tegernsee 48←
Teignmouth 22↘
Teil, le 52←
Teinestölen 26↗
Teisko 87→
Teiuş 92↘
Tejares 64↘
Tejo, river 68→
Tekirdağ 93↗
Telciu 92←
Telese 57↙
Telford 22↗
Telfs 48←
Telgte 36←
Telish 93←
Tellejokk 85↘
Teltow 37→
Tembleque 70←
Temerin 90↗
Temmes 86→
Tempelhof, airport 37→
Tempio Pausánia 61↖
Temple 19↙
Temple Bar 22↖
Temple Ewell 23→
Temple Sowerby 21←
Templemore 25←
Templepatrick 24↖
Templin 37↗
Tempo 24→
Tempsford 23↖
Ten Boer 35↗
Tenala 87↘
Tenay 46↙
Tenbury Wells 22↗
Tenby 22←
Tenebrón 69↖
Tence 52↖
Tende 53←
Tenhult 30↗
Tenterden 23→
Tentugal 68↖
Teo 63←
Tepasto 86←
Tepelenë 94↖
Teplice 41↗
Ter Apel 36↖
Teramo 57↖
Terborg 35→
Teregova 92↘
Terena 68↘
Terešov 41→
Terespol 89→
Terézin 41→
Tergnier 38→
Terheijden 35←
Termens 67←
Termini 60←
Térmoli 57→
Terndrup 32←
Terneuzen 35↙
Ternhill 21↙
Terni 56↗
Ternitz 49→
Terracina 57↙
Terralba 61←
Terras do Bouro 63←
Terrasson 51↖
Terriente 71↖
Terroba 65→
Terrugem 68↙
Tertenia 61→
Teruel 71↖
Terval 93↖
Tervo 87←
Tervola 86↙
Tervuren 39↖
Tessin 33↗
Teste-de-Buch, la 50←
Tessy 43←
Tét 90←
Tetbury 22→
Teterow 33↗
Tetovo 91↗
Tettnang 47→
Teulada, *Italy* 61↙
Teulada, *Spain* 71↘
Teuva 87↙
Tevansjö 84↗
Teviothead 20↗
Tewkesbury 22→
Thale 37↙
Thalerhof, airport 49←
Thalfang 39→
Thalheim 41↗
Thalmässing 41↙
Thalwil 12↖
Thame 23←
Thames, river 23←
Thamshamn 84↙
Thann 46↗
Thannhausen 47↗
Thaon-les-Vosges 46↗
Tharsis 72→
Thásos 94↗
Thaxted 23→
Theale 23←
Themar 40→
Thénezay 44↘
Thenon 51↖
Theóktiston 96→
Theológos, *Greece* 96↖
Theológos, *Greece* 96→
Théoule 53↙
Thèrmon 96↙
Thermopylai, anc. site 96←
Thespiaí 96←
Thessaloníki 94↗
Thetford 23↗
Theux 39←
Thiaucourt 39↙
Thiberville 43→
Thiede 37←
Thiene 54↗
Thiers 45↘
Thiesi 61←
Thilliers, les 38↙
Thillot, le 46↗
Thionville 39↘
Thíra 95↙
Thirsk 21←
Thisted 32↙
Thísvi 96←
Thívai 96←
Thiviers 51↖
Thizy 46↙
Thoissey 46↙
Thoknia 96↘
Tholen 35↙
Thomastown 25←
Thônes 46↘
Thonon-les-Bains 46↘
Thorens-Glières 12↘
Thorigny-sur-Oreuse 45↗
Thornaby-on-Tees 21←
Thornbury 22→
Thornby 23↖
Thornhill, *Scotland* 18↘
Thornhill, *Scotland* 20↗
Thorne 21↙
Thorney 23↖
Thornton 20→
Thorrington 23→
Thouars 44→
Thouría 96↘
Thrapston 23↖
Threekingham 21↘
Threshfield 21←
Thueyts 52←
Thuin 38→
Thuir 67↗
Thum 41↗
Thumby 33←
Thun 47↙
Thürkow 33↗
Thurles 25←
Thurlestone 22↙
Thurnscoe 21↙
Thursby 20→
Thurso 19↖
Thury-Harcourt 43←
Thusis 47↘
Thyatera, anc. site 95↗
Thyborøn 32↘
Tibro 28↘
Tichfield 23↙
Ticino, prov. 47↙
Tickhill 21↙
Tidaholm 30↗
Tidan 28↘
Tidersrum 31↖
Tidworth 23←
Tiębas 65→
Tiefencastel 47↘
Tiel 35→
Tielt 38↗
Tienen 39↖
Tiengen 47←
Tierga 65↘
Tiermas 66↖
Tierp 29↗
Tighnabruaich 18↘
Tihany 90←
Tijesno 55↘
Tíjola 74←
Tilburg 35→
Tilbury 23→
Tilchâtel 46←
Tileagd 92↙
Tillberga 29←
Tillicoultry 18↘
Tílos 95↙
Tilshead 22→
Timahoe 25←
Timbákion 94↘
Timfristos 96↙
Timişoara 90↗
Timmell 30↗
Timmernabb 31←
Timmersdala 28↘
Timmervik 28↙
Timolin 25↖
Timrå 84↗
Tinahely 25↖
Tinchebray 43←
Tineo 64↖
Tingewick 23←
Tinglev 33↙
Tingsryd 30→
Tingstade 31↗
Tingvatn 27→
Tingvoll 84↘
Tinn = Atrå
Tinnoset 27↖
Tinntorp 28←
Tínos 95←
Tintagel 22↙
Tintern 22→
Tintigny 39←
Tione di Trento 47↘
Tipperary 25←
Tiptree 23→
Tirana = Tiranë
Tiranë 91↗
Tirano 47↘
Tiraspol 92↖
Tire 95→
Tîrgovişte 92→
Tîrgu Frumos 92↖
Tîrgu Jiu 92↘
Tîrgu Mureş 92→
Tîrgu Neamţ 92←
Tîrgu Ocna 92→
Tîrgu Secuesc 92→
Tirilye 95↗
Tírins 96→
Tiriolo 59←
Tírnavos 94↗
Tirol, prov. 48←
Tirrénia 54↙
Tirschenreuth 41←
Tirstrup 32→
Tiszafüred 90↖
Titaguas 71←
Tithoréa 96←
Titisee 47↖
Titograd 91↖
Titova 55→
Titov Veles 93↘
Titovo Užice 91↖
Titran 84↙
Tittelsnes 27↙
Tittling 41↘
Tittmoning 48↗
Titu 92→
Titz 39↗
Tiuccia 62←
Tived 28↘
Tiverton 22↘
Tivisa 67↙
Tívoli 56→
Tjällmo 29↙
Tjåmotis 85→
Tjernmoen 28↖
Tjölling 27↖
Tjönnefoss 27←
Tjörnekalv 30↖
Tjötta 85↙
Tkon 55↘
Tobarra 71↙
Tobercurry 24→
Tobermore 24↖
Tobermory 18↙
Tobblach 48↘
Tobru 26↗
Tocha 68↖
Töckfors 28←
Todal 84↘
Toddington 23←
Todi 56↗
Tödi, mt. 47←
Todmorden 21↙
Todtmoos 47←
Todtnau 47↖
Tofta, *Sweden* 30←
Tofta, *Sweden* 31↗
Toftlund 33↙
Tohmajarvi 87↖
Toijala 87→
Toivakka 87←
Tokajo 90↖
Toksovo 87↗
Tolentino 55↙
Toledo 69→
Tolfa 56→
Tolg 30→
Tolga 84→
Tollarp 30↘
Tollsjö 30↖
Tolmezzo 48↘
Tolna 90→
Tolo 96→
Tolob 19→
Tolós 96→
Tolosa, *Spain* 65↗
Tolosa, *Portugal* 68→
Tolve 58→
Tomar 68←
Tomala 87↘
Tomashevka 89→
Tomashpol 92↖
Tomaszów Lubelski 89↘
Tomaszów Mazowiecka 89←
Tomatin 18→
Tomdoun 18→
Tomelilla 30↘
Tomelloso 70↘
Tomgraney 25←
Tomich 18→
Tomintoul 19←
Tommerup 33←
Tomnavoulin 19←
Tomta 29←
Tomter 28←
Tona 67→
Tonbridge 23→
Tondal 27←
Tondela 68↗
Tønder 33↙
Tongeren 39↖
Tongue 18↗
Tönjum 26→
Tonnay-Charente 44↙
Tonneins 50→
Tonnerre 45↗
Tönnet 28↗
Tonning 33↘
Tonsberg 27↖
Tonstad 27↘
Toome 24↖
Toomyvara 25←
Topcliffe 21←
Toplita 92←
Topola 91↖
Toholampi 87←
Topolčane 91↗
Topol'čany 90←
Topólia 96←
Topolovení 92→
Topolovgrad 93→
Topraisar 93↖
Topsham 22↘
Toques 63↗
Tor Vaiánica 56→
Torá 67←
Torbalı 95←
Torbay 22↘
Torberget 84→
Torcello 54↗
Torcross 22↙
Torcy 38↙
Tordera 67→
Tordesillas 64↘
Töre 85↘
Töreboda 28↘
Torekov 30←
Torelló 67→
Toreno 64←
Torgås 28↗
Torgau 37↘
Torhamn 31↙
Torhout 38↗
Torigni 43←
Torija 70↗
Toril 71↖
Torino 53↖
Torino di Sangro 57←
Torla 66↗
Törmänen 86←
Tornavacas 69↖
Torneträsk 85→
Tornevalla 29↙
Tornio 86→
Toro 64↘
Törökszentmiklós 90↖
Torp 84↗
Torpa, *Sweden* 30↗
Torpa, *Sweden* 30→
Torpe 26↗
Torphins 19←
Torpo 26→
Torpoint 22↙
Torpshammar 84↗
Torquay 22↘
Torquemada 65←
Torralba de Calatrava 70↙
Torrão 68↙
Torre Annunziata 57↙
Torre Baja 71←
Torre Cardela 74←
Torre de Dona Chama 63↘
Torre de Juan Abad 70↘
Torre de Miguel Sesmero 68↘
Torre de Moncorvo 63↘
Torre del Campo 73→
Torre del Greco 57↙
Torre del Mar 73↘
Torre Mileto 58↙
Torre Pacheco 75←
Torre Pedrera 11↗
Torre-Péllice 53←
Torreblanca 71↗
Torreblascopedro 73↗
Torrecampo 69↘
Torrecilla de la Orden 64↘
Torrecilla en Cameros 65→
Torrecillas de la Tiesa 69←
Torredembarra 67↙
Torredonjimeno 73→
Torrejón de Ardoz 70↖
Torrejon de la Calzada 70←
Torrejón el Rubio 69←
Torrejoncillo 69←
Torrelaguna 65↙
Torrelapaja 65↘
Torrelavega 65↖
Torrelobatón 64↘
Torrelodones 69↗
Torremaggiore 58↙
Torremegia 69↙
Torremocha 69←
Torremolinos 73↘
Torrente 71←
Torrente de Cinca 66↘
Torrenueva 14↘
Torreorgaz 69←
Torreperogil 74←
Torres 73→
Torres de Albánchez 70↘
Torres Novas 68←
Torres Vedras 68↙
Torresandino 65←
Torrevieja 75←
Torrico 69→
Torridon 18→
Torriglia 53→
Torrijo 65↘
Torrijos 69→
Tørring 32↘
Torroal 68↙
Torroella de Montgri 67→
Torrox 73↘
Torsåken 29↖
Torsåker 84↗
Torsång 29↖
Torsås 31←
Torsborg 84→
Torsby 28↗
Torseter 28↖
Torshatta 29←
Torsken 85↖
Torslanda 30↖
Torsnes 26↘
Torsö 28↘
Torsvåg 85↗
Tortellá 67→
Tórtoles de Esqueva 65←
Tortona 53→
Tortosa 66↘
Tortozendo 68→
Toruń 89↖
Torup 30←
Torviscon 74↙
Torvsjö 84↖
Torvund 26↘
Torzym 88→
Toscaig 18←
Toscana, prov. 54↙
Tosen 84←
Tossá 67→
Tossåsen 84→
Tösse 28↙
Tostedt 36↗
Toszek 89↙
Totana 74→
Tôtes 38←
Tótkomlos 90↗
Tötlandsvik 27↙
Totnes 22↙
Tottenham 23←
Totton 23↙
Toucy 45→
Touët 53↙
Toul 39↙
Toulon 52↘
Toulon-sur-Arroux 45→
Toulouse 51↙
Touquet-Paris-Plage, le 83←
Tour-du-Pin, la 52↗
Tour-Fondue, la 10←
Touraine, reg. 42↘
Tourcoing 38↗
Tourkoléka 96→
Tournai 38↗
Tournan 38↘
Tournay 66↗
Tournon 52↖
Tournon-d'Agenais 51←
Tournon-St-Martin 44→
Tournus 46↙
Tourouvre 43→
Tours 44→
Tourves 52↘
Toury 45↖
Toužim 41→
Tovdal 27→
Tow Law 21←
Towcester 23↖
Towyn 22↖
Töysä 87←
Trabazos 64←
Traban-Trarbach 39→
Tradate 53↗
Tradet 30↗
Trafaria 68↙
Tragacete 70↗
Traganón 96↘
Tragwein 49↖
Traian 92↗
Traiguera 71↗
Traisen 49↗
Traiskirchen 49↗
Tralee 25↙
Tramagal 68←
Tramaríglio 61↖
Tramelan 46→
Tramore 25→
Tranås, *Sweden* 30↗
Tranås, *Sweden* 30↘
Tranche, la 44↙
Trancoso 64↙
Tranebjerg 32→
Tranekær 33←
Tranemo 30↗
Tranent 19↙
Tranhult 30↗
Trani 58←
Tranöy 85↖
Transtrand 28↗
Transylvania, reg. 92↘
Transylvanian Alps = Carpati Meridionali
Tranvik 29→
Trápani 60↙
Traryd 30→
Träskholm 86↘
Träslövsläge 30←
Trasmiras 63→
Trassem 39→
Traun 49↖
Traunstein 48→
Travad 28↘
Travemunde 33→
Travnik 91←
Trawsfynydd Llanuwchllyn 20↘
Trayas, le 53↙
Trbovlje 49↙
Trebbin 37→
Třebenice 41→
Trébeurden 42←
Třebíč 88↘
Trebinje 91←
Trebisacce 58→

Trebišov 90↖
Trebujena 73↙
Trecastle 22←
Trecate 53↗
Tredegar 22→
Treen 22↘
Trefnant 20↘
Tregaron 22↖
Trégastel 42←
Tregony 22↙
Tréguier 42→
Treherbert 22→
Trehörna 30↗
Trehörningsjö 84↖
Treib 47←
Treignac 51↖
Treis 39→
Trekanten 31←
Trelleborg 30↙
Trélon 38→
Tremadoc 20↘
Tremblade, la 44↙
Tremezzina 47↘
Třemošna 41→
Tremp 67←
Trenance 22↙
Trenčín 90←
Trendelburg 36↘
Trengereid 26↘
Trent, river 21↘
Trentino-Alto Adige, prov. 48↙
Trento 48↙
Tréport, le 38←
Trescore Balneario 54↖
Tresenda 47↘
Tresjuncos 70→
Trespaderne 65↖
Tresta 19↗
Tretower 22→
Trets 52↘
Tretten 26↖
Trettin 84→
Treuburg = Olecko
Treuchtlingen 40↘
Treuenbrietzen 37→
Trevélez 74↙
Treviglio 53↗
Treviño 65→
Treviso 54↗
Třevoň 88↘
Trévoux 46↙
Treysa 40↗
Trezzo 53↗
Trhové Sviny 49↖
Triabo 31←
Triánda 95↘
Tribanj 55→
Triberg 47↖
Tribsees 33↗
Tricárico 58→
Tricase 58↗
Trickett's Cross 22↘
Trie 66↗
Trieben 49←
Trier 39→
Trieste 55↖
Tríglia 96↗
Trigueros 72→
Tríkkala 94↖
Tríklinos 96↙
Tríkorfon 96↙
Trillo 70↗
Trim 24↗
Trimouille, la 45↙
Trinafour 18→
Trindade 63↘
Tring 23←
Trinité-Porhoet, la 42↘
Trino 53↗
Tríodhos 96↘
Tripi 96→
Tripití 96↘
Trípolis 96→
Triponzo 56↗
Tripótama 96↘
Tripótamos 96↘
Triptis 41↖
Trístenon 96↙
Trittau 33→
Trnava 90←
Trödje 29↖
Trogen 47→
Trogir 91↙
Trögstad 28←
Troia 58↙
Troina 60↖
Troisdorf 39↗
Troizín 96→
Trollhättan 28↙
Tromsö 85↗
Tromsöysund 85↗
Tronche, la 52↗
Trondenes 85←
Trondheim 84←
Trönödal 84↗
Trontveit 27←
Troon 20↗
Trópaia 96↘
Tropea 59↘
Trosa 29↘
Trossingen 47↖
Trostberg 48↗
Trostyanets 92↖
Troutbeck 20→
Trouville 43↗
Trovåg 27↙
Trowbridge 22→
Trowse Newton 23↗
Troy, anc. site 95↖
Troyan 93←
Troyes 45↗
Trstena 89↙
Trstenik 91↖
Trubia 64↗
Truchas 64←
Trujillo 69←
Trumpet 22→
Trumpington 23↗
Trun 47↙
Truro 22↙
Truşeşti 92←
Trutnov 88↘
Tryserum 31↖
Trysil 84→
Trzebiatow 88↗
Trzebnica 89←
Trzemeszno 89←
Tsamandás 94↖
Tsar Kaloyan 93←
Tsiótion 94↗
Tsitália 96→
Tsoukaléika, *Greece* 96↙
Tsoukaléika, *Greece* 96↘
Tuam 24↘
Tubbergen 36←
Tübingen 47↖
Tučepi 16→
Tuchan 67↗
Tuchola 89↖
Tuchów 89↘
Tudanca 65↖
Tuddal 27←
Tudela 65→
Tudela de Duero 64↘
Tuéjar 71←
Tuffe 44↗
Tulcea 92↗
Tulchin 92↖
Tulla 25↙
Tullamore 25←
Tulle 51↖
Tullinge 29→
Tulln 49↗
Tullow 25↘
Tulovo 93←
Tulsk 24→
Tumba 29→
Tumby 21↘
Tumleberg 28↙
Tummel Bridge 18→
Tun 28↙
Tuna, *Sweden* 29↙
Tuna, *Sweden* 29→
Tuna, *Sweden* 31↖
Tuna, *Sweden* 84↗
Tunbridge Wells = Royal Tunbridge Wells
Tundradalsseter 26↙
Tungelsta 29↘
Tungenes 27↙
Tungestölen 26←
Tunnhovd 26→
Tunstall, *England* 21↙
Tunstall, *England* 23↗
Tuomioja 86→
Tupos 86→
Turcifal 68↙
Turda 92↘
Turégan 65↙
Turek 89←
Tîrgovishte 93←
Turgutlu 95←
Turis 71←
Türje 90←
Turka 89↘
Türkheim 47↗
Turku 87↘
Turleque 70←
Turnberry 20↖
Turnhout 39↖
Turnov 88↘
Turnu Măgurele 93←
Turnu Roşu 92→
Turnu Severin 92↘
Turón 74↙
Turriff 19←
Turtagro 26←
Turtola 86↙
Turvey 23↖
Tuscánia 56↗
Tuscany = Toscana
Tusilovic 55↗
Tuşnad 92→
Tuszyn 89←
Tutiyarvi 86↘
Tutrakan 93←
Tuttlingen 47↖
Tuulos 87→
Tuupovarra 87↖
Tuusmemi 87↖
Tuxford 21↘
Tuy 63←
Tuzla 91←
Tvååker 30←
Tvärålund 87↙
Tvarån 86↘
Tveit 27↙
Tverra 27↘
Tving 31←
Tvinne 26↘
Tvrdonice 49↗
Tvŭrdîtsa 93←
Tweedmouth 19↙
Tweedsmuir 20↗
Twimberg 49←
Twistringen 36→
Two Bridges 22↙
Twycross 23↖
Twyford, *England* 23↙
Twyford, *England* 23←
Twynholm 20↗
Tyachevo 92↙
Tychowo 88↗
Tychy 89↙
Tydal 84→
Tyedstrand 27↗
Tyfors 28↗
Tygelsiö 30↙
Tyin 26←
Tyinholmen 26←
Tylldal 26↖
Tylösand 30←
Tymbark 89↙
Tynderö 84↗
Tyndrum 18↘
Tynemouth 21↖
Tyngsjö 28↗
Tynset 84→
Tyringe 30↘
Tyristrand 26↗
Tyrol = Tirol
Tyrrelspass 25←
Tysdal 27↙
Tysfjord 85←
Tysnes 27↙
Tysse 26↘
Tyssedal 26↘
Tysvær 27↙
Tyvoll 84→

Ualand 27↘
Ubach 39↖
Ubbedissen 36→
Ubeda 74←
Ubergsmoen 27→
Überlingen 47←
Ubierna 65←
Ubrique 73↙
Uceda 65↙
Ucero 65↙
Uchte 36→
Uckermünde-haff 88↗
Uckfield 23↘
Udbina 55→
Udbyhøj 32→
Uddebo 30↗
Uddevalla 28↙
Uden 35→
Udenäs 28↘
Udine 48↘
Uelsen 36←
Uelzen 37↖
Uerzell 40→
Uetersen 33↘
Uettingen 40→
Uetze 37←
Uffenheim 40→
Ugerløse 33↖
Uggelheden 28↖
Ugljar 74↙
Ugine 52↗
Ugljane 91←
Uherské Hradište 89↙
Uherský Brod 89↙
Uhingen 40↘
Uhlfeld 40→
Uig, *Scotland* 18↖
Uig, *Scotland* 18←
Uithuizen 55↗
Ujazd 89↙
Ujpest 90←
Ujścíe 88↗
Ukhta 86↗
Ukna 31↖
Ula 95→
Ulbjerg 32↘
Ulceby Cross 21↘
Ulcinj 91↗
Uleåborg = Oulu
Ulefoss 27↖
Uleila del Campo 74→
Ulfborg 32↘
Uljma 90↗
Ulladal 27↙
Ullånger 84↗
Ullapool 18↗
Ullared 30←
Ullatti 85→
Ullava 87←
Ulldecona 71↗
Ullsfjord 85↗
Ulm 47↗
Ulmen 39→
Ulmeni 92→
Ulog 91←
Ulricehamn 30↗
Ulrika 31↖
Ulriksberg 28↗
Ulriksfors 84↖
Ulsjöl 29←
Ulstein 84↘
Ultvängstorp 30↗
Ulubey 95→
Ulverston 20→
Ulvik 26→
Ulvila 87↘
Ulvshyttan 29↖
Ulzio 53↖
Umag 55↖
Umberleigh 22↙
Umbértide 54↘
Umbria, prov. 56↗
Umbukta 85↙
Umeå 87↙
Umfors 85↙
Umgransele 84↖
Umnås 85↙
Umurbey 95↖
Unari 86↙
Uncastillo 66←
Underåsen 27→
Undersvik 84↗
Undredal 26→
Uněšov 41→
Ungeny 92↖
Ungerdorf 49↘
Unhais da Serra 68↗
Unhošt 41→
Uniejów 89←
Unirea 92↘
Unna 36↙
Unquera 64↗
Unt Stinkenbrunn 49↗
Unterägeri 47←
Unterdigisheim 47↖
Untergarching 48↗
Untergröningen 40↘
Unterhaching 48↖
Unterluss 37←
Unterpullendorf 49→
Unterwasser 47→
Unterweissenbach 49↖
Unterweitersdorf 49↖
Unterzolling 48↖
Untorp 84→
Upavon 23←
Upper Austria = Oberösterreich
Upper Chapel 22↗
Upper Tean 21↙
Uppgränna 30↗
Upphärad 28↙
Uppheim 26↘
Uppingham 23↖
Uppsala 29→
Uppsala Väsby 29→
Uppsjohytta 26↗
Upton upon Severn 22→
Upwell 23↗
Urach 47↗
Uras 61←
Urbánia 54↘
Urbino 54↘
Urda 70←
Urdos 66↗
Uriage 52↗
Urjala 87→
Urk 35↗
Urla 95←
Urlaţi 92→
Urlingford 25←
Urmston 21↙
Urnäsch 12↗
Urnes 26←
Uroševac 91↖
Urrea de Gaén 66↘
Urshult 30→
Ursviken 86↘
Urziceni 92→
Usagre 69↙
Uşak 95→
Used 65↘
Useldange 39←
Useras 71→
Usingen 40←
Usk 22→
Uskedal 27↙
Usküdar 93↗
Uslar 36↘
Ussat-les-Bains 67↖
Ussel 51↖
Ustaoset 26→
Ustaritz 66↖
Uster 47←
Ustí nad Labem 41↗
Ustilug 89→
Ustovo 93→
Ustrzyki Dolne 89↘
Utajärvi 86→
Utåker 27↙
Utebo 66←
Utersum 33↙
Utiel 71←
Utifällen 86↘
Utne 26↘
Utrecht 35→
Utrera 73←
Utrillas 66↙
Utsjoki 86↖
Uttendorf 48↗
Uttersberg 29←
Uttoxeter 21↙
Utvik 26↙
Uukuniem 87↗
Uurainen 87←
Uusi Värtsilä 87↖
Uusikaarlepyy = Nykarleby
Uusikaupunki 87↘
Uvana 28↗
Uvdal 26→
Uxbridge 23←
Uyeasound 19↗
Uzès 52←
Uzhgorod 92↙
Uzhok 89↘
Uzunköprü 93→

Vä, *Sweden* 30↘
Vä, *Sweden* 30↘
Vaajakoski 87←
Vaala 86→
Vaalimaa 87↗
Vaaraslahti 87←
Vaasa 87↙
Vaattojärvi 86↙
Vabre 51↘
Vác 90←
Vacha 40↗
Väckelsång 30→
Vad 29←
Väddö 29→
Väderstad 28↘
Vadfoss 27↗
Vadheim 26↙
Vadje 28↗
Vado Ligure 53→
Vadsbro 29↙
Vadsö 86↗
Vadstena 28↘
Vaduz 47→
Værøy 85←
Vagå 26↘
Vagan 85←
Vage 26↘
Vaggeryd 30↗
Vägjöfors 28↗
Vagnhärad 29↘
Vagos 68↖
Vähakyrö 87↙
Vaignes 44↗
Vaihingen 40↙
Vaikijaur 85↘
Vailly, *France* 38↘
Vailly, *France* 45←
Vainikkala 87↗
Vairano Scalo 57↙
Vaison 52→
Vaite 46←
Vakarel 93↙
Vakern 28↗
Vákhlia 96↘
Vaksdal 26↘
Val d'Esquières 53↙
Val-d'Isère 53↖
Val-Suzon 46←
Våladålen 84→
Valadares 63↙
Valado 68←
Valais, prov. 47↙
Valandovo 93↘
Valasske Klobouky 89↙
Valasske Meziříči 89↙
Valberg, *France* 53←
Valberg, *Norway* 85←
Valberg, *Sweden* 28→
Valbondione 47↘
Valcarlos 66↖
Valdagno 54↗
Valdahon, le 46→
Valdealgorfa 66↘
Valdecaballeros 69→
Valdefuentes 69←
Valdeganga 71↙
Valdelacasa de Tajo 69←
Valdemaluque 65↙
Valdemarsvik 31↖
Valdemorillo 69↗
Valdemoro 70←
Valdemoro Sierra 71←
Valdeolivas 70↗
Valdepeñas 70↙
Valdepeñas de Jaén 73→
Valderas 64→
Valderrobres 66↘
Valderrueda 64↗
Valdeverdeja 69←
Valdobbiádene 54↗
Valdoviño 63↗
Vale de Cambra 63↙
Vale de Prazeres 68→
Valea lui Mihai 92↙
Valea Viseului 92←
Valebö 27↖
Valenca do Minho 63←
Valençay 45←
Valence, *France* 50↘
Valence, *France* 52→
Valence-d'Agen 51←
Valence-d'Albigeois 51↘
Valencia 71→
Valencia de Alcántara 68→
Valencia de Don Juan 64→
Valencia de las Torres 69↙
Valencia del Mombuey 72↗
Valencia del Ventoso 72↗
Valenciennes 38→
Vălenii de Munte 92→
Valenza 53↗
Valenzuela 73→
Våler 28↖
Valescure 53↙
Valevag 27↙
Valflaunès 52↙
Valguarnera Caropepe 60→
Valhelhas 68↗
Valimí 96←
Valíra 96↘
Valjevo 91↖
Valjok 86↖
Valkeakoski 87→
Valkeala 87→
Valkenswaard 35↘
Vall 31↗
Vall d'Alba 71→
Vall de Uxó 71→
Valla 29↙
Valladolid 64↘
Vallata 58↙
Vallauris 53↙
Valldal 26←
Valldalsseter 27←
Valldemosa 75←
Valle 27←
Valle d'Aosta, prov. 53↖
Valle de Cabuérniga 65↖
Valle de la Serena 69↙
Valle de los Caidos 69↗
Vallecas 70↖
Vallelado 65↙
Vallelunga Pratameno 60→
Vallen, *Sweden* 84↖

Villardefrades 64↘
Villarejo de Fuentes 70→
Villarejo de Salvanés 70→
Villares del Saz 70→
Villargordo 73→
Villargordo del Cabriel 71↗
Villarluengo 66↘
Villarquemado 71↘
Villarramiel 64→
Villarrasa 72→
Villarreal 71→
Villarreal de Alava 65↗
Villarrín de Campos 64→
Villarrobledo 70→
Villarroya de la Sierra 65↘
Villarroya de los Pinares 71↘
Villarrubia 73→
Villarrubia de Santiago 70→
Villarrubia de los Ojos 70→
Villars 46↘
Villars-les-Dombes 46↙
Villarta de los Montes 69↘
Villarta de San Juan 70↙
Villasana de Mena 65↖
Villasandino 65←
Villasarracino 64→
Villasequilla de Yepes 70←
Villasimíus 61↘
Villasor 61↙
Villastar 71↖
Villatobas 70←
Villatoya 71←
Villava 65↗
Villaverde de Guadalimar 70↘
Villaverde de Trucios 65↖
Villaverde del Rio 73←
Villaviciosa 64↙
Villaviciosa de Córdoba 73↙
Villaviciosa de Odon 69↗
Villavieja de Yeltes 64↙
Villé 46↗
Villeblanca 72→
Villebois-Lavalette 50↙
Villecroze 52↘
Villedieu-les-Poêles 43←
Villefagnan 44↙
Villefort 52←
Villefranche, France 46↙
Villefranche, France 53↙
Villefranche-de-Lauragais 67↙
Villefranche-de-Rouergue 51←
Villefranche-du-Périgord 51←
Villefranche-sur-Cher 45←
Villel 71↖
Villemayor de Gállego 66→
Villemur 51↙
Villena 71↙
Villenauxe 45↙
Villeneuve, Italy 53↖
Villeneuve, Switzerland 46↘
Villeneuve-d'Aveyron 51←
Villeneuve-de-Marsan 50↙
Villeneuve l'Archevêque 45↙
Villeneuve-les-Avignon 52↙
Villeneuve-St-Georges 38↙
Villeneuve-sur-Lot 51←
Villeneuve-sur-Yonne 45↗
Villeparisis 38↙
Villeréal 51←
Villers 43↘
Villers-Bocage 43←
Villers-Cotterets 38↘
Villersexel 46↗
Villerville 43↗
Villia 96←
Villiers-St-Georges 38↘
Villingen 47↖
Villoldo 64→
Villoria 64↙
Villotta 54↙
Vilnes 26↙
Vilppula 87↙
Vilsbiburg 48↗
Vilshofen 41↘
Vilshult 30↗
Vilsund 32↙
Vilusi 91←
Vilvestre 63↘
Vilvoorde 39↘
Vimercate 53↗
Vimianzo 63↘
Vimieiro 68↘
Vimioso 63↘
Vimmerby 31↘
Vimo 28↙
Vimoutiers 43→
Vimpeli 87↙
Vimperk 41↘
Vinaixa 67↙
Vinaroz 71↗

Vinay 52↙
Vinberg 30←
Vindafjorden 27↘
Vindelgransele 84↖
Vindeln 86↘
Vinderup 32↘
Vinga 90↙
Vingåker 29↙
Vinhais 63→
Vinje, Norway 26↘
Vinje, Norway 27←
Vinjeora 84↘
Vinjevingen 27←
Vinju Mare 93↙
Vinkovci 90→
Vinnelys 85↙
Vinninga 28↘
Vinogradov 92↙
Vinon 52↘
Vinslöv 30↙
Vinstra 26↘
Vintjarn 29↙
Vintl 48↙
Vintrosa 28→
Vinuesa 65→
Viöl 33↗
Vipiteno 48←
Virancık 95↗
Vire 43←
Vireux 39←
Virfurile 92↘
Virginia 24→
Virje 90↙
Virolahti 87↗
Virovitica 90→
Virpazar 91↗
Vírra 96↖
Virrat 87←
Virsbo 29↙
Virserum 31↖
Virtasalmi 87↖
Virton 39↙
Vis 57↙
Visbek 36↙
Visby 31↗
Visé 39↖
Višegrad 91↖
Viserba 11↗
Viserbella 11↗
Viseu 68↗
Viseu de Sus 92←
Visiedo 66↙
Vişina Veche 93←
Viskafors 30↙
Viskinge 32→
Vislanda 30→
Visnes 27↙
Visnum 28↘
Viso, Monte, mt. 53←
Viso del Marqués 70↙
Visoko 91←
Visp 47↙
Vissefjärda 31↙
Visselhövede 36↙
Vissenbjerg 33←
Visso 53↗
Vissoie 47↙
Vistabella del Maestrazgo 71↗
Viste 33↙
Vistula = Wisła
Viterbo 56↗
Vitigudino 64↙
Vitína 96↙
Vitoria 65↗
Vitré 44↙
Vitrey 46↙
Vitry-le-François 39↙
Vitsand 28↙
Vitsi = Vérnon
Vittangi 85→
Vitteaux 46←
Vittel 46↙
Vittoria 60↙
Vittório Veneto 54↗
Vittsjö 30↙
Vitvattnet 85↘
Viuf 33↙
Vivario 62↙
Vivel del Rio Martin 66↙
Viveli 26→
Viver 71←
Vivero 63↙
Viverols 52↖
Viveros 70↘
Vivier-sur-Mer, le 10↗
Viviez 51←
Vivunki 85↗
Vize 93↗
Vizíkion 96↘

Vizille 52↙
Viziru 92↗
Vizovice 89↙
Vizzavona 62←
Vizzini 60↙
Vlaardingen 35←
Vladičin Han 93↙
Vladimirci 91↖
Vlakhérna 96→
Vlákhoi 96↘
Vlakhokerasiá 96→
Vlakhomándra 96↙
Vlakhópoulon 96↙
Vlakhorráfti 96↘
Vlasenica 91←
Vlašim 88↙
Vlaslei 55→
Vlasotince 93↙
Vlássi 96↖
Vlčany 90↙
Vlieland 35↖
Vlissingen 35↙
Vlorë 91↙
Vlotho 36→
Vöcklabruck 48↙
Voditsa 93←
Vodnjan 55←
Voe 19↙
Voghera 53↗
Vohburg 41↙
Vohenstrauss 41←
Vöhringen 47↗
Void 39↙
Voiron 52↗
Voislova 92↙
Voitsberg 49←
Voitsdorf 49↗
Vojens 33↙
Vojnic 55↙
Vojvodina, prov. 90↗
Vöklamarkt 48↗
Volary 41↘
Volda 26↙
Volders 48←
Volgsele 84↖
Volgsjöfors 84↖
Volissós 95←
Volkach 40→
Völkermarkt 49↙
Völklingen 39↘
Volkmarsen 36↘
Vollenhove 35↙
Vollsbu 26↘
Vollset 26←
Vollum 26↖
Vólos 94↙
Volovọ 92↙
Volterra 54↙
Voltri 53→
Volturara Appula 58↙
Volvic 45↙
Volyně 41↘
Vomvokoú 96↙
Vónitsa 96↙
Voorschoten 35←
Voorst 35→
Voorthuizen 35→
Võra 87↙
Vorarlberg, prov. 47→
Vorbasse 33↙
Vorden 35→
Vordernberg 49←
Vorderriss 48←
Vordhónia 96↙
Vordingborg 33↗
Voreppe 52↗
Vorey 51↗
Vormsund 28↘
Vorupør 32↙
Vosges, mts. 46↗
Voss 26↘
Vossevangen = Voss
Votice 88↙
Vouillé 44↘
Voúla 96↙
Voulgarélion 94↙
Voulte, la 52←
Voulx 45↗
Voúnargon 96↘
Vouvray 44→
Vouzela 68↙
Vouziers 39↙
Voves 45↙
Voxna 84↙
Vrå, Denmark 32←
Vrå, Sweden 30→
Vråble 90↖
Vrakháti 96→
Vrakhnéika 96↙
Vråliosen 27←

Vrana 55→
Vranganiótika 96↘
Vranja 55↘
Vranje 93↙
Vratsa 93↙
Vrbnik 55→
Vrbovec 93↙
Vrbovsko 55↙
Vreden 36←
Vrena 69→
Vrésthena 96↘
Vreta Kloster 29↙
Vretstorp 28↘
Vrgorac 91←
Vrh 55→
Vrhnika 49↙
Vries 35↙
Vriezenveen 35→
Vrigstad 30↗
Vríses 96↘
Vrísis 96↖
Vrlika 91↖
Vrondádhes 95←
Vrondamas 94→
Vršac 90↗
Vrsar 55↙
Vrútky 89↙
Vsetin 89↙
Vsevolozhskay 87↗
Vučitrn 91↖
Vukova Gorica 55↙
Vukovar 90↗
Vulpera 47→
Vuodas 85↙
Vuokatti 87↘
Vuoksenniska 87↗
Vuolijok 87↗
Vuollerim 85↘
Vuostimo 86↙
Vuotso 86←
Vūrbitsa 93←
Vust 32↙
Vyartsilya 87↖
Vyborg 87↗
Východocesky 88↘
Vysoké Mýto 88↘
Vysokoye 89↗
Vysotsk 87↗
Vyšší Brod 49↘

Waabs 33←
Waalwijk 35→
Wąbrzezno 89↘
Wachtendonk 39↗
Waddesdon 23←
Waddington, England 21↘
Waddington, Rep. of Ireland 25↘
Wadebridge 22↙
Wädenswil 47←
Wadern 39↙
Wadhurst 23→
Wagenfeld 36→
Wageningen 35↙
Waging 48↗
Wahn, airport 39↙
Waiblingen 40↘
Waibstadt 40↙
Waidhofen 49↘
Wainfleet All Saints 21↘
Wakefield 21↙
Walachia, reg. 92→
Wałbrzych 88↘
Walchensee 48←
Wałcz 88↗
Wald, Austria 48→
Wald, Switzerland 47←
Waldbröl 40↙
Waldeck 40↘
Waldenbuch 40↙
Waldenburg, Poland = Wałbrzych
Waldenburg, W. Germany 41↖
Waldfenster 40↙
Waldfischbach 39↘
Waldheim 41↗
Waldkappel 40↗
Waldkirch 47↖
Waldkirchen 41↘
Waldmünchen 41↘
Waldsassen 41↘
Waldshut 47↘
Walenstadt 47→
Walford 22↗
Wall 22↗
Wallasey 20↘
Walldürn 40↘
Wallingford 23←
Walls 19↙
Wallsbüll 33↙
Wallsend 21↖
Walmer 23→

Walsall 22↗
Walsrode 36→
Walsum 36↙
Waltershausen 40↗
Waltham 23↗
Waltham Abbey 23→
Walton-le-Dale 21↙
Walton on the Naze 23→
Waltrop 36↙
Wanderup 33↙
Wanfried 40↗
Wangen 47↖
Wangersen 36↗
Wangs 12↗
Wanlockhead 20↗
Wanne-Eickel 36↙
Wansford 23↗
Wantage 23←
Wanzleben 37←
Warboys 23↗
Warburg 36↘
Wardenburg 36↖
Ware 23←
Wareham 22↘
Waremme 39↘
Waren 33↗
Warendorf 36←
Warin 33↗
Warka 89↘
Warley 22↗
Warmington 23↖
Warminster 22↗
Warmwell 22↘
Warnemunde 33↗
Warrenpoint 24↙
Warrington 21↙
Warsaw = Warszawa
Warstade 36↗
Warstein 36↘
Warszawa 89↘
Warszow 88↗
Warth 47↗
Warwick, England 21↖
Warwick, England 23←
Washington, England 21↖
Washington, England 23↙
Wasselonne 39↙
Wassen 47↙
Wassenaar 35←
Wasseralfingen 40↘
Wasserauen 47→
Wasserburg 48↗
Wassertrüdingen 40↘
Wassy 46↘
Wast Water, lake 20↙
Wasungen 40↗
Watchet 22↙
Watenstedt 37←
Waterbeach 23↗
Waterford 25→
Wateringhouse 19↘
Waterloo 38↙
Waterloo Cross 22↘
Waterville 25↘
Watford 23←
Watlington 23←
Watten 19↙
Wattens 48←
Wattenscheid 39↗
Watton 23↗
Watton at Stone 23←
Wattwil 47←
Wavre 39↘
Wearhead 21←
Wedel 33↘
Wedmore 22↗
Wednesbury 22↗
Wednesfield 22↗
Weedon 23↙
Weener 36↘
Weert 35↘
Weesen 47←
Weesp 35←
Weeze 35→
Weferlingen 37←
Wegberg 39↗
Wegeringhausen 40↖
Weggis 47←
Węgorzewo 89↗
Węgorzyno 88↗
Węgrów 89↘
Wehr 47↘
Weichshofen 41↙
Weida 41↘
Weiden 41↘
Weidenau 40↙
Weikersheim 40↘
Weil der Stadt 40↙
Weilburg 40←
Weilheim 48←

Weimar 41↖
Weinfelden 47←
Weingarten, W. Germany 40↙
Weingarten, W. Germany 47↗
Weinheim 40←
Weinsberg 40↘
Weissbach 47↗
Weissenbach, Austria 47→
Weissenbach, Austria 48→
Weissenburg 41↙
Weissenfels 41↘
Weissenhorn 47↗
Weissenstadt 41←
Weisskirchen 49←
Weisstannen 47↖
Weitendorf 33↗
Weitersfeld 49↗
Weitra 49↖
Weiz 49↙
Welle 36↗
Wellesbourne 23↖
Wellingborough 23↗
Wellington, England 22↗
Wellington, England 22↘
Wells, England 22→
Wells, Rep. of Ireland 25↖
Wells next the Sea 23↗
Welney 23↗
Wels 49↗
Welsh Corner 20↘
Welshpool 22↗
Welsickendorf 37→
Welwyn 23←
Welwyn Garden City 23←
Welzheim 40↘
Wem 22↗
Wemding 40↘
Wemperhardt 39←
Wenden 40↘
Wendisch-Baggendorf 33↗
Wendling 23↗
Wendover 23←
Wenduine 38↗
Wengen 47↙
Wennington 21←
Wensley 21↗
Weppersdorf 49→
Werben 37←
Werbig 37↗
Werbomont 39←
Werdau 41↘
Werden 39↗
Werdohl 40↗
Werl 36↗
Werlte 36←
Wermelskirchen 39↗
Wernberg 41←
Werne 36↗
Werneuchen 37→
Wernigerode 37↙
Wertheim 40↘
Wertingen 40↙
Weseke 36←
Wesel 36↙
Wesenberg 37↗
Weser, river 36→
Wesselburen 33↖
Wesseling 39↗
West Auckland 21↖
West Bridgford 21↙
West Bromwich 22↗
West Calder 19↙
West End 23↙
West Haddon 23↗
West Ham 23←
West Harptree 22→
West Kilbride 20↖
West Linton 19↙
West Malling 23→
West Meon 23↙
West Mersea 23↗
West Sandwick 19↗
West Tanfield 21↙
West Terschelling 35↙
West Wittering 23↙
West Woodburn 21↘
Westbury 22→
Westerbork 35↗
Westerburg 39↗
Westerham 23↗
Westerland 33↙
Westerstede 36↘
Westgate 23→
Weston super Mare 22→
Westport 24↘
Westruther 19↙
Westward Ho! 22←
Wetheral 21↖
Wetherby 21←
Wetter 39↗

Wetteren 38↗
Wettin 37↙
Wettingen 47←
Wettringen 36←
Wetwang 21→
Wetzikon 47←
Wetzlar 40↖
Wexford 25↖
Weybridge 23←
Weyer 49←
Weyhill 23←
Weymouth 22↘
Whaley Bridge 21↙
Whalley 21↙
Wharfe 21←
Wheatley 23←
Wheddon Cross 22→
Wherwell 23←
Whicham 20→
Whiddon Down 22↙
Whipsnade 23←
Whiston 21↙
Whitburn 18↘
Whitby 21→
Whitchurch, *England* 21↙
Whitchurch, *England* 22→
Whitchurch, *England* 23←
Whitchurch, *Wales* 22→
Whiteabbey 24↖
Whitebridge 18→
Whitehall 19↗
Whitehaven 20→
Whitehead 24↖
Whiteparish 23↙
Whithorn 20→
Whitland 22←
Whitney 22↗
Whitstable 23→
Whitstone 22↙
Whittingham 21↖
Whittington 20↘
Whittlesey 23↖
Wiblingen 47↗
Wick 19↖
Wickford 23→
Wickham 23↙
Wickham Market 23↗
Wicklow 25↖
Widdrington 21↖
Widecombe-in-the-Moor 22↙
Widford 23→
Widmerpool Cross Roads 21↙
Widnes 21↙
Wiebelskirchen 39↘
Więcbork 89↖
Wiedenbrück 36↘
Wiehe 41↖
Wiehl 40↖
Wielbark 89↗
Wieliczka 89↙
Wielun 89←
Wien 49↗
Wiener Neustadt 49→
Wierden 35→
Wieruszów 89←
Wies 48←
Wiesbaden 40←
Wieselburg 49↖
Wiesenburg 37→
Wiesloch 40↙
Wigan 21↙
Wigston Magna 23↖
Wigton 20→
Wigtown 20↗
Wijhe 35→
Wil 47←
Wildbad 40↙
Wildberg 37↗
Wildeshausen 36↗
Wildhaus 47→
Wildon 49↘
Wildschönau = Auffach/Niederau
Wilhelmsburg, *Austria* 49↗
Wilhelmsburg, *W. Germany* 36↗
Wilhelmshaven 36↖
Wilhermstadt 40↘
Wilkau-Hasslau 41↖
Willand 22↘
Willemstad 35←
Willersley 22↗
Willesden 23←
Williton 22→
Wilmslow 21↙
Wilnsdorf 40↖
Wilsdruff 41↗
Wilsontown 19↙
Wilster 33↘
Wilton 22→
Wiltz 39←
Wimbledon 23←
Wimborne Minster 22↘
Wimereux 38↖
Wincanton 22↘
Winchcomb 22→
Winchelsea 23↘
Winchester 23←
Windecken 40←
Windermere 20→
Windischgarsten 49←
Windsbach 40↘
Windsheim 40↘
Windsor 23←
Windygates 19↙
Wing 23←
Wingham 23→
Winkleigh 22↙
Winklern 48↘
Winnenden 40↘
Winningen 39→
Winschoten 36↖
Winsen 37↖
Winsford 21↙
Winslow 23←
Winsum 35↗
Winterbach 49↖
Winterberg 40↖
Winterbourne Abbas 22↘
Wintermoor 36↗
Winterswijk 35→
Winterthur 47←
Winterton 21↘
Wintzenheim 46↗
Wipperfürth 40↖
Wirksworth 21↙
Wisbech 23↗
Wishaw 18↘
Wisła 89↙
Wisła, river 89→
Wismar 33↗
Wissant 38↖
Wissembourg 40↙
Wissen 39↗
Witham 23→
Witheridge 22↙
Withernsea 21↘
Witney 23←
Witten 39↗
Wittenberg 37→
Wittenberge 37↖
Wittenburg 37↖
Wittensee 33→
Wittersham 23↘
Wittingen 37←
Wittlich 39→
Wittmund 36↖
Wittstock 37↗
Witzenhausen 40↗
Wiveliscombe 22↘
Władysławowo 89↖
Wleń 88→
Włocławek 89←
Włodawa 89→
Woburn 23←
Wodzisław 89↙
Woerden 35←
Woerth 39↘
Wohlen 47←
Woking 23←
Wola 89←
Wolfach 47↖
Wolfegg 47↗
Wolfen 37↘
Wolfenbüttel 37←
Wolfenschiessen 12↖
Wolfhagen 40↗
Wolfratshausen 48↖
Wolfsberg 49↙
Wolfsburg 37←
Wolgast 88↗
Wolhusen 47←
Wolin 88↗
Wolkersdorf 49↗
Wollaston 23↖
Wöllstein 40←
Wolmirstedt 37←
Wolsingham 21←
Wolsztyn 88→
Woltersdorf 37↖
Wolthausen 36→
Wolvega 35↗
Wolverhampton 22↗
Wolverton 23↖
Wolvey 23↖
Wolviston 21←
Wombwell 21↙
Wonersh 23←
Woodbridge 23↗
Woodenbridge 25↖
Woodford, *England* 23→
Woodford, *Rep. of Ireland* 25↘
Woodhall Spa 21↘
Woodhead 21↙
Woodstock 23←
Woodtown 24↗
Woodville 23↖
Woofferton 22↗
Wookey Hole 22→
Wool 22↘
Woolacombe 22←
Wooler 21↖
Woolsthorpe 21↘
Woolwich 23→
Woore 21↙
Wootton 21↘
Wootton Basset 22→
Worb 47←
Worbis 37↙
Worcester 22↗
Wörgl 48←
Workingham 23←
Workington 20→
Worksop 21↙
Workum 35↗
Wörlitz 37↘
Wormbridge 22→
Wormhoudt 38↖
Wormit 19↙
Worms 40←
Wornitz 40↘
Wörrstadt 40←
Wörth, *W. Germany* 40↙
Wörth, *W. Germany* 40→
Wörth, *W. Germany* 41↙
Wörther See, lake 49↙
Worthing 23↙
Wortley 21↙
Wössingen 40↙
Wotton under Edge 22→
Woudenberg 35→
Wragby 21↘
Wrentham 23↗
Wrexham 20↘
Wriezen 88→
Wrocław 89←
Wronki 88→
Wrotham 23→
Wrotham Heath 23→
Wroughton 22→
Wroxham 23↗
Września 89←
Wschowa 88→
Wulfen 36↙
Wülfrath 39↗
Wünnenberg 36↘
Wunsiedel 41←
Wunstorf 36→
Wuppertal 39↗
Würnsdorf 49↖
Würselen 39↖
Wurzach 47↗
Wurzbach 41←
Würzburg 40→
Wurzen 37↘
Wusterhausen 37↗
Wuustwezel 39↖
Wych Cross 23→
Wye 23→
Wyk 33↙
Wymondham 23↗
Wyrzysk 89↖
Wysokie Maz 89↗
Wyszków 89→
Wyszogrod 89←
Wythall Heath 22↗
Xanten 35→
Xánthi 93→
Xanthus, anc. site 95↘
Xeriés 96↘
Xertigny 46↗
Xidheika 96→
Xilókastron 96←
Xinía 96←
Xirokhóri 96↘
Xironomí 96←
Yablonov 92←
Yakoruda 93↘
Yalakdere 95↗
Yalova 95↗
Yambol 93←
Yampol 92↖
Yanguas 65→
Yannina = Ioánnina
Yarcombe 22↘
Yaremcha 92←
Yarm 21←
Yarmouth 23↙
Yaryshev 92↖
Yasinya 92←
Yatağan 95→
Yátova 71←
Yatton 22→
Yavorov 89↘
Ybbs 49↖
Ydby 32↘
Yderby 32→
Ydrefors 31↖
Yealmpton 22↙
Yecla 71↙
Yedintsy 92↖
Yelverton 22↙
Yeméni 96←
Yenice 95↖
Yeniköy 95↗
Yenino 89↖
Yenişehir 95↗
Yeoryítsi 96→
Yeovil 22↘
Yepes 70←
Yeráki, anc. site 96→
Yerville 38←
Yesa 66↖
Yeşilova 95→
Yeste 70↘
Yetholm 21↖
Yetminster 22↘
Yiáltra 96←
Yiannádhes 96↘
Yiannitsá, *Greece* 93↘
Yiannitsá, *Greece* 94←
Yiannítsi 96↖
Yiannitsoú 96←
Yihärmä 87↙
Yimnón 96↖
Yíthion 94↘
Yllämaa 87↗
Yläne 87↘
Ylikiminki 86→
Yli-Ii 86→
Ylimarkku 87↙
Ylistaro 87↙
Ylitornio 86↙
Ylivieska 87←
Ylöjärvi 87→
Yngsjö 30↘
York 21←
Youghal 25→
Yoxford 23↗
Ypäjä 87↘
Yport 43↗
Ypenburg, airport 35←
Yppari 86→
Yrke 27↙
Ysby 30→
Yspytty Ystwyth 22↖
Yssingeaux 52↖
Ystad 30↘
Ystradgynlais 22←
Ytre Arna 26↘
Ytterhogdal 84↗
Ytterlännäs 84↗
Yttermalung 28↗
Yttersjön 86↘
Yuncos 69→
Yunquera de Henares 70↗
Yverdon 46↘
Yvetot 43↗
Yvonand 46→
Yxe 29←
Yxerum 29↙
Zaandam 35←
Zabari 91↖
Zabkowice 88↘
Zabłudów 89↗
Zabno 49↘
Zabalotov 92←
Zábřeh 88↘
Zabrze 89↙
Zadar 55→
Zafra 69↙
Zaga 48↘
Zagań 88→
Zagorá 94↗
Zagorje 49↙
Zagreb 49↘
Zagubica 93↙
Zahara 73↙
Zahna 37→
Záhorská Bystrica 49↗
Zaidín 66→
Zaječar 93↙
Zákány 49↘
Zákha 96↘
Zakháro 96↘
Zákinthos 96↘
Zakopane 89↙
Zakroczym 89←
Zákros 95↙
Zalaegerzeg 49↘
Zalalövo 49↘
Zalamea de la Serena 69↙
Zalamea la Real 72→
Zalău 92↙
Zaleshchiki 92←
Zalewo 89↖
Zalla 65↖
Zaluznica 55→
Zamberk 88↘
Zambrana 65→
Zambrów 89↗
Zamora 64↘
Zamość 89↘
Zandvoort 35←
Zapponeta 58←
Zarafóna 96→
Zaragoza 66←
Zarauz 65↗
Zarkon 94↗
Zărneşti 92→
Żarnów 89←
Zarrentin 37↖
Zary 88→
Zarza Capilla 69↘
Zarza la Mayor 68→
Zarzadilla de Totana 74→
Zas 63↖
Zasieki 88→
Zastavna 92←
Zate 41→
Zawiercie 89↙
Zazvic 55↘
Zbąszyń 88→
Žďár 88↘
Zdice 41→
Zdihovo 55↗
Zdrelac 55↘
Zduńska 89←
Zealand = Sjælland
Zebreira 68→
Zechlin 37↗
Zegar 55→
Zehdenick 37↗
Zeist 35→
Zeitlarn 48↗
Zeitz 41↖
Zele 38↗
Zelechów 89→
Zelenogorsk 87↗
Zelezny Brod 88↘
Zelhem 35→
Zeliezovce 90←
Zélion 96←
Zell 39→
Zell am See 48→
Zell am Ziller 48←
Zella-Mehlis 40↗
Zellingen 40→
Zeltweg 49←
Zelzate 38↗
Zemenón 96→
Zemun 90↗
Zemunik 55→
Zenica 91←
Zennor 22↘
Zepče 91←
Zerbst 37→
Zergan 91↗
Zermatt 47↙
Zernez 47↘
Zeulenroda 41↖
Zeven 36↗
Zevenaar 35→
Zevgolatió, *Greece* 96→
Zevgolatió, *Greece* 96↘
Zevgolatión 96→
Zeytindağ 95↖
Zgierz 89←
Zgorzelec 88→
Zhabe 92←
Zheleznodorozhnyy 89↗
Ziar-nad-Hronom 90←
Zicava 62←
Ziegenhain 40↗
Zielona Góra 88→
Zierenberg 40↗
Zierikzee 35←
Zierzow 37↖
Ziesar 37→
Zilina 89↙
Zimnicea 93←
Zingst 33↗
Zirikia 96←
Zirl 48←
Zirndorf 41↙
Zirovnica 91↗
Zistersdorf 49↗
Zítsa 94↖
Zittau 88→
Zlatitsa 93←
Zlatni Piassatsi = Golden Sands
Złocieniec 88↗
Złoczew 89←
Zlonice 41→
Zlutice 41→
Zmigród 88→
Zminj 55↖
Znamensk 89↗
Znin 89←
Znojmo 49↗
Zofingen 47←
Zollhaus 40←
Zollikofen 46→
Zóni 96→
Zonza 62←
Zörbig 37↘
Zorita 69←
Zorleni 92↗
Zorneding 48↖
Zory 89↙
Zossen 37→
Zrenjanin 90↗
Zruč 88↘
Zsáka 90↖
Zschopau 41↗
Zucaina 71←
Zuéra 66→
Zufre 72→
Zug 47←
Zugspitze, mt. 48←
Zúhéros 73→
Zuidhorn 35↗
Zuidlaren 35↗
Zújar 74←
Zülpich 39↗
Zumárraga 65↗
Zumaya 65↗
Zundert 35↙
Zuoz 47↘
Zupanja 90→
Zurich, *Netherlands* 35↗
Zürich, *Switzerland* 47←
Zürich-See, lake 47←
Zurndorf 49↗
Zürs 47→
Zurzach 47←
Zusmarshausen 47↗
Züsow 33↗
Zuta Lokva 55→
Zutphen 35→
Zuzemberk 49↙
Zvezdets 93↗
Zvolen 90←
Zvornik 91↖
Zwartsluis 35↗
Zweelo 35↗
Zweibrücken 39↘
Zweisimmen 46↘
Zwenkau 41↖
Zwettl 49↖
Zwickau 41↖
Zwiefalten 47↗
Zwiesel 41↘
Zwischenahn 36↖
Zwoleń 89→
Zwolle 35→
Zwönitz 41↗
Zyrardów 89←
Zywiec 89↙

Cities (along the diagonal, in order): Amsterdam, Antwerpen, Athinai, Barcelona, Basel, Beograd, Biarritz, Bolzano, Bordeaux, Boulogne, Brindisi, Bruxelles, Calais, Cherbourg, Dieppe, Dunkerque, Esbjerg, Firenze, Frankfurt, Genève, Genova, Gibraltar, Hamburg, Hannover, Helsinki, Hoek van H...

156	3022	1538	780	1843	1241	1117	1056	394	2153	203	360	827	483	320	760	1439	462	920	1265	2459	476	447	1664	
	2971	1365	629	1791	1085	1044	900	238	2018	47	204	671	325	164	916	1287	394	769	1114	2303	546	480	1820	
		3313	2567	1180	3486	2211	3301	3188	219	2963	3175	3339	3165	3125	3260	1117	2576	2610	2388	4485	2977	2816	4165	
			1056	2134	581	1334	650	1292	2102	1318	1326	1294	1199	1312	2201	1213	1403	803	925	1172	2018	1757	3106	
97				1405	1143	407	861	684	1389	550	705	838	665	682	1149	690	351	253	517	2227	866	705	2053	
1878	1846				2306	986	2121	2008	1696	1789	1992	2160	1955	1955	2081	1078	1397	1431	1209	3273	1798	1637	2985	
956	848	2059				1487	185	996	2218	1038	1030	817	846	1032	1962	1407	1350	890	1120	1218	1678	1518	2866	
485	391	1595	656				1302	1217	1096	998	1191	1341	1062	1080	1368	402	671	597	410	2507	1085	924	2272	
1145	1113	733	1326	873				811	2033	853	845	644	661	846	1777	1334	1165	705	1061	1403	1493	1332	2681	
771	674	2166	361	710	1433				2076	225	34	426	150	74	1154	1413	623	766	1204	2190	871	718	2087	
694	649	1374	829	253	613	924				1923	2061	2206	2033	2057	2446	885	1724	1326	1176	3273	2163	2002	3351	
656	559	2051	404	535	1318	115	809				204	583	299	169	875	1241	386	677	1067	2256	597	491	1867	
245	148	1981	803	425	1227	619	756	504				460	183	40	1085	1379	591	747	1184	2224	714	684	2024	
1338	1254	136	1306	863	1054	1378	681	1263	1290				303	471	1398	1529	927	853	1291	2047	1115	1096	2444	
126	29	1841	819	342	1112	645	620	530	140	1195				191	1241	1342	686	706	1144	2065	958	805	2145	
224	127	1973	824	428	1238	640	740	525	21	1281	127				1080	1387	549	731	1168	2211	669	650	1857	
514	417	2075	804	521	1342	508	833	400	265	1371	362	286				1770	798	1402	1646	3219	283	444	1167	
300	202	1967	745	413	1215	526	660	411	93	1263	186	114	188				1028	724	286	2483	1487	1326	2675	
199	102	1942	815	424	1215	641	671	526	46	1278	105	25	293	119				603	867	2575	515	354	1703	
472	569	2026	1368	714	1293	1219	850	1104	717	1520	544	674	869	771	671				470	1975	1118	958	2306	
894	800	694	754	429	670	874	250	829	878	550	771	857	950	834	862	1100				2097	1363	1202	2551	
287	245	1601	872	218	868	839	417	724	387	1071	240	367	576	426	341	496	639				2897	2736	4084	
572	478	1622	499	157	889	553	371	438	476	824	421	464	530	439	454	871	450	375				161	1188	
786	692	1484	575	321	751	696	255	659	748	731	663	736	802	711	726	1023	178	539	292				1349	
1528	1431	2787	728	1384	2034	757	1558	872	1361	2034	1402	1382	1272	1283	1374	2000	1481	1600	1227	1303				
296	339	1850	1192	538	1117	1043	674	928	541	1344	371	444	693	595	416	176	924	320	695	847	1800			
278	298	1750	1092	438	1017	943	574	828	446	1244	305	425	681	500	404	276	824	220	595	747	1700	100		
1034	1131	2588	1930	1276	1855	1781	1412	1666	1279	2082	1160	1258	1519	1333	1154	725	1662	1058	1433	1585	2538	738	838	
46	78	1883	926	449	1150	752	696	637	226	1302	107	205	454	293	180	518	890	282	556	811	1509	342	287	1080
612	575	1368	828	238	635	882	74	767	648	748	546	639	759	651	647	776	324	314	329	323	1556	543	443	1281	625
482	579	2036	1378	724	1303	1254	860	1139	662	1534	600	706	960	781	602	173	1110	506	881	1033	1986	186	286	552	528
162	130	1716	931	330	983	773	532	658	268	1183	128	254	488	312	229	462	748	115	437	640	1514	286	186	1024	167
369	261	1994	758	440	1261	535	687	420	162	1290	255	183	141	69	197	823	858	508	466	738	1292	647	547	1385	287
264	167	1941	836	424	1208	600	671	485	19	1269	159	40	253	74	65	736	873	399	445	717	1357	524	473	1262	253
1421	1324	2816	810	1360	2083	650	1578	765	1269	2028	1295	1275	1135	1176	1291	1869	1524	1507	1203	1345	420	1660	1593	2398	1417
231	164	1710	722	207	977	680	485	565	252	1060	135	263	418	280	238	585	636	186	314	528	1437	409	309	1147	231
555	458	1711	401	255	978	455	460	340	454	922	429	459	490	396	449	906	497	408	98	319	1129	720	620	1458	536
1094	997	2454	395	1051	1702	323	1224	438	942	1648	968	963	837	849	964	1542	1149	1159	894	970	434	1366	1266	2104	1075
754	657	1780	324	448	1068	445	507	410	653	989	628	658	684	595	648	1105	474	607	264	247	1052	919	819	1657	735
671	604	1418	630	233	685	748	167	633	656	620	575	669	751	643	657	935	196	451	204	88	1358	769	669	1507	682
524	492	1354	848	241	621	914	170	796	598	844	464	607	721	594	587	680	420	247	367	419	1594	500	400	1238	529
1171	1104	399	1043	733	974	1164	554	1127	1190	263	1075	1169	1251	1143	1157	1404	304	951	754	462	1771	1224	1124	1962	1182
833	736	1614	429	450	881	566	384	531	732	823	707	737	768	674	727	1164	308	668	293	130	1157	977	877	1715	814
167	70	1911	828	412	1178	654	690	539	78	1290	70	57	325	151	32	585	841	310	473	733	1411	409	368	1147	148
659	680	2213	1555	901	1480	1406	1037	1291	904	1707	731	861	1056	958	858	187	1287	683	1058	1210	2163	363	463	456	705
303	206	1864	642	310	1131	468	556	353	161	1160	177	174	211	103	173	751	739	362	319	591	1225	545	475	1283	284
1035	968	508	907	597	838	1028	418	991	1054	399	939	1033	1115	1007	1021	1268	168	815	618	332	1635	1088	988	1826	1046
506	409	2104	723	550	1156	434	796	319	298	1365	393	319	121	220	326	924	927	602	533	762	1191	748	696	1486	487
610	606	1268	928	327	535	982	174	867	684	848	550	693	807	680	673	766	423	333	429	423	1656	586	486	1324	615
886	983	2440	1782	1128	1707	1658	1264	1518	1066	1938	1004	1110	1364	1085	1006	577	1514	910	1285	1437	2415	590	686	148	932
370	303	1576	742	88	843	796	334	681	376	941	274	385	499	372	377	626	517	130	245	409	1470	450	350	1188	381
706	639	1503	561	312	770	663	252	548	637	697	582	625	698	600	615	1020	281	482	161	111	1289	812	712	1550	717
856	819	1138	905	482	405	1042	208	913	892	643	790	883	1003	895	893	1015	265	582	484	346	1633	835	735	1573	864
820	782	1240	803	411	507	940	134	811	837	541	722	816	929	793	804	984	163	551	382	244	1531	804	704	1542	818
754	712	1238	1120	507	505	1174	366	1059	864	921	730	858	987	860	853	894	543	467	633	615	1848	718	618	1456	749

MILES